BIOLOGY
Exploring the Science of Life

Gustavo Loret de Mola

McGraw Hill **Wright Group**

The **McGraw·Hill** Companies

Gustavo Loret de Mola is a science educator with more than 35 years of teaching and administrative experience at the middle school, high school, and college level. He holds a B.S. degree in education from the University of Miami, an M.S. degree in science from Nova University, and an Ed.D. in Educational Leadership from Nova University. Dr. Loret de Mola taught high school science for 11 years before serving as a project manager in science and a middle school assistant principal for Dade County Public Schools. Dr. Loret de Mola then served 20 years as District Science Supervisor for Dade County Public Schools. As a committee member, Dr. Loret de Mola helped develop the Teacher Certification Tests in Science for the State of Florida. He also served as State Chairperson for the State of Florida Life Sciences Instructional Materials Council. Currently, Dr. Loret de Mola continues to work in the sciences as adjunct professor at the University of Miami, where he teaches science and science methodology courses for graduate students in education.

Series Consultants

Richard Audet, EdD
  Roger Williams University

Matthew Marino, PhD
  Washington State University

Barbara Scott, MD
  Los Angeles Unified School District

Lisa Soll, BS
  San Antonio Independent School District

Content Reviewers

Tamara Kirshtein, MAT, Charleston, South Carolina
Sonja Oliveri, MA, Chicago, Illinois
Surey Rios, BS, Miami, Florida
Ann Shioji, MA, Long Beach, California
Gary Yoham, EdD, Miami, Florida

ELL Consultants

Mary Smith, MA, Merced, California
Brian Silva, Long Beach, California

Laboratory Reviewer

Garrett Hall, Pleasant Hill, Iowa

Laboratory Safety Consultant

Jeff Vogt, MED, West Virginia University at Parkersburg

About the Cover

Photo credits are on pages 499–500.

www.WrightGroup.com

Wright Group

Send all inquiries to:
Wright Group/McGraw-Hill
P.O. Box 812960
Chicago, IL 60681

ISBN 978-0-07-704536-4
MHID 0-07-704536-X

1 2 3 4 5 6 7 8 9 YAK 13 12 11 10 09 08 07

Contents

How to Use the Book

This book contains many features to help you learn science. Becoming familiar with these features will help you get more from this book.

Before You Read

Units: Units start with a list of the chapters and major science questions.

Chapters: Chapters start with a **Key Concept** that introduces the big scientific idea of the chapter. Introductions and **Think About** link the science concepts of the chapter to everyday life.

Lessons: Lessons start with **Learning Goals** and **Vocabulary** to give you an idea of the key points of the lesson.

As You Read

SciLinks: Codes let you use the Internet to learn more about the chapter topic.

Reading Activities: Before You Read asks you to answer questions, fill out charts, and make predictions about the lesson. **As You Read** keeps you on the right track while you are in the middle of the lesson. **After You Read** brings together what you learned throughout the lesson and asks questions about the lesson.

Science Activities: Explore It! activities offer small, quick, and fun experiments to do at your desk or at home. **Explain It!** activities give you an opportunity to explain scientific ideas in writing. **Extend It!** activities offer science topics that you can research to learn more about.

Special Features: Did You Know? boxes are full of interesting facts relating to the lesson. **People in Science** tells the stories of some key figures in science. **Science Connection** tells how the lesson topics relate to other subjects or other branches of science. **Figure It Out** questions will help you understand the pictures and their captions.

After You Read

Chapter Summaries: Each chapter ends with a summary of key points, vocabulary, and a set of questions that will help you prepare for the chapter test.

Beyond the Book: The **Student Workbook** reviews the key science concepts and vocabulary terms related to each chapter. The **Laboratory Manual** is designed to give you a hands-on science experience. The **Student CD-ROM** lets you explore the text and the science concepts in greater detail.

Flip through the book and examine all of its features. Get a good feel for how the features work, and have fun learning science!

Unit 1

Learning About Living Things

Studying Life Science

KEY CONCEPT Science is a process of observation and investigation through which information about the natural world is learned.

Did you know that communities teeming with living things can be found in warm, shallow waters around the world? This home to many creatures is known as a coral reef. Do you wonder what a coral reef is? Are you interested in knowing if a reef is alive, if it can grow, or what living things can be found in it? If you want to ask any of these questions, you are acting like a scientist. In other words, scientists ask questions about the world and try to find answers to those questions.

Think About Acting Like a Scientist

You probably ask many questions in the course of a day. Each time you learn the answer to a question, you acquire new information. Some answers may even lead you to ask more questions. Think about an animal or plant you find interesting.

- In your Science Notebook, draw a picture of the animal or plant. Then write three questions you have about it.

- Write one or two sentences explaining how you can find answers to each of your questions.

www.scilinks.org
Scientific Method
Code: WGB01

Before You Read

Make predictions about what you think is the definition of science. Preview the Learning Goals and the headings in this lesson. Write at least five descriptive words and examples in your Science Notebook to describe your predictions.

There is a sunflower plant growing in your garden. Unfortunately, the plant does not look healthy. You try giving it more water. Then you try giving it less water. Perhaps you place it in direct sunlight, or you move it into the shade. After each action, you study the plant's condition and decide whether your action has helped. It may surprise you to know that when you follow this type of organized process of learning about the plant, you are acting like a scientist.

The Nature of Science

What comes to mind when you think about science? Perhaps you imagine a set of books packed with information. Maybe you think of a laboratory filled with colored liquids in a variety of containers. Although these images are related to science, they do not completely describe the true nature of science. **Science** is an organized method of using evidence to propose explanations for events in the natural world. Science is both the process of gaining knowledge and the resulting body of knowledge.

Explain It!

Scientists from many different fields usually work together. Choose an example of a scientific task, such as sending astronauts into space or forming images of structures inside the body. In your Science Notebook, explain how the input of different scientists would be needed to complete the task.

Figure 1.1 Trying to learn more about living things, such as these sunflowers, is one goal of science.

Branches of Science

Imagine being a scientist. Are you wearing a lab coat and mixing substances together? Are you looking at images of the newest spot on Jupiter? Perhaps you are scuba diving to see animals that live in the ocean. Scientists do all of these things and many more. The specific description of a scientist's work depends on the topics that scientist investigates. The topics of scientific study are divided into three main branches: life science, Earth and space science, and physical science.

Life Science   Scientists who study life science work with living things. This might include the more obvious life forms, such as animals and plants, or the living things that are so small they cannot be seen with the unaided eye. In addition, life scientists investigate how living things interact with one another and obtain the things they need from their environments.

Earth and Space Science   Scientists who study Earth and space science examine the characteristics of planet Earth. They study Earth's water, land, and air, as well as weather events and natural disasters. Earth and space scientists also consider the process through which Earth developed, how it changes, and where it fits into the universe.

Physical Science   Scientists who study physical science investigate topics such as motion, forces, matter, energy, sound, light, electricity, and magnetism. Physical scientists also study the interactions that occur between different types of particles and different types of matter. Even roller coaster designers need to know a lot about physical science to do their job well.

Figure 1.2 Jane Goodall, who studies chimpanzees, is a life scientist.

As You Read

In your Science Notebook, make a chart with three columns. Label the columns *Life Science, Earth and Space Science*, and *Physical Science*. What are the general topics studied in each branch? What are two specific topics that are studied in each branch? Write your answers in the chart.

Figure 1.3 The weather on Earth and the flow of water over a mountain are some of the topics studied by Earth and space scientists.

Figure 1.4 The way in which objects move and the properties of light are two of the many topics studied by physical scientists.

Figure It Out

1. Which type of life scientist would most likely conduct an experiment to find out how to produce tomato plants that yield a greater number of tomatoes?
2. How might the work of a plant taxonomist and an ecologist be related?

Major Fields of Life Science

Branch of Life Science	The Study of	Branch of Life Science	The Study of
botany	plants	taxonomy	how living things can be organized
zoology	animals	ecology	how living things interact with one another and with their environments
genetics	how traits are passed from one generation to the next	microbiology	living things too small to be seen with the unaided eye
anatomy	the structures of living things	medicine	diagnosing, treating, and preventing disease

Figure 1.5 Life scientists specialize in certain fields.

Branches of Life Science

Each branch of science is further divided into more specific groups. This textbook explores the world of life science. Not all life scientists study the same things. **Figure 1.5** describes some of the major fields of life science. These fields are discussed throughout this textbook.

Did You Know?

The names of many fields of science end with the suffix *-ology*. When this suffix is added to a word, it means "the study of." The name *zoology* means "the study of animals."

Who Studies Science?

Perhaps you are thinking that science is only for scientists in laboratories. How wrong you are! Many people working in other fields study science or use the work of scientists.

For example, people who work in government need an understanding of many science topics in order to make useful decisions. They must have knowledge of how hurricanes and tornadoes behave in order to prepare safety and response plans. They should understand how living things interact with their environments in order to make decisions about water conservation. Almost every job imaginable requires some understanding of scientific topics.

After You Read

1. Using the notes in your Science Notebook, describe how science can be both an action and a thing.
2. Explain which type of scientist is most likely to study weather patterns.
3. How might the work of a botanist be related to some of the foods you eat?
4. The work of a microbiologist differs from that of a zoologist. Predict how the two scientists might work on different research that is somehow related.

1.2 How Is Science Studied?

Before You Read

Create a sequence chart in your Science Notebook. Imagine your teacher giving you a topic to research. You are not familiar with the topic, but your teacher would like you to prepare a report on it. In your sequence chart, write the steps that you might follow to learn about the topic.

The process scientists use to attempt to answer questions about the natural world is generally known as the **scientific method**. Although there is no single series of steps that is always followed in the scientific method, all scientific investigations use an orderly approach.

Making Observations

The process of science often begins with an observation. An **observation** is information you gather by using your senses. Your senses involve seeing, hearing, smelling, tasting, and touching. Maybe you *see* that the leaves of a tree have turned orange. It could be that you *hear* a frog in a swampy forest. Perhaps you *smell* the fragrance of a flower. Can you think of other examples of observations that use your senses?

Asking Questions

Suppose you step outside one morning and smell smoke in the air. The first thing you might do is ask yourself what is burning. The scientific process happens in a similar way. A scientist asks a question about an observation. A scientific question is one that can be studied through observation, testing, and analysis.

Figure 1.6 A scientist studying whales can make observations by looking at the size, shape, and color of whales. The scientist might also listen to the sounds whales make to communicate.

As You Read

Review the sequence chart you created in your Science Notebook. Make changes or additions so that your chart describes the steps of the scientific method.

How does the hypothesis relate the dependent variable to the independent variable?

Developing a Hypothesis

When wondering about the source of the smoke, you guess that it might be a brush fire in a nearby forest. This is your hypothesis. A **hypothesis** (hi PAH thuh sus, plural: hypotheses) is a proposed explanation for an observation. A hypothesis is not a random guess. It must be based on observations, previous knowledge, and research.

A good hypothesis is proposed in a way that it can be tested to find out if it is true. Some hypotheses are tested by making more observations. Others are tested with experiments. An **experiment** is an investigation in which information is collected under controlled conditions.

Designing an Experiment

All experiments involve **variables**, which are factors that can vary, or change. For example, suppose you want to conduct an experiment to find out if the temperature of turtle eggs affects whether the hatching turtles are male or female. Variables include the amount of moisture in the eggs' environment and the temperature of the eggs. If you allow both of the variables to change throughout the experiment, you will not be able to determine which variable affected the resulting turtles.

Instead, you must do a **controlled experiment**, or an experiment in which only one variable at a time is allowed to change. The variable that changes during the experiment is called the **independent variable**. In the turtle experiment, temperature is the independent variable. Something that is independent does not rely on other factors in order to vary.

The variable that is observed to find out if it also changes as a result of a changing independent variable is called the **dependent variable**. Something that is dependent relies on another factor. The gender of the hatching turtles is the dependent variable. All other variables must remain unchanged.

Turtle Eggs Exposed to Different Temperatures

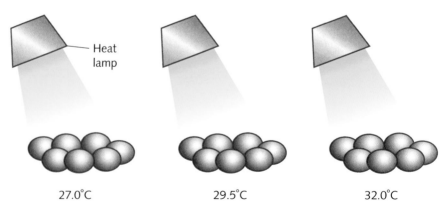

Heat lamp

27.0°C   29.5°C   32.0°C

Figure 1.7 Every variable must remain unchanged, or constant, except for the independent variable. What is the independent variable in this experiment?

Control Group Some experiments have two groups. In the **control group**, all the variables are kept the same. In the **experimental group**, the independent variable is allowed to change or made to change. The control group is used for comparison. The control group shows what would have happened if nothing were to change.

Not every experiment can be controlled, however. Imagine a scientist trying to find out how changes to wetlands affect a population of local birds. The scientist would need to observe the birds in their natural habitat without disturbing them. Although this type of research must be conducted in the field, it still requires the same methods of questioning and observation as an experiment conducted in a laboratory.

Explore It!

Design an experiment to find out how the amount of water a plant receives affects the number of tomatoes it produces. Propose a hypothesis. Identify the independent and dependent variables. Describe the control and experimental groups. Share your experiment with a partner.

PEOPLE IN SCIENCE Ignaz Philipp Semmelweis 1818–1865

It may seem obvious to you that washing your hands is a good way to get rid of germs, but that wasn't always obvious. Long before people knew that germs existed, a doctor named Ignaz Philipp Semmelweis, working at Vienna General Hospital in Austria, made an important observation. He noticed that a large number of women were dying of a condition known as puerperal (pyhew ER pe ral) fever shortly after delivering their babies. Doctors of the time thought the illness could not be prevented. However, Semmelweis noticed something very interesting. The hospital had two identical clinics for delivering babies. In one clinic, roughly 13 percent of new mothers died from puerperal fever. In the other clinic, the number of deaths was only about 2 percent.

Semmelweis recognized that the only difference between the clinics was the people who worked in them. The clinic with the higher number of deaths was staffed by medical students. The other was staffed by people called midwives, who were trained to deliver babies. Semmelweis asked the question, "Is there something the medical students do that causes more women to develop the disease?"

At about the same time, a friend of Semmelweis's performed an examination of the body of a person who had died. During the procedure, the friend accidentally cut his finger. Shortly after, he died of symptoms similar to those of women who died of puerperal fever. Semmelweis hypothesized that the medical students were carrying some type of particles from the dead bodies they were studying to the women in the hospital.

To test his hypothesis, Semmelweis began requiring doctors to wash their hands in a cleaning solution before delivering babies. The death rate in the medical students' clinic immediately dropped. Semmelweis concluded that some agent was indeed causing disease.

It took a while for Semmelweis's conclusions to be accepted by other doctors. His work, however, not only saved many lives, it also laid the groundwork for later scientists to discover the microscopic germs that cause disease.

Safety   An important aspect of every scientific experiment is working safely. Scientists and students alike must take precautions to protect themselves from possible dangers. In the experiments you will be doing as you study life science, you will see **safety symbols** that warn you about possible dangers. The chart in **Figure 1.8** describes what you should do when you see some of the major safety symbols. Appendix A contains more information about laboratory safety.

Most important, always read the instructions for every experiment in advance. Make sure you understand what you need to do. Tell your teacher if you have questions or if there is an accident in the laboratory.

Lab Safety Symbols

	This symbol appears when a danger exists for cuts or punctures caused by the use of sharp objects.
	This symbol appears when substances used could stain or burn clothing.
	This symbol appears when a danger to the eyes exists. Safety goggles should be worn when this symbol appears.
	This symbol appears when chemicals used can cause burns or are poisonous if absorbed through the skin.

Figure 1.8 These and other safety symbols alert you to follow safety procedures when doing a laboratory experiment.

Collecting and Analyzing Data

The information obtained through observation is called **data** (singular: datum). Some experiments produce huge amounts of data. To make sense of all the information, scientists organize data into forms that are easier to read and analyze. One way to organize data is in a table. **Figure 1.9** shows a possible data table resulting from the turtle experiment.

Figure It Out

1. What are the two variables represented in this table?

2. What trend do the data show?

Turtle Egg Data

Temperature of Eggs	Percentage of Males	Percentage of Females
27.0°C	68.4	31.6
29.5°C	47.8	52.2
32.0°C	29.3	70.7

Figure 1.9 The data for this experiment consist of numbers that can be organized in a table. A table has rows and columns with headings that describe the information.

Another way to organize data is to create a graph. Line graphs, such as the one in **Figure 1.10a**, are best for data that change continuously, such as the size of a population of bacteria. Bar graphs, such as the one in **Figure 1.10b**, are useful for comparing data, such as the data from the turtle experiment. Circle graphs, such as the one in **Figure 1.10c**, are best for analyzing data that are divided into parts of a whole, such as the elements in the human body.

Drawing Conclusions

Once the experiment has been conducted and the data have been collected, a scientist tries to determine if the hypothesis is supported. A **conclusion** is a statement that uses evidence from an experiment to indicate whether or not the hypothesis is supported.

A conclusion is not necessarily the end of the investigation. If the hypothesis is not supported, the scientist might develop a new hypothesis and design a new experiment to test it. If the hypothesis is supported, the scientist needs to repeat the experiment many times to make sure that the conclusion is valid.

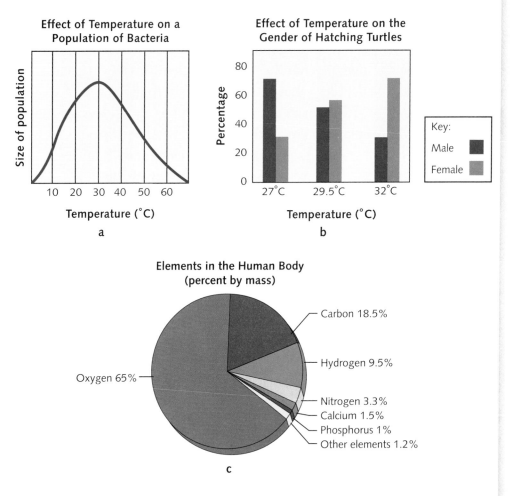

Figure 1.10 Data can be presented in a variety of formats, including **a)** a line graph, **b)** a bar graph, and **c)** a circle graph.

Communicating the Results

A scientist also must share his or her experimental procedures and results with the scientific community. In this way, other scientists can repeat the experiment to confirm the results. Throughout an experiment, the scientist must keep careful records that describe everything about the research. These records should include not only the data but also information about the experimental design, possible sources of error, unexpected results, and any remaining questions.

Scientists share their results with one another by communicating through written reports and journal articles, as well as through oral presentations. Sharing information in this way not only adds to the body of scientific knowledge, it also gives other scientists an opportunity to repeat the experiment and confirm its results. The results of an investigation can only be considered valid if they are continually achieved during repeated trials of the same procedure.

Scientific Theories

If a hypothesis is supported by many separate investigations over a long period of time, it may lead to the development of, or become part of, a theory. A **theory** is an explanation for a broad range of observations that is supported by a body of scientific evidence. A theory is not the same as a scientific law. A scientific law is a rule that describes an observed pattern in nature.

Scientists use theories to make predictions about new situations. A **prediction** is a statement that suggests what a person thinks will happen in the future based on past experience and evidence.

A theory is not considered to be absolute truth, but it is supported by a vast amount of scientific evidence. Scientists continually analyze the strengths and weaknesses of a theory. If new evidence is discovered, a theory can be revised or replaced.

Figure 1.11 One theory you will learn about later in this textbook suggests how some living things change over time and how others, such as saber-toothed cats, disappear completely. The skull of a saber-toothed cat is shown above.

After You Read

1. How is a hypothesis related to an observation?

2. How might a hypothesis become part of a theory?

3. What is the difference between the independent and dependent variables in a controlled experiment?

4. You see a symbol of an apron in a laboratory procedure. What does this symbol tell you about the experiment you will be conducting?

5. Look at the sequence chart you completed. In a well-developed paragraph, explain how the last step might lead back to the first step.

1.3 Measurements in Science

Before You Read

Look at the headings in the lesson to find the measurements discussed. In your Science Notebook, create a chart with three columns. Label one column *Quantity*, the next column *Units*, and the third column *Examples*. Include a row for each measurement you find.

It's a fly ball to center field. You run to catch the baseball. You've got it! Now all you have to do is throw it to home plate to catch the runner. Can you throw the ball that far? Can you get it there in time? How far and how fast you throw the ball are examples of measurements. A **measurement** is a quantity, dimension, or amount. Scientists rely on measurements to gather and analyze data.

Units of Measurement

Would you say that it takes you 15 to get to school? Probably not. You would probably say that it takes you 15 minutes to get to school. That description would help another person know how long it takes for you to get from your home to school. Minutes are an example of units. A unit is a precise quantity that does not change. All measurements must be described with the appropriate unit.

During the 1700s, people used all sorts of different units to describe measurements. To describe the length of an object, a person could use any one of more than ten different units. In addition, because many of the units were based on body parts such as feet and hands, the size of a unit could vary from one person to another. The variety of units not only made it confusing to talk about measurements, it also made it difficult to buy and sell goods.

To put an end to the confusion, the French government asked the Academy of Sciences to develop a standard system of measurement. The Academy produced a system, known as the metric system, based on the number 10. This makes it very simple to change a unit of one magnitude to a unit of a different magnitude by dividing or multiplying by 10. You can learn more about the metric system in Appendix B.

In 1960, some elements of the metric system became part of the **International System of Units (SI)**. This system of measurement is used by most countries throughout the world and in almost all scientific activities.

Learning Goals

- Identify the SI units of several scientific measurements.
- Recognize how prefixes indicate the magnitude of a unit.
- Describe how common tools of measurement are used.

New Vocabulary

measurement
International System of Units (SI)
mass
volume
meniscus
derived unit
density

Figure 1.12 The distance from the outfield to home plate is one measurement on a baseball field. What are some other measurements related to a baseball game?

Length

When you describe how far you throw a ball, how long a table is, or how tall a building is, you are describing length. The base SI unit of length is the meter (m). A baseball bat is about one meter long. An average doorknob is about one meter above the floor.

A meter is divided into smaller units called centimeters (cm) and millimeters (mm). The prefix *centi-* means "one-hundredth." There are 100 centimeters in a meter. The prefix *milli-* means "one-thousandth." There are 1,000 millimeters in a meter. A millimeter is about the thickness of one dime.

For longer lengths, scientists use a unit called a kilometer (km). The prefix *kilo-* means "one thousand." There are 1,000 meters in a kilometer. The distance from San Diego, California, to New York, New York, is 3,909 km. The tallest mountain in the world, Mt. Everest, has a height of almost 9 km.

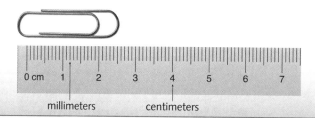

Figure 1.14 Length can be measured with a metric ruler. Line up one end of an object with the first mark on the ruler. Find the line closest to the other end of the object. The longer lines show centimeters, and the shorter lines show millimeters.

Mass

The amount of matter in an object is that object's **mass**. The base unit of mass in the SI system is the kilogram (kg). Recall that the prefix *kilo-* means "one thousand." A kilogram consists of one thousand smaller units called grams (g). The mass of a paper clip is about one gram. Even smaller masses are measured in milligrams (mg). There are 1,000 milligrams in one gram. A tool called a balance is used to measure mass.

CONNECTION: Math

All number systems have a base number. The number system you use every day has a base of 10. This makes it convenient to multiply or divide by multiples of 10. To multiply a number by a multiple of 10, move the decimal point to the right the same number of places as there are zeroes. For example, to multiply a number by 100, move the decimal point two places to the right. You can follow a similar process to divide by multiples of 10. However, move the decimal point to the left instead.

Metric Prefixes

Prefix	Meaning
kilo-	1,000
hecto-	100
deka-	10
base unit	1
deci-	0.1 (one-tenth)
centi-	0.01 (one-hundredth)
milli-	0.001 (one-thousandth)

Figure 1.13 You can use the prefix of an SI unit to figure out how large or small a measurement is. How many millimeters are in one kilometer?

Extend It!

The United States is the only major industrialized nation in the world that does not use the metric system. Work in a small group to research the history of the metric system. Prepare a poster or time line showing the major developments in the system and the laws governing it. Then separate into teams and debate the topic of whether the United States should adopt this system.

Volume

Have you ever tried to pour all the liquid from a large container into a smaller container? If so, you probably made a mess! That's because the liquid takes up a certain amount of space. If the smaller container does not have enough space, the liquid spills over the sides.

The amount of space an object takes up is its **volume**. How volume is measured depends on the sample being measured. The sample being measured also determines the units used for the measurement. The three basic types of samples are liquids, regular solids, and irregular solids.

Finding the Volume of a Liquid The volume of a liquid, such as water, is measured in a unit called a liter (L). A liter is not an SI unit, but because it is important and widely used, it is accepted for use with the SI. You may have seen one-liter and two-liter bottles of drinks at the grocery store. Smaller volumes can be measured in milliliters. Recall that the prefix *milli-* means "one-thousandth." There are 1,000 milliliters in a liter.

One way to measure the volume of a liquid is by pouring the liquid into a container that has measured markings on it. Perhaps you have poured milk into a measuring cup to find a specific amount for a recipe. Scientists often use a graduated cylinder to measure liquid volumes.

Figure 1.15 shows a liquid being measured in a graduated cylinder marked in one-milliliter intervals. You can see that the top surface of the liquid is slightly curved. The curve is known as the **meniscus** (meh NIHS cus). To find the volume of the liquid, you must read the measurement of the marking at the bottom of the meniscus.

Finding the Volume of a Regular Solid A regular solid is an object that has matching shapes for sides. A number cube and a shoebox are examples of regular solids. You can find the volume of a regular solid by multiplying its length by its width by its height.

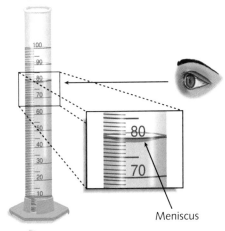

Meniscus

Figure 1.15 According to the graduated cylinder, what is the volume of this liquid?

$$\text{volume} = \text{length} \times \text{width} \times \text{height}$$

If the measurements are made in centimeters, the unit of volume is cm³ (cubic centimeters: cm × cm × cm). If the measurements are made in meters, the unit of volume is m³ (cubic meters: m × m × m). A unit that consists of more than one base unit is called a **derived unit**. Volume is a derived unit.

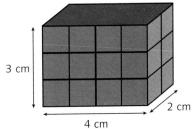

3 cm
4 cm
2 cm

Figure 1.16 The volume of this solid is 4 cm × 2 cm × 3 cm = 24 cm³.

As You Read

In your chart of measurements, fill in the second column with the units in which each measurement can be described. Provide at least two examples of each measurement in the third column.

🔍 Explore It!

Find three regular solids in your home or classroom. Examples include a book, an eraser, and a tissue box. Work with a partner to find the volume of each solid.

Finding the Volume of an Irregular Solid   Not all solids have regular shapes. A stone or a shell, for example, has an irregular shape. One method for finding the volume of an irregular solid is to place the object in water. Once you place the object in water, the level of the water will rise. If you subtract the original water level from the new water level, you will find the volume of the solid.

Figure 1.17 The water rises from the 25-mL mark to the 31-mL mark after the stone is placed in it. Therefore, the volume of the stone is 6 mL.

Density

If you place a table tennis ball in water, it will float. If you place a golf ball in water, it will sink. Even though the balls have about the same volume, they have different masses. The amount of mass in a given volume is an object's **density**. Density can be affected by temperature. For example, the density of water is different at room temperature than at 32°F. This is because the particles are farther apart at higher temperatures and closer together at lower temperatures. As a result, the same amount of mass can take up more or less space depending on its temperature.

You can find the density of an object by dividing its mass by its volume.

> **density = mass/volume**

Density is a derived unit. The units of density are made up of a unit of mass and a unit of volume. If mass is measured in grams and volume is measured in cubic centimeters, the unit of density is grams per cubic centimeter (g/cm^3). If volume is measured in milliliters, the unit of density is grams per milliliter (g/mL).

Sample Densities	
Substance	**Density (g/cm³)**
air	0.001293
aluminum	2.7
gold	19.3
mercury	13.6
water	1.00

Figure 1.18 This table lists the densities of several common substances at room temperature.

Temperature

Before going outside, many people check the air temperature so they know what to wear. Most people think temperature is a measure of heat. To a scientist, temperature is the measure of the average kinetic energy of the particles in an object. Temperature can be measured on different scales. One temperature scale commonly used in science is the Celsius scale. On this scale, temperature is measured in degrees Celsius. Water freezes at 0°C and boils at 100°C.

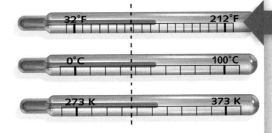

Figure 1.19 This diagram compares the Fahrenheit, Celsius, and Kelvin scales.

The SI base unit of temperature is the kelvin. Units on the Kelvin scale are the same size as those on the Celsius scale. However, each value on the Kelvin scale is 273 degrees more than on the Celsius scale.

The scale with which you may be most familiar is the Fahrenheit scale. On this scale, water freezes at 32°F and boils at 212°F. Fahrenheit measurements are generally not used in science.

A thermometer is used to measure temperature. To use a thermometer, place the device in the substance being measured. Allow time for the liquid in the thermometer to move up or down. Then read the number next to the top of the liquid. Some thermometers are digital and display a number once the temperature of the substance is recorded. Scientists often use probes that automatically make temperature measurements and create a table or graph of the data when attached to a computer.

Time

Time describes how long an event takes to occur. The second (s) is the SI base unit of time. Short periods of time are measured in parts of one second, such as milliseconds (ms). There are 1,000 milliseconds in a second. Longer periods of time are measured in multiples of a second, such as minutes and hours. There are 60 seconds in one minute and 60 minutes in one hour. Time is measured using such instruments as clocks, watches, and stopwatches.

CONNECTION:
Social Studies

Before clocks and watches were invented, people used the Sun to tell time. A sundial is a device that forms a shadow of a stick. As Earth rotates on its axis, the shadow moves and changes in length. By designing a sundial so that adjustments can be made for the location on Earth and the time of year, a person can keep an accurate record of time.

After You Read

1. Which type of measurement can be made using a graduated cylinder? Describe how to use this tool.

2. How many milliseconds are in five seconds?

3. Review your chart of measurements. Which units are base units in the SI system? Which units are derived units? Explain your answers.

Summary

KEY CONCEPTS

1.1 What Is Science?

- Science is a process of observation and investigation through which information about the natural world is learned.
- The branches of science are life science, Earth and space science, and physical science.
- Life scientists focus on different fields such as botany, zoology, ecology, and microbiology.

1.2 How Is Science Studied?

- The scientific method is an organized process of obtaining evidence to learn about the natural world.
- Scientists make observations, ask questions, develop hypotheses, conduct experiments, analyze data, and draw conclusions.
- The steps of the scientific method are not always followed in the same order or in the same way.
- A hypothesis is a proposed answer to a scientific question.

1.3 Measurements in Science

- SI base units include the meter for length, the kilogram for mass, the kelvin for temperature, and the second for time. The liter, for volume, is a non-SI unit.
- Derived units are made up of two or more base units. Volume is a derived unit. Density is a derived unit made up of a unit of mass divided by a unit of volume.

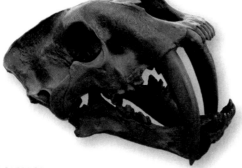

VOCABULARY REVIEW

Write each term in a complete sentence, or write a paragraph relating several terms.

1.1
science, p. 4

1.2
scientific method, p. 7
observation, p. 7
hypothesis, p. 8
experiment, p. 8
variable, p. 8
controlled experiment, p. 8
independent variable, p. 8
dependent variable, p. 8
control group, p. 9
experimental group, p. 9
safety symbol, p. 10
data, p. 10
conclusion, p. 11
theory, p. 12
prediction, p. 12

1.3
measurement, p. 13
International System of Units (SI), p. 13
mass, p. 14
volume, p. 15
meniscus, p. 15
derived unit, p. 15
density, p. 16

PREPARE FOR CHAPTER TEST

To prepare for the chapter test, create a question from each Learning Goal. Use the information in your Science Notebook to answer each question. Then use these answers to write a well-developed essay about the chapter. Use the Key Concept on the first page of this chapter as your topic sentence.

True or False
If the statement is true, write "true." If it is false, change the underlined word or words to make the statement true.

1. The amount of salt in ocean water is a topic of study in <u>Earth and space</u> science.

2. A <u>conclusion</u> is an educated guess that attempts to explain an observation.

3. The variable that a scientist changes during an experiment is called the <u>dependent</u> variable.

4. The facts, figures, and other evidence obtained during an experiment make up the <u>hypothesis</u>.

5. A graduated cylinder is used to measure the <u>length</u> of an object.

6. The <u>mass</u> of an object is measured with a balance.

Short Answer
Answer each of the following in a sentence or brief paragraph.

7. Describe how the three branches of science are alike. Describe how they are different.

8. Explain why a hypothesis needs to be testable.

9. What are the two types of variables in a controlled experiment? Distinguish between them.

10. Discuss why using a common system of measurement is useful to scientists.

11. A student pours 14.2 mL of water into a graduated cylinder. She then places a stone in the cylinder. The meniscus of the water rises to 18.8 mL. Calculate the volume of the stone.

Critical Thinking
Use what you have learned in this chapter to answer each of the following.

12. **Infer** Plants take in carbon dioxide and release oxygen during photosynthesis. A researcher places ten plants in separate containers. Nine of the plants receive different levels of carbon dioxide. After a period of time, the researcher measures the amount of oxygen produced by each plant. Propose a possible hypothesis for this experiment.

13. **Compare and Contrast** Compare the figures. Which object has a greater volume?

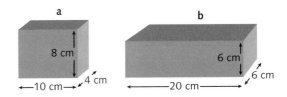

14. **Sequence** List and describe the steps you could follow to find the density of a marble.

Standardized Test Question
Choose the letter of the response that correctly answers the question.

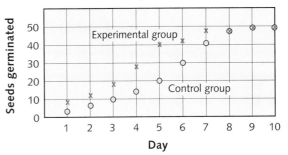

The Effect of Temperature on Germination

15. A team of scientists measured the number of seeds that germinated (grew into plants). A control group was kept at 18°C and an experimental group at 25°C. The graph shows their data. According to the data, what is the independent variable in the experiment?

   A. the number of seeds they studied

   B. the number of seeds that germinated

   C. the temperature of the seeds

   D. the type of seeds they used

Test-Taking Tip

Quickly look over the entire test so you know how to use your time wisely. Allow more time for sections of the test that look the most difficult.

Chapter

2

The Nature of Life

KEY CONCEPT Living things share the same characteristics and have similar needs.

High on a mountaintop, water collects and begins to flow downhill. Animals from all around come to drink from the brook and find food. Along the brook, a variety of trees and other plants grow.

Although the water flows from one place to another, grows in size, and changes shape, it is not alive. What is the difference between living and nonliving things? Scientists use specific characteristics to define the exact nature of life.

Think About Living and Nonliving Things

You may take for granted that you know the difference between living and nonliving things. Think about the characteristics you use when deciding if something is living or nonliving.

- In your Science Notebook, make a T-chart. Label one column *Living* and the other *Nonliving*. List at least ten things from your daily life in each column.

- At the bottom of each column, write some of the characteristics of the items in the column. For example, do all of the living things breathe or eat? Review your choices as you read the chapter.

www.scilinks.org
Characteristics of Living Things
Code: WGB02

2.1 Characteristics of Living Things

Learning Goals

- Identify the characteristics of living things.
- Classify items as living or nonliving.

New Vocabulary

organism
cell
stimulus
response
homeostasis
adaptation
reproduction
species

Before You Read

Create a concept map in your Science Notebook. Write and circle the title *Living Things*. Draw five circles surrounding the title. Draw lines connecting each circle to the title. Without writing in the circles, predict some of the characteristics that you will use to fill them in as you read the lesson.

Polar bears search for food in the icy waters near the North Pole. Toucans fly among the branches of the rain forests of Brazil. Outside your window, a grasshopper might munch on a leaf. Wherever you look, you will find that Earth's environments are filled with living things. Another name for a living thing is an **organism**.

Many organisms, such as those in **Figure 2.1**, are easy to recognize as living. But often, deciding whether something is living or nonliving is not that simple. For example, the Spanish moss hanging from the tree in **Figure 2.2** does not do many of the things an animal does. The moss, however, is very much alive.

Scientists use several characteristics to describe something as living. Many nonliving things can display *some* of these characteristics. However, living things display *all* of these characteristics.

Figure 2.2 Spanish moss is an air plant that grows from tree branches in warm, humid climates. It does not have roots like other plants. Instead, it absorbs what it needs from the air.

Figure 2.1 Bears, birds, and insects are examples of living things. What are some of the characteristics of these organisms?

Read the headings listing the characteristics of living things. Write the headings in the circles of your concept map.

How do these characteristics compare with your predictions?

Living Things Are Made Up of Cells

What does a giant elephant have in common with a tiny mouse? They are both made up of smaller units called cells. A **cell** is a structure that contains all of the materials needed for life. Each cell is separated from its surroundings by a barrier called a cell membrane. A cell is the smallest unit of an organism that can be considered alive. All living things are made up of cells.

Figure 2.3 All living things, large and small, are made up of cells. Larger organisms, such as elephants, have more cells than do smaller organisms, such as field mice.

Some organisms are made up of only one cell. These organisms, called unicellular organisms, carry out all of their functions in just one cell. Multicellular organisms are made up of more than one cell. Most of the organisms with which you are familiar are multicellular. Some of these organisms are made up of millions or even trillions of cells. In many of these organisms, different types of cells perform specific functions. Groups of cells work together to keep an organism alive.

Figure It Out

1. What tool do scientists use to see cells?

2. An adult human is estimated to have one hundred trillion cells. How is this number expressed?

Figure 2.4 This bird is made up of many cells. Some of these cells are blood cells. Together, the bird's cells carry out all the processes needed for the bird to survive.

Living Things Respond to Their Environment

Has a doctor ever shined a light into your eyes during a checkup? The doctor was checking to see if the dark center of each eye, the pupil, would change in size as a result of the light. A condition or event in your surroundings, such as the doctor's light, is called a **stimulus** (STIHM yuh lus, plural: stimuli). A change that happens because of a stimulus, such as a change in the size of your pupils, is known as a **response**. All living things can sense stimuli and respond to them.

Light is just one type of stimulus. Other stimuli include darkness, chemicals, sounds, tastes, and touch. You might think that only animals can respond to stimuli. This is not the case. Although plants cannot get up and move, they can change the way in which they grow in response to stimuli. Many plants can grow toward sunlight and water, for example. Others, such as the plant in **Figure 2.5**, bend to curl around objects that they touch.

Not all stimuli are external, or outside living things. Living things also respond to stimuli that occur inside them. They do so in order to regulate their internal surroundings and maintain conditions necessary for survival. The process by which living things respond to stimuli in ways that allow them to maintain and balance internal conditions necessary for life is called **homeostasis** (hoh mee oh STAY sus). Through homeostasis, living things maintain characteristic internal temperature ranges and the correct amounts of water and minerals in their cells. Homeostasis is a characteristic of all living things.

Living Things Can Adapt

Organisms depend on certain traits to help them survive. The hedgehog in **Figure 2.6** has stiff spines that protect it from predators, or animals that might eat it. Any trait that helps an organism survive in its environment is called an **adaptation**. Other adaptations help organisms attract mates, obtain food, build homes, and survive in harsh weather conditions.

An adaptation such as the hedgehog's stiff spines does not develop during an organism's life. Instead, adaptations develop over many generations as conditions in the environment change. In order to survive, living things must be able to adapt.

Living Things Reproduce

Living things make more organisms like themselves in a process called **reproduction**. A group of organisms that has similar physical characteristics and can mate to produce offspring capable of reproducing their own offspring is called a **species** (SPEE sheez). Lions, tigers, and panthers are different species of cats.

Figure 2.5 This climbing plant has parts called tendrils that coil when they touch solid objects.

Explore It!

Your body responds to physical activity by changing your heart rate. This is the number of times your heart beats each minute. Design and conduct an activity that compares the heart rate before and after an action such as jumping rope or climbing stairs.

Figure 2.6 When this hedgehog feels threatened, it rolls into a ball. This causes the spines to point outward. How does this adaptation help the hedgehog survive?

Reproduction is essential to the continuation of a species. This means that while not every single organism has to reproduce, members of a species must be able to reproduce so that the species continues to exist on Earth. For example, not every single sea turtle will produce offspring. However, many sea turtles must reproduce or the species will eventually become extinct, or die out.

Living Things Grow and Develop

All living things grow and develop at some point in their lives. For unicellular organisms, growth usually means that the cell becomes larger. Multicellular organisms go through a more complex process of development. They each begin as a single cell that divides into two cells. Each new cell then divides again. This process continues over and over again to produce the many cells of the organism.

In addition, some organisms go through periods of dramatic change. **Figure 2.8** shows the development of a frog. Even humans experience periods of rapid growth at different points in their lives.

Figure 2.7 A female sea turtle lays eggs in a hole she digs on the beach. When the young turtles hatch, they dig their way out and then use clues from the environment to find their way back to the ocean. When they grow up, they will produce another generation of turtles so the species can survive.

Figure It Out

1. What is the first stage in a frog's life?

2. Describe how a frog's place in its environment changes as the frog grows and develops.

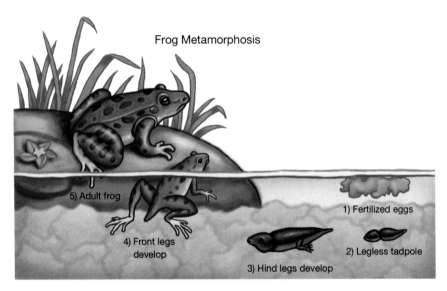

Frog Metamorphosis

5) Adult frog
4) Front legs develop
3) Hind legs develop
2) Legless tadpole
1) Fertilized eggs

Figure 2.8 An adult frog looks and acts very different from the way it does in its younger form.

After You Read

1. Review your concept map. Use it to list five characteristics of all living things.

2. What type of stimulus might make you feel hungry? What might your response be?

3. Why can an organism be considered living even if it never reproduces?

4. A raindrop grows and becomes larger. It is made up of smaller units, and it moves from a cloud to the ground. Why isn't a raindrop considered living, while a leaf that it falls on is?

The Needs of Living Things

Before You Read

In your Science Notebook, make a list of things you would like to have. Classify the items on your list as things you want and things you truly need. Label each item in the list accordingly. Write a few sentences describing the difference between a want and a need.

Learning Goals

- Examine the needs of living things.
- Recognize the importance of water to life on Earth.

New Vocabulary

energy
photosynthesis
autotroph
heterotroph
solvent
endotherm
ectotherm

You have decided to grow a tomato plant. Do you pour the seeds into an empty pot and put the pot in a closet? Not if you ever hope to see tomatoes. Instead, you must add soil to the pot, give the seeds water, and place the pot in sunlight.

Like the tomato plant, all living things have certain needs that must be met in order for them to survive. While the needs of every organism are not exactly the same, they fall into several general categories.

Energy

Living cells need energy for their life processes. **Energy** is the ability to cause change or do work. Many organisms obtain the energy they need from sunlight. Plants, some bacteria, and most algae get energy in this way. In a process called **photosynthesis** (foh toh SIHN thuh sus), these organisms change the energy of sunlight into energy that can be stored. At the same time, they take nonliving matter from the air, soil, and water. Plants store matter and energy in the substances that form their bodies. Organisms that capture matter and energy from their surroundings are called **autotrophs** (AW tuh trohfs). Autotrophs are also called producers.

When organisms that conduct photosynthesis need energy, they break down certain substances in their bodies to supply energy to their cells. If organisms that conduct photosynthesis, such as plants, are eaten as food, the stored energy is passed on. The animals in **Figure 2.10** obtain stored matter and energy, or food, when they eat grasses. Like plants, these animals use some of the energy and store the rest. Organisms that do not feed on plants get matter and energy by eating organisms that do eat plants. Organisms that get matter and energy by eating plants or other organisms are called **heterotrophs** (HE tuh roh trohfs). Heterotrophs are also called consumers.

Figure 2.9 Tomato plants have needs that must be met in order for them to grow and produce tomatoes. What are some of the needs of tomato plants?

Figure 2.10 Grazing animals, such as this gazelle and zebra, get energy by eating plants. This energy is passed along to animals that eat them, such as lions. What are organisms such as gazelles, zebras, and lions called? Why?

Figure 2.11 Roughly three-quarters of Earth's surface is covered by water. Life on Earth depends on liquid water.

Water

Did you know that humans can survive for weeks without food but only days without water? The reason is that water makes up about 60 percent of the human body. Water is just as important to most other living things.

Water is involved in almost all of the chemical reactions and other processes that take place in living things. Water is important because it acts as a solvent. A **solvent** (SAHL vunt) is a substance in which other substances dissolve, or break apart. Water is known as the universal solvent because it can dissolve so many substances.

Because water can dissolve other substances, it can carry gases, nutrients that come from food, and waste products throughout an organism. For example, water is part of the digestive fluids. These fluids help break down food into nutrients the body can use. As part of blood, water helps carry the nutrients to cells where they are needed. The cells release waste products into the blood. In this way, water helps carry away waste products so they can be removed from the body.

Water is also important in maintaining human body temperature. When a body becomes warm, it releases a fluid called perspiration. As perspiration evaporates from the skin, liquid water changes into a gas. In the process, heat is released from the skin. The result is that the body cools off.

Figure 2.12 The human body needs to remain within a certain temperature range. When the body becomes too warm, it perspires. This causes the body's temperature to fall back within the proper range. This is an example of homeostasis.

Earth is home to millions of different types of living things. Astrobiologists and exobiologists, or scientists who search for signs of life beyond Earth, subscribe to a basic principle: follow the water. This means that if liquid water is necessary for living things to exist, the only way to find life is to find water.

Astrobiologists describe a habitable zone as the region of space in which a planet could support life because it can have liquid water. Earth is within this habitable zone. So is Mars. To date, scientists have not found liquid water on Mars, but they have found evidence that liquid water existed in the recent past or still exists underground. Such evidence includes the presence of dry riverbeds, flood plains, and gullies.

Scientists are also searching for water in other places, such as the moons of planets. One of Jupiter's moons, Europa, shows signs that it may have a deep ocean of liquid water covered by ice.

Does the presence of liquid water prove that life exists elsewhere in the universe? No one knows for sure, but astrobiologists hope to find out in time.

Temperature

Organisms need to live within a certain temperature range. Some organisms, such as birds and kangaroos, maintain a constant internal temperature. When the environment is too cold for these organisms, they can change the Sun's energy stored in food into thermal energy. This type of energy keeps them warm. An organism that produces thermal energy to maintain its temperature is called an **endotherm**.

If these organisms become too warm, they can remove thermal energy from their bodies. Recall that water in perspiration helps maintain temperature in this way.

Perhaps you have seen a lizard basking on a rock. Lizards do this because they are ectotherms. An **ectotherm** does not maintain a constant temperature. Instead, it takes on the temperature of the environment. When ectotherms become cool, they might bask in the sunlight to absorb energy from it. If they become too warm, they move to a shady area, open their mouths, or even burrow into the soil to stay cool.

As You Read

In your Science Notebook, choose an organism and draw its outline. You might include a picture of the organism, if possible. Around the picture, draw and label the things the organism needs to obtain from its environment in order to survive.

Is this organism an autotroph or a heterotroph? Is it an endotherm or an ectotherm? How do you know?

Figure It Out

1. How can you tell that the spider takes on the temperature of its environment?

2. Which of the organisms would be classified as an ectotherm and which as an endotherm?

Figure 2.13 The red and orange colors of the bird show that it can stay warmer than its environment. The spider cannot.

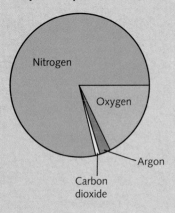

Air

You need to breathe in order to survive. Each time you breathe in, or inhale, you take in oxygen from the air. Your cells use oxygen to release the energy they need. When you breathe out, or exhale, you release carbon dioxide. Plants use carbon dioxide from the air to conduct photosynthesis. Recall that photosynthesis is the process through which plants store matter and the Sun's energy in their bodies. Because so many organisms depend on plants for food, it is important to all organisms that plants obtain the carbon dioxide they need for photosynthesis.

Even organisms that live in water use gases from the air. Fish, for example, absorb oxygen and release carbon dioxide. Unlike animals on land, however, they exchange gases directly with the water.

Space

All organisms need a place to live. The place must provide all of the things the organism needs to survive. In addition, it should provide the organism with protection from natural events, such as storms, and from other organisms.

For some organisms, such as the Florida panther in **Figure 2.14**, the place must be quite large. A male panther's place will cover an area of about 500 square kilometers. Meanwhile, the dust mite in **Figure 2.14** can fulfill all its needs on the tip of an eyelash.

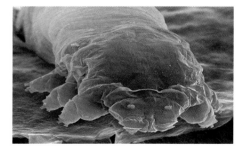

Figure 2.14 All living things need a place to live. This Florida panther *(left)* travels across a large space, whereas this mite *(right)* might never leave an eyelash.

After You Read

1. What is the original source of energy in a salad you eat?

2. What are three ways in which water is essential to living things?

3. How is a snake different from an ostrich in terms of body temperature?

4. Review your list of wants and needs. Based on what you learned about the needs of living things, would you still classify the same items as needs? How are your true needs basically the same as those of the tomato plant described at the beginning of the lesson?

2.3 Chemistry of Living Things

Before You Read

Look at the four main headings within this lesson. Create a chart in your Science Notebook with four columns. Write one heading in each column. Then write a sentence summarizing what you know about each one.

Everything around you is made up of matter, including living things. **Matter** is defined as anything that has mass and takes up space. Every type of matter is made up of smaller units called **atoms**. The atoms of each element, such as carbon, oxygen, and hydrogen, are different from every other type of atom.

Atoms bond, or join together, to form molecules. The chemical processes in living things involve organic molecules. Scientifically, the term **organic** refers to molecules that contain bonds between carbon and hydrogen atoms. There are four main classes of organic molecules: carbohydrates, proteins, lipids, and nucleic acids.

Carbohydrates

You may have heard that breads, pasta, and fruits contain carbohydrates. A **carbohydrate** (kar boh HI drayt) is a molecule made up of carbon, oxygen, and hydrogen. Carbohydrates provide energy to the cells of an organism.

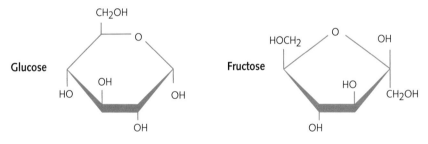

Figure 2.15 Two carbohydrates are glucose and fructose. They are made up of the same atoms arranged in different ways.

The simplest type of carbohydrate is called a monosaccharide (mah nuh SA kuh ride), or simple sugar. Glucose, fructose, and galactose are all simple sugars. Two monosaccharide molecules bonded together form a disaccharide. You may have sprinkled the disaccharide known as sucrose on your oatmeal. Sucrose, or table sugar, is formed when glucose and fructose link together.

CONNECTION: Health

The human body needs to obtain six kinds of nutrients from food: carbohydrates, fats, proteins, minerals, vitamins, and water. These nutrients are required in different amounts for proper health. The Food and Drug Administration has created a Food Pyramid to compare these amounts.

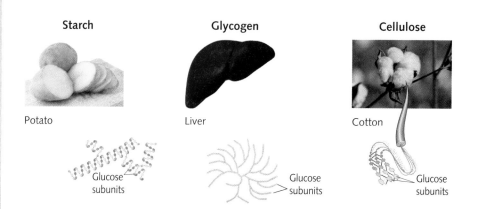

Starch

Potato

Glucose subunits

Glycogen

Liver

Glucose subunits

Cellulose

Cotton

Glucose subunits

Figure 2.16 Starch, found in structures such as potatoes, stores energy for plant cells. Glycogen stores energy in the livers of mammals, such as humans. Cellulose makes up the cell walls of plant cells and fibers such as cotton.

As You Read

In the chart you made in your Science Notebook, summarize the information you learn about each type of organic molecule. Correct any errors you might have included in the information you wrote before reading the lesson.

Which type of organic molecule serves as an energy source for cells?

Figure It Out

1. What are three foods that contain large amounts of lipids?

2. Determine how many carbon, hydrogen, and oxygen atoms are in a glycerol molecule.

Chains of monosaccharides form polysaccharides (pah lee SA kuh ridez). Starch, cellulose, and glycogen are polysaccharides made up of chains of glucose. Examples of where these carbohydrates can be found are shown in **Figure 2.16**.

Lipids

Many foods contain fats and oils. These are examples of **lipids**, which are molecules made up mostly of carbon and hydrogen. Lipids also contain small amounts of oxygen. Many lipids are made up of a glycerol molecule and fatty acids. A fatty acid is a long chain of carbon and hydrogen. Examples of food that contain lipids are shown in **Figure 2.17**.

Lipids are said to be **insoluble** (ihn SAHL yuh bul), which means that they do not dissolve, or break apart, in water. This is one reason why lipids are the main component of the membranes that surround living cells. Lipids are also important to cells because they store energy and provide insulation.

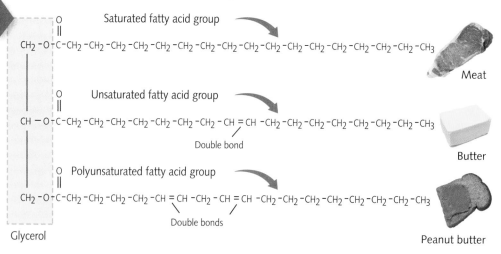

Figure 2.17 A lipid is saturated if each carbon atom in its fatty acid chains is joined to the adjoining carbon atoms by single bonds. A lipid is unsaturated if there is at least one double bond between two carbon atoms. What is a polyunsaturated lipid?

Proteins

A **protein** is an organic molecule made up of carbon, hydrogen, oxygen, and nitrogen. Proteins occasionally contain sulfur, as well. They are made up of building blocks called **amino acids**. Amino acids, such as the one in **Figure 2.18**, join together to form chains.

There are about 20 different amino acids that can join together in a huge number of combinations. The order and arrangement of the amino acids determine the nature of the protein. Some proteins act as the building and binding materials of living things. Collagen, for example, forms tendons, ligaments, and cartilage. Other proteins take part in chemical reactions and transport materials throughout an organism. Others cause actions, such as contraction of muscles, to occur.

Nucleic Acids

Living things tend to resemble their parents because of information passed from one generation of cells to the next. This information, known as genetic information, is stored in nucleic acids. A **nucleic** (new KLAY ihk) **acid** is a large molecule made up of smaller units called nucleotides.

One important nucleic acid is deoxyribonucleic (dee AHK sih rib oh noo klay ihk) acid, or **DNA**. Genetic information is stored in DNA. This information makes up the instructions used to form all of a cell's proteins. Another nucleic acid is ribonucleic acid (RNA). The information stored in DNA is translated by RNA and used to direct the production of proteins.

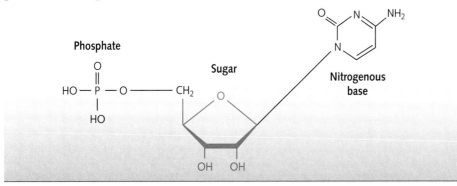

Figure 2.20 Each nucleotide consists of a phosphate group, a simple sugar, and a nitrogenous base.

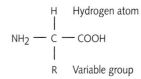

Figure 2.18 An amino acid has an amino group (–NH$_2$) and a carboxyl (organic acid) group (–COOH) attached to the same carbon atom. The rest of the amino acid, or R-group, makes each kind different from the others.

Figure 2.19 Proteins make up structures such as hair, nails, horns, and hoofs.

Extend It!

Nutrition labels on food products list the amounts of fats, carbohydrates, and proteins in each serving of the food. Compare the nutrition labels for several food products. Identify the types of foods that are sources of each organic molecule. Record your findings in your Science Notebook.

After You Read

1. Why might a runner eat carbohydrates before a race?

2. What are three functions of lipids in living things? Of proteins?

3. Use the information in your Science Notebook to write a well-developed paragraph about the subunits of each type of organic molecule.

KEY CONCEPTS

2.1 Characteristics of Living Things

- All living things are made up of cells, respond to stimuli, adapt over time, are capable of reproduction, and grow and develop.

- The process by which living things respond to stimuli in ways that allow them to maintain internal conditions necessary for life is called homeostasis.

- The characteristics of living things can be used to classify objects as living or nonliving.

2.2 The Needs of Living Things

- All living things need to obtain energy, take in water, maintain a proper temperature, exchange gases in air, and have a space in which to live.

- Organisms that capture energy from their surroundings are called autotrophs. Organisms that eat other organisms to get energy to survive are called heterotrophs.

- Water, the universal solvent, is essential to living things because it dissolves substances involved in chemical reactions, carries materials, and helps maintain body temperature.

2.3 Chemistry of Living Things

- Carbohydrates provide energy to all cells and support to plant cells. They can be simple sugars or combinations of simple sugars.

- Lipids store energy, provide insulation, and are the main component of cell membranes. They are often made up of glycerol molecules and fatty acids.

- Proteins play important roles in an organism, such as providing structure, taking part in chemical reactions, and transporting materials. They are made up of chains of amino acids.

- Nucleic acids carry genetic information from one generation to the next. They are made up of nucleotides.

VOCABULARY REVIEW

Write each term in a complete sentence, or write a paragraph relating several terms.

2.1
organism, p. 21
cell, p. 22
stimulus, p. 23
response, p. 23
homeostasis, p. 23
adaptation, p. 23
reproduction, p. 23
species, p. 23

2.2
energy, p. 25
photosynthesis, p. 25
autotroph, p. 25
heterotroph, p. 25
solvent, p. 26
endotherm, p. 27
ectotherm, p. 27

2.3
matter, p. 29
atom, p. 29
organic, p. 29
carbohydrate, p. 29
lipid, p. 30
insoluble, p. 30
protein, p. 31
amino acid, p. 31
nucleic acid, p. 31
DNA, p. 31

PREPARE FOR CHAPTER TEST

To prepare for the chapter test, create a question from each Learning Goal. Use the information in your Science Notebook to answer each question. Then use these answers to write a well-developed essay about the chapter. Use the Key Concept on the first page of this chapter as your topic sentence.

True or False

If the statement is true, write "true." If it is false, change the underlined word or words to make the statement true.

1. A <u>molecule</u> is the smallest unit that can be considered alive.

2. A living thing can respond to a <u>stimulus</u> in its environment.

3. During the process of <u>adaptation</u>, living things make more organisms like themselves.

4. A rattlesnake is an <u>endotherm</u> because its temperature depends on its environment.

5. <u>Lipid</u> molecules are made up of chains of amino acids.

6. A molecule of <u>DNA</u> carries instructions a cell can use to make proteins.

Short Answer

Answer each of the following in a sentence or brief paragraph.

7. What does it mean to say that living things respond to both their external and internal environments?

8. The ability to change color is an adaptation that helps a chameleon survive in its environment. Analyze how this adaptation increases the chameleon's ability to survive.

9. Water is known as the universal solvent. Summarize why this makes water essential to life on Earth.

10. Explain the relationship between autotrophs and heterotrophs. Include definitions of the terms in your answer.

11. Describe proteins and DNA in your own words. Then relate DNA to proteins.

Critical Thinking

Use what you have learned in this chapter to answer each of the following.

12. **Compare and Contrast** Identify ways in which unicellular organisms are similar to and different from multicellular organisms.

13. **Give Examples** The fire alarm at school is an example of a stimulus. Tell how you and your classmates respond to this stimulus. Then give another example of a stimulus in your environment and your response to it.

14. **Relate** How are carbohydrates and lipids related to the need for energy in the body?

Standardized Test Question

Choose the letter of the response that correctly answers the question.

15. The diagrams show the same plant over the course of ten days. What is the plant doing?

　A. changing into a different species

　B. obtaining more space in which to live

　C. passing along its traits

　D. responding to the stimulus of light

Test-Taking Tip

Read each question carefully. Think about what is asked. If you are not sure, read the question again. If you are still not sure, go on to the next question. Sometimes, a later question will help you remember information you need to answer a question you skipped. Remember to check all of your answers before turning in the test.

Chapter 3

Cell Structure, Function, and Processes

KEY CONCEPT Cells are the basic units of structure and function in living things.

The players on your soccer team work to kick the ball into the other team's goal and keep the ball out of your goal. When each player performs his or her job correctly, the team functions efficiently. Your team is part of a league, and your league is part of state and national organizations.

The cells that make up living things are organized in a similar way. Each cell is made up of individual parts that work together. For many organisms, cells are then organized into different levels and groups.

Think About Identifying Groups

Without realizing it, you are part of a group that is part of a larger group.

- Think about a group to which you belong. Consider how the group is part of a larger group.

- In your Science Notebook, draw a diagram showing how the groups are related. Write a few sentences describing your role in the group and why it is important for the largest group to be divided into smaller groups.

NSTA

SCiLINKS
THE WORLD'S A CLICK AWAY

www.scilinks.org
Cell Features **Code: WGB03**

3.1 Cells: Basic Units of Life

Learning Goals

- Identify the roles played by different scientists in the development of the cell theory.
- Recognize how spontaneous generation came to be disproved over time.

New Vocabulary

microorganism
spontaneous generation
cell theory

Recall Vocabulary

cell (p. 22)

Before You Read

It is often helpful to summarize information about people and events. In your Science Notebook, draw a horizontal time line. Label the left end of the line *1600* and the right end of the line *1900*. Make marks to show each interval of 50 years on the time line. Skim through the lesson to look for any important years that are mentioned. Write these years in the appropriate places on the time line.

Biologists now know that cells are the basic units of life. As you may recall from Chapter 2, a cell is a structure that contains all of the materials needed for life. Recognizing the importance of cells, however, was not an easy task. After all, most cells are so small that they cannot be seen with the unaided eye. Discovering the structure and function of cells involved the work of many scientists as well as the development of the right scientific tools.

Robert Hooke

In the 1600s, a scientist named Robert Hooke designed one of the first compound microscopes. A microscope is a tool that a scientist can use to look at very small objects. In 1665, Hooke used his microscope to study a slice of cork. He observed that the cork was made up of tiny, empty chambers. These chambers reminded Hooke of rooms in a monastery called cells. He therefore gave this name to the chambers he observed in the cork. **Figure 3.1** shows what Hooke may have seen when observing cells. It also shows the microscope he used to see them.

Anton van Leeuwenhoek

The cork that Hooke observed was no longer living. Because of this, the chambers that Hooke saw were not living cells. They were only parts of cells left behind. The first person to observe living cells was Anton van Leeuwenhoek (LAY vun hook). At about the same time that Hooke made his observations, Leeuwenhoek designed and used a simple microscope to study pond water. In 1678, he described how the water was filled with tiny living organisms. He named them *animacules*, which means "tiny animals." Tiny living things that can be seen only through a microscope are now known as **microorganisms** (mi kroh OR guh nih zumz).

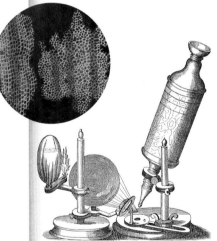

Figure 3.1 Robert Hooke used a compound microscope similar to the one shown here. It uses two lenses to magnify an object. Among the many things Hooke observed through his microscope was a slice of cork. He made drawings to represent the chambers he saw. What did he name these boxlike chambers?

Figure 3.2 Matthias Schleiden and Theodor Schwann used observations made using microscopes to conclude that all plants and animals are made up of cells.

As You Read

On the time line you drew in your Science Notebook, record the name of each scientist mentioned near the year when that person made an important discovery. Then write a sentence summarizing what that person discovered.

How did the discoveries of Schleiden and Schwann relate to the observations of Hooke and Leeuwenhoek?

Schleiden and Schwann

In 1838, almost 200 years after Robert Hooke first observed microscopic cells, a German botanist named Matthias Schleiden (SHLI dun) studied a variety of plants. By this time, the microscope had been improved. It could be used to see even smaller objects in greater detail. Schleiden used the microscope to study the composition of plants.

After many careful observations, Schleiden concluded that all plants are made of cells. He went on to suggest that plants grow because new cells are produced. Schleiden's friend, German biologist Theodor Schwann, reached the same conclusion about animals the next year.

Spontaneous Generation

By the mid-1800s, scientists had learned a great deal about cells. They knew that cells were the basic units of plants and animals, and they knew about some structures inside cells. What they did not know was where cells came from. That was part of a much larger question that had been debated for centuries: Can living things arise from nonliving matter?

The belief that living things come from nonliving matter is known as **spontaneous generation**. As early as the fifth century B.C., philosophers concluded that living things developed from nonliving elements in nature. Today, this idea may seem unrealistic based on what you know. Yet these conclusions were based on people's observations. Some of them are summarized in **Figure 3.3**.

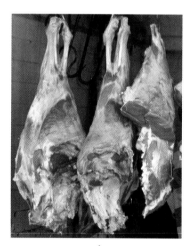

a   b   c

Figure 3.3 a) Farmers of medieval Europe stored grain in barns with thatched roofs. In time, the grain became moldy and overrun with mice. People concluded that mice came from moldy grain. **b)** Before refrigerators were invented, butchers hung meat from hooks. In warm weather, maggots would appear on the meat. People concluded that rotting meat produced maggots. **c)** After the Nile River floods its banks each spring, frogs appear in the mud left behind. Years ago, people observed this and concluded that mud gave rise to frogs.

Redi Critics first questioned the idea of spontaneous generation in 1668. The Italian physician Francesco Redi believed that maggots did not suddenly appear on rotting meat. He proposed, instead, that maggots come from eggs laid by flies that land on the meat.

To test his hypothesis, Redi set out jars similar to those shown in **Figure 3.4**. Some of the jars were open to air, some were completely sealed, and some were covered with cheesecloth. Air can pass through cheesecloth, but flies cannot. After a period of time, Redi observed that maggots appeared only in the jars that could be reached by flies.

Needham Despite Redi's conclusions, most people continued to believe in spontaneous generation. In fact, even Redi believed it could occur under certain conditions. It was around this time that Leeuwenhoek first observed living cells. People used his discovery to support the idea of spontaneous generation. After all, the microscopic world of living things that seemed to suddenly appear had to come from somewhere.

In 1745, an English clergyman named John Needham designed an experiment to test the idea. He boiled chicken broth to kill any microorganisms in it. He then put the broth into a flask, sealed it, and waited. In time, tiny organisms appeared in the broth. Needham concluded that spontaneous generation did occur.

Spallanzani An Italian priest named Lazzaro Spallanzani (spah lahn ZAH nee) was not convinced by Needham's experiment. He suggested that the microorganisms came from the air after the broth was boiled but before the flask was sealed. To test his hypothesis, he placed the chicken broth in a flask, sealed the flask, and then removed the air. Then he boiled the broth. This time, no microorganisms grew in the broth. He concluded that spontaneous generation did not occur. Many critics, however, argued that Spallanzani only proved that spontaneous generation does not occur without air.

Explain It!

"No scientist works in isolation." In your Science Notebook, write a paragraph explaining this statement. As examples, use the overlapping work of scientists studying cells and others studying spontaneous generation.

Figure It Out

1. What was the control group in Redi's experiment?

2. Did Redi's observation support his hypothesis? Explain.

Figure 3.4 In this experiment conducted by Francesco Redi, the spontaneous generation of maggots from rotting meat was tested.

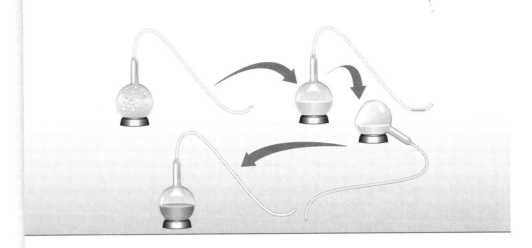

Figure 3.5 Louis Pasteur designed an experiment to settle the issue of spontaneous generation. He used flasks that would allow air to pass through but not microorganisms.

Pasteur In 1859, a French chemist named Louis Pasteur (pas TUHR) designed a variation of the experiments done by Needham and Spallanzani. As shown in **Figure 3.5**, he bent the neck of a flask into an S-shape. This made it possible for air to reach the broth inside. However, any microorganisms in the air would get stuck in the neck of the flask and be unable to reach the broth. As in Spallanzani's experiment, no microorganisms grew in the broth. This time, however, Pasteur knew that spontaneous generation did not occur even in the presence of air. To confirm that microorganisms came from the air, he then tilted the flasks. Microorganisms suddenly appeared in the broth. In this way, Pasteur proved that microorganisms exist in the air and do not arise from nonliving matter.

The Cell Theory

Around the same time that Pasteur was conducting his experiments, a German physician named Rudolf Virchow was studying cells. Like Pasteur, he did not support the idea of spontaneous generation. In 1855, Virchow presented a theory that cells are produced only by other cells.

This last piece of information on cells helped complete what is now known as the cell theory. The **cell theory** states:

- All living things are made up of cells.
- Cells are the basic units of structure and function in living things.
- New cells are produced from existing cells.

CONNECTION: History

Inventors were very busy during the years that these scientists were studying cells. For example, the thermometer was invented in 1724, the lightning rod in 1752, and a better steam engine in 1769. The battery was introduced in 1799, the steam locomotive in 1814, and the telegraph in 1837.

After You Read

1. How were Leeuwenhoek's observations different from those of Hooke?
2. What contribution to the cell theory was made by each scientist studied here.
3. According to the summaries on your time line, how did research into spontaneous generation affect the development of the cell theory?

3.2 The Parts of a Cell

Learning Goals

- Identify the parts of animal cells and plant cells.
- Differentiate between prokaryotic cells and eukaryotic cells.

New Vocabulary

organelle
nucleus
cytoplasm
ribosome
endoplasmic reticulum
Golgi apparatus
mitochondrion
lysosome
vacuole
chloroplast
cell membrane
cell wall
eukaryotic cell
prokaryotic cell

Recall Vocabulary

DNA (p. 31)
photosynthesis (p. 25)

Before You Read

Review the Learning Goals and the vocabulary for this lesson. Many of these terms describe parts of a cell. In your Science Notebook, create a concept map with the word *Cell* in the center. Draw several smaller circles around the word *Cell*. Think about the functions that a cell needs to carry out.

You pull a box of cereal from the kitchen cabinet and head to the table. As your stomach rumbles, you pour some cereal into a bowl. The cereal you're about to eat was processed in a factory. In a factory, many different people and machines work together to manufacture a product. In a cereal factory, ingredients such as grains, sugar, and flour are taken into the factory. They are mixed together and processed to produce cereal. The finished cereal is then placed into packages that are sent out to stores.

In many ways, living cells are like factories that produce goods. They take in raw materials, use them to build products such as proteins, package the products, and transport them to different parts of the cell or to other cells. The different jobs are performed by structures within the cell called **organelles**. The cells of animals and plants share most of the same kinds of organelles and other cell parts. **Figure 3.6** on page 40 shows the structure of a cell from an animal and a cell from a plant.

Nucleus

The **nucleus** (NEW klee us) can be described as the control center of the cell. Almost all of the DNA in a cell is contained in the nucleus. Recall from Chapter 2 that DNA is an organic molecule that stores the instructions for making proteins and other molecules needed by the cell.

The nucleus is surrounded by two thin membranes that make up the nuclear envelope. Pores, or small openings, in the nuclear envelope allow specific materials to move between the nucleus and the rest of the cell.

Cytoplasm

The cell is generally divided into two parts. One part is the nucleus. The part of the cell outside the nucleus is the **cytoplasm**. The cytoplasm contains the organelles as well as chemicals needed by the cell. It also contains the cytoskeleton, which is a network of fibers that gives the cell its shape and anchors organelles in place.

1. What is the largest and most obvious structure in each cell?

2. Identify some differences between these two cells.

Animal Cell

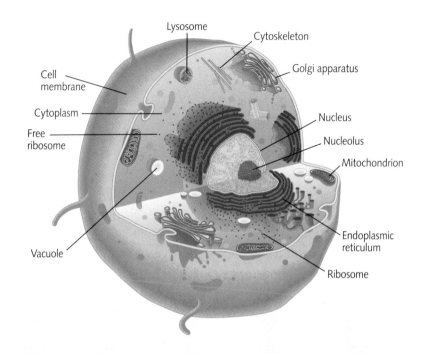

Plant Cell

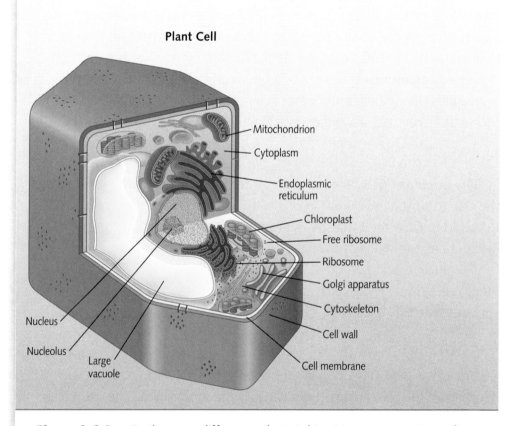

Figure 3.6 Despite the many differences that might exist among organisms, they are all made up of cells that share many common features. This figure shows the structures of a typical animal cell and a typical plant cell.

Ribosomes

The proteins that are made according to instructions in DNA are assembled on **ribosomes** (RI buh sohmz). A ribosome is made up of the nucleic acid RNA and protein. Ribosomes are formed within a structure known as the nucleolus, which is in the nucleus of the cell. Unlike most other organelles, ribosomes are not surrounded by membranes.

Endoplasmic Reticulum

The cell contains a system of membranes and sacs known as the **endoplasmic reticulum** (en duh PLAZ mihk • rih TIHK yuh lum), or ER. The endoplasmic reticulum acts like a highway along which molecules can move from one part of the cell to another.

The part of the ER that is involved in the production of proteins has ribosomes along its surface. As a result, this type of ER is known as rough ER. It is common in cells that make large amounts of proteins. The other part of the ER is known as smooth ER because it does not have ribosomes. This type of ER is involved in regulating processes in cells.

As You Read

Write the name of one cell part in each smaller circle of your concept map. Add circles as needed. Connect each smaller circle to the circle labeled *Cell* with a line. On each line, write a few words describing the function of that cell part.

Which part of the cell is like the power supply of a factory?

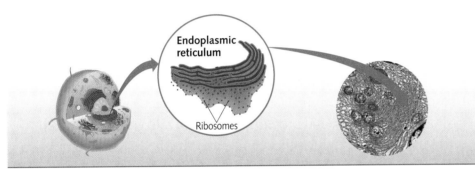

Figure 3.7 The endoplasmic reticulum (ER) is a series of folded membranes. Chemical processes necessary to the cell occur in and on the ER.

Golgi Apparatus

Proteins produced in rough ER are passed along to an organelle known as the **Golgi apparatus**. The Golgi apparatus consists of a stack of membranes. Its job is to modify and package proteins and other molecules so they can either be stored in the cell or sent outside the cell.

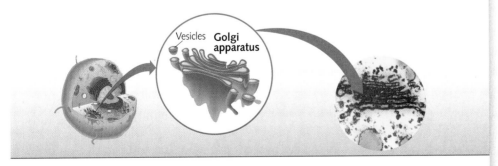

Figure 3.8 The Golgi apparatus looks like a flattened stack of pancakes. What is the role of this organelle?

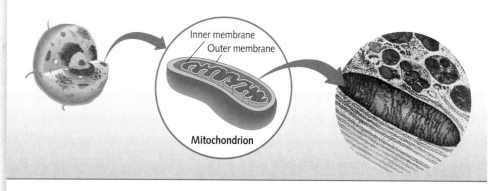

Extend It!

Think about the ways in which a cell is like a factory. Work with a partner to create a poster that relates each organelle in a cell to a job in a factory. Use both illustrations and descriptions. Then present your poster to the class.

Figure 3.9 Each mitochondrion resembles a seed with a folded inner membrane. The folded shape of the membrane allows a large membrane to fit in a smaller space.

Mitochondria

All cells need energy. Organelles called **mitochondria** (mi tuh KAHN dree uh, singular: mitochondrion) convert energy stored in organic molecules into compounds the cell can use. The greater the energy needs of a cell, the more mitochondria that cell will have.

Lysosomes

The cleanup crew of a cell consists of **lysosomes** (LI suh sohmz). These are small organelles filled with enzymes. An enzyme is a protein that speeds up the rate of a chemical reaction. The enzymes enable the lysosomes to digest, or break down, organic molecules such as carbohydrates, lipids, and proteins. The larger molecules are broken down into small molecules that can be used by the cell. Lysosomes also digest old organelles that are no longer useful to the cell.

Vacuoles

A saclike organelle that stores materials for the cell is known as a **vacuole** (VAK yuh wohl). In some cells, vacuoles store water, salts, carbohydrates, and proteins. Most plant cells have one large central vacuole. The pressure of the liquid in this vacuole helps to structurally support the plant.

Not every cell contains vacuoles. For example, they are found in only some animal cells. For some unicellular organisms, the vacuole plays an important role in motion. The organism pumps water out of the cell in order to move forward.

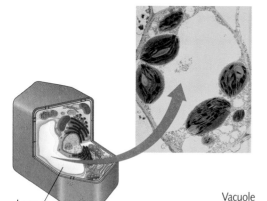

Large vacuole

Vacuole

Figure 3.10 There is usually one large vacuole in a plant cell. Animal cells may contain many smaller vacuoles.

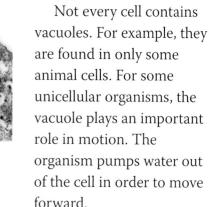

Chloroplasts

Plants and some other organisms contain **chloroplasts** (KLOR uh plasts). These organelles are the sites of photosynthesis. Recall from Chapter 2 that during photosynthesis, an organism captures the energy of sunlight and converts it into energy that is stored in organic molecules.

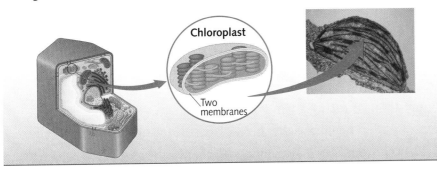

Figure 3.11 Structures within chloroplasts resemble stacks of coins. Chloroplasts trap sunlight and use it to produce organic molecules.

Cell Membrane

At the outer edge of a cell is a thin barrier called the **cell membrane**. A cell must be able to take in nutrients and dispose of wastes. The cell membrane controls how these substances pass into and out of a cell.

Cell Wall

In some cells, such as plant cells, an additional barrier is located outside the cell membrane. The **cell wall** is a rigid outer layer that supports the cell and protects it from harm. Pores in the cell wall allow materials to pass into and out of the cell.

Types of Cells

The animal cells and plant cells you just read about are eukaryotic (yew KAR ee ah tik) cells. Some other organisms, such as bacteria, are prokaryotic (proh KAR ree oht ik) cells. What makes the cells different?

A **eukaryotic cell** is larger and more complex than a **prokaryotic cell**. A eukaryotic cell has a nucleus, and a prokaryotic cell does not. In addition, many of the organelles in a eukaryotic cell are enclosed in their own membranes. A prokaryotic cell does not contain membrane-bound organelles. **Figure 3.12** shows both types of cells.

CONNECTION: Art

Many of the photographs you may see of cells appear in color. The colors do not represent the actual colors of the cell and cell parts. Instead, they are used to make the cells easier to study and to highlight different parts of the cells. Draw and label your own version of an animal cell. Use different colors to identify the various organelles.

Prokaryotic Cell

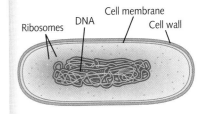

Eukaryotic Cell

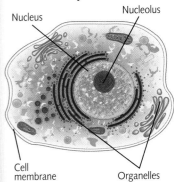

Figure 3.12 Some parts of prokaryotic cells and eukaryotic cells are shown here.

After You Read

1. What is the role of ribosomes in a cell?
2. List three ways in which a eukaryotic cell is different from a prokaryotic cell.
3. Identify which parts in the diagram you drew in your Science Notebook are found in plant cells but not in animal cells.

Before You Read

In your Science Notebook, draw a T-chart. Label one side *Passive Transport* and the other side *Active Transport*. In your own words, define the terms *passive* and *active*.

Have you ever noticed that after being in a bathtub or swimming pool for a long time, your fingers and toes become wrinkled? Fingers and toes become wrinkled because skin cells at the surface absorb water. The outer skin cells are tightly attached to the layers below. Having nowhere to go, the skin wrinkles to make up for the swelling.

Just as water can pass through the outer membrane of a skin cell, so too can other materials. This enables the cell to obtain materials it needs from its environment and to get rid of wastes.

Concentration of Solutions

The materials that enter and leave a cell are dissolved in the liquids on either side of the cell membrane. A liquid containing dissolved materials is known as a **solution**. A solution is a mixture of two or more substances. The substance in a solution that does the dissolving is known as the solvent. The substance that is dissolved is known as the solute. As discussed in Chapter 2, water is known as the universal solvent because it can dissolve so many substances.

The **concentration** of a solution compares the amount of substance dissolved to the amount of substance doing the dissolving. For example, suppose you add a pinch of powdered drink mix to a glass of water. The drink will not be very flavorful because the concentration of drink mix is low. However, if you add a large spoonful of drink mix to the water, the drink will taste much sweeter. This is because the concentration is high.

Figure 3.13 The concentration of a solution depends on how much of one substance is dissolved in another. The more drink mix that is dissolved in water, the sweeter and more flavorful the drink will taste.

Diffusion

As with all particles, the particles in a solution are in constant motion. As they move, they collide with each other. This causes them to spread out. They tend to move from an area where they are more crowded to an area where they are less crowded. The movement of particles from an area of higher concentration to an area of lower concentration is known as **diffusion** (dih FYEW zhun).

Diffusion is one way that particles enter and leave a cell. Suppose, for example, that the concentration of a substance is different on either side of the cell membrane, as in **Figure 3.14**. If the substance can cross the cell membrane, its particles will move toward the area where it is less concentrated.

Particles will continue to diffuse until their concentration is the same on both sides of the cell membrane. The system is then said to be at **equilibrium** (ee kwuh LIH bree um). Once equilibrium is reached, particles continue to move across the cell membrane in both directions. However, the same number of particles moves in each direction, so there is no additional change in concentration.

The difference in concentration between two regions is known as the concentration gradient. Diffusion does not require the cell to use energy because it depends on the random movements of particles to move substances along the concentration gradient. The movement of materials into or out of the cell without the cell's use of energy is called **passive transport**.

Diffusion Across a Cell Membrane

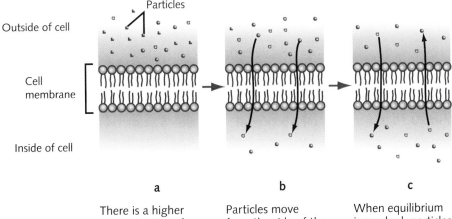

a There is a higher concentration of particles on one side of the membrane.

b Particles move from the side of the membrane with a higher concentration to the side of the membrane with a lower concentration.

c When equilibrium is reached, particles continue to move across the membrane in both directions.

Figure It Out

1. In Part a, on which side of the cell membrane are the particles more concentrated?

2. Describe what occurs when the system reaches equilibrium.

Figure 3.14 During diffusion, particles move from areas of higher concentration to areas of lower concentration.

Osmosis

Not all particles can diffuse across a cell membrane. Most membranes are **selectively permeable membranes**, which means that some particles can pass across them while others cannot. A membrane is said to be permeable to substances that can pass across it and impermeable to those that cannot.

Perhaps the most important substance that passes through the cell membrane is water. Water molecules pass through a selectively permeable membrane by a type of diffusion known as **osmosis** (ahs MOH sus). During osmosis, water molecules move from a place of higher concentration of water to a place of lower concentration of water—either into or out of the cell.

When there is a difference in the concentration of a solution outside a cell compared with the cytoplasm inside the cell, osmosis will occur. If the concentration of solute particles outside the cell is higher than the concentration inside the cytoplasm, water diffuses out of the cell, causing the cell to shrivel. If the concentration of solute particles inside the cytoplasm is higher than the concentration outside the cell, water diffuses into the cell. This causes the cell to swell.

As You Read

List examples of passive and active transport described in the lesson in the T-chart you created in your Science Notebook.

Which form of transport requires the cell to use energy?

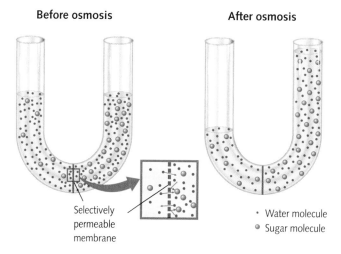

Before osmosis   **After osmosis**

Selectively permeable membrane

• Water molecule
◦ Sugar molecule

Figure 3.15 The barrier at the bottom of this U-tube is a selectively permeable membrane. It allows water molecules to pass through it, but not sugar molecules. How do the numbers of water and sugar molecules on each side of the membrane change after osmosis?

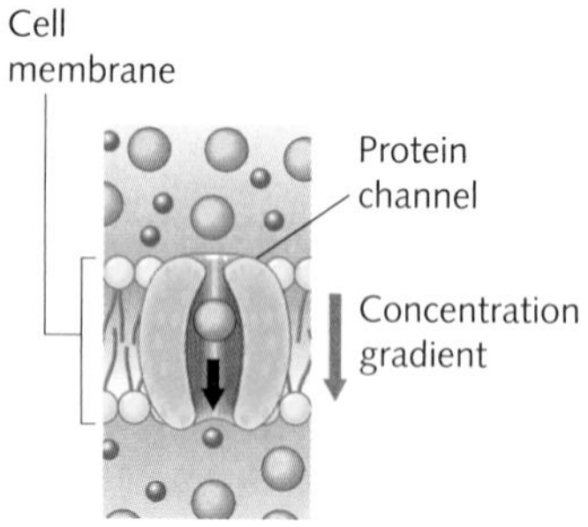

Facilitated Diffusion

Another type of diffusion is known as **facilitated diffusion**. During this process, a few molecules pass through the cell membrane more easily than might be expected. The reason is that cell membranes have pathways in them known as protein channels. The channels allow specific types of molecules to pass into or out of the cell. Protein channels in red blood cells, for example, allow only glucose to pass through.

Cell membrane

Protein channel

Concentration gradient

Figure 3.16 Protein channels help certain particles move across the cell membrane. The process is still diffusion and occurs only if there is a difference in concentration.

Active Transport

Some materials must be moved against the concentration gradient by **active transport**. This type of movement requires the cell to use energy.

Molecule Transport   Like facilitated transport, active transport can use proteins in the membrane to move substances across the membrane. For example, sodium ions (Na^+) are pumped out of the cell and potassium ions (K^+) are pumped into the cell by specific carrier proteins. The cell must use energy to pump these ions across the membrane.

Endocytosis and Exocytosis   Some large molecules are transported by movements of the cell membrane itself. During endocytosis (en duh sy TOH sus), the cell membrane surrounds a substance and brings it into the cell. In an opposite process known as exocytosis (ek soh si TOH sus), the cell membrane releases substances from the cell.

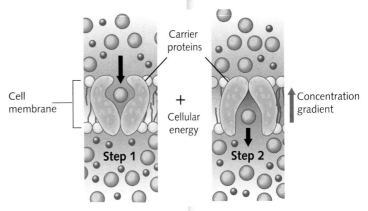

Figure 3.17 In active transport, carrier proteins can pick up and transport ions and molecules across the cell membrane. Why does this process require energy?

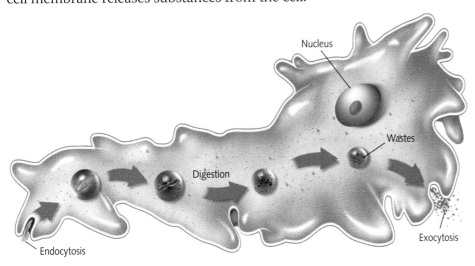

Figure 3.18 Some cells take in or release materials through movements of the cell membrane that use energy. Endocytosis and exocytosis are forms of active transport.

Explore It!

Obtain four raisins. Soak two in water overnight. Set the other two aside. Blot the water-soaked raisins dry. Be careful not to break the skins. Examine both sets of raisins. Carefully poke the ends of each raisin with a toothpick. Squeeze each raisin over a paper towel. How can you account for any differences in the raisins?

After You Read

1. How is diffusion related to the concentration of solutions inside and outside the cell?
2. Compare and contrast osmosis and facilitated diffusion.
3. What is active transport? Give two examples of how it occurs.
4. In a well-developed paragraph, compare active transport and passive transport in terms of the cell's use of energy. Use examples from the T-chart in your Science Notebook in your answer.

Learning Goals

- Trace the flow of energy that occurs during photosynthesis.

- Examine the process that occurs during cellular respiration.

- Identify the circumstances under which fermentation occurs.

New Vocabulary

producer
cellular respiration
aerobic
fermentation
anaerobic

Recall Vocabulary

energy (p. 25)
autotroph (p. 25)
heterotroph (p. 25)

Did You Know?

Most environments on land depend on organisms that conduct photosynthesis. However, some organisms rely on another method of capturing matter and energy. Deep beneath the surface of the ocean, near hot vents in the ocean floor, some organisms produce organic molecules from chemicals released into the ocean water. This process is known as chemosynthesis.

Before You Read

Think about what you do when you compare and contrast. Then select a plant and an animal. In your Science Notebook, create a chart to compare and contrast the two organisms. Write words or draw pictures in your chart.

Have you ever felt as if you were filled with energy? Or maybe you have felt drained of energy after a long day. Recall from Chapter 2 that energy is the ability to cause change or do work. Cells need energy to perform their functions. Where does that energy come from? Almost all of the energy used by living things comes from the Sun.

Photosynthesis

Light, such as sunlight, is one type of energy. There are other types of energy as well. A different type of energy known as chemical energy is stored in organic substances such as foods and fuels. Each form of energy can be changed into another form.

The process by which some organisms capture light energy and change it into chemical energy is called photosynthesis. Organisms that conduct photosynthesis are called **producers** because they make, or produce, food for other organisms. All autotrophs, or organisms that capture matter and energy from their surroundings, are producers.

Recall from Lesson 2 that plant cells contain chloroplasts. Within chloroplasts is a pigment known as chlorophyll. It is this pigment that gives plants their green color and allows plants to capture the energy of sunlight. As shown in **Figure 3.19**, this energy is required for a process that uses carbon dioxide gas from the air and water to produce organic molecules, such as carbohydrates, and oxygen gas.

Some of the energy stored in organic molecules is used by the plant's cells. Any remaining energy can be stored in the form of carbohydrates, such as starch and sugar, and lipids.

$$\overset{\text{Light}}{\underset{\text{energy}}{}}$$
$$6CO_2 \ + \ 6H_2O \ \rightarrow \ C_6H_{12}O_6 \ + \ 6O_2$$

Carbon dioxide — Water — Simple sugar — Oxygen

Figure 3.19 The process of photosynthesis can be summarized by this equation. It shows that six molecules of carbon dioxide and six molecules of water are needed to form one simple sugar molecule and six molecules of oxygen.

Cellular Respiration

Plants get energy from sunlight. Where do you get energy? You, along with other organisms that do not conduct photosynthesis, get energy from the foods you eat. Such organisms are called heterotrophs.

Organisms must release the energy stored in organic molecules. One way they do this is through a process known as cellular respiration. During **cellular respiration**, glucose and other organic molecules are broken down to release energy in the presence of oxygen. Cellular respiration occurs in the mitochondria of eukaryotic cells.

The process of cellular respiration is summarized in **Figure 3.20**. Despite the simple equation, cellular respiration does not take place in one simple step. Instead, the process occurs in a series of steps. In each step, cells can trap bits of energy and change that energy into a form the cell can use.

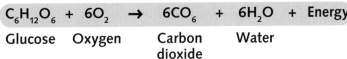

$$C_6H_{12}O_6 + 6O_2 \rightarrow 6CO_6 + 6H_2O + Energy$$

Glucose   Oxygen   Carbon   Water
                   dioxide

Figure 3.20 The complete process of cellular respiration is summarized by this equation. It shows that during cellular respiration, organisms use organic compounds, such as glucose, and oxygen to produce carbon dioxide, water, and energy.

As You Read

Extend the chart you made for comparing and contrasting plants and animals so that you can now compare and contrast photosynthesis and cellular respiration.

How are the two processes similar? How are they different?

Figure It Out

1. How many molecules of oxygen are produced for each molecule of glucose that is broken down?

2. Explain where the energy released during cellular respiration came from.

⊕ CONNECTION: Physics

Everything you do involves energy. Whether you are riding a skateboard, getting dressed, or reading a book, you are using energy. Energy is all around you in different forms. For example, nuclear energy powers the Sun and other stars. Light, also known as electromagnetic energy, is then given off by the Sun. That energy is captured by plants and changed into chemical energy.

Perhaps you are recognizing the fact that each form of energy can be converted into another form. What you may not know, however, is that no energy is destroyed or created in the process. The total amount of energy in a system remains the same. This fundamental concept of physics is known as the law of conservation of energy. According to this law, energy is neither created nor destroyed by ordinary processes.

Does that mean that the amount of energy in the universe is always exactly the same? Well, not exactly. In 1905, Albert Einstein proposed the idea that in some situations, matter can change into energy and energy can change into matter. Therefore, the total amount of matter *and* energy in the universe remains constant.

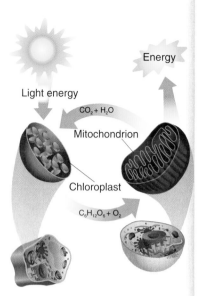

Figure 3.21 Photosynthesis and cellular respiration are related in a continuous process. Both plants and animals conduct cellular respiration, but only plants conduct photosynthesis.

Figure 3.22 Many foods, such as these loaves of bread, are produced using a type of fermentation. Why is fermentation an anaerobic process?

A Continuous Cycle

Did you notice anything familiar about the equation for cellular respiration? It is the opposite of the equation describing photosynthesis. In other words, the materials needed for cellular respiration are the materials produced during photosynthesis. The materials produced during cellular respiration are the materials needed for photosynthesis.

Fermentation

Cellular respiration is said to be an **aerobic** (er ROH bihk) process because it requires oxygen. When cells cannot get the oxygen they need, they use a process called **fermentation** (fur mun TAY shun) to release energy stored in organic molecules. This process is said to be **anaerobic** (an uh ROH bihk) because it does not require oxygen. Fermentation does not provide cells with as much energy as cellular respiration does.

There are two types of fermentation: lactic acid fermentation and alcoholic fermentation.

Lactic Acid Fermentation   During this type of fermentation, cells produce a compound called lactic acid. This type of fermentation is used to manufacture certain foods, such as yogurt and cheese. It also occurs in your muscles during strenuous exercise.

Alcoholic Fermentation   During alcoholic fermentation, some single-celled organisms produce ethyl alcohol. This type of fermentation is used to produce bread and beverages such as wine and beer.

CONNECTION: Nutrition

Fermentation has been used to preserve foods throughout history. One advantage of fermented foods is that they are partially broken down already, which makes them easier to digest. In addition, important enzymes, vitamins, and other nutrients are preserved because they are not broken down by heat.

After You Read

1. What changes in energy occur during photosynthesis?

2. Why do organisms conduct cellular respiration?

3. How are the raw materials and products of photosynthesis related to those of cellular respiration?

4. Extend the chart in your Science Notebook once more. This time, compare and contrast cellular respiration and fermentation. How are these processes alike? How are they different?

3.5 Cell Growth and Development

Before You Read

Preview the lesson by looking at the pictures and reading the headings and the Learning Goals. In your Science Notebook, write a paragraph describing what you think the lesson is about based on your preview.

In Chapter 2, you learned that some organisms are made up of one cell and others are made up of more than one cell. Cells of multicellular organisms vary in their structure and function.

Cell Shape

The shape of a cell is often related to its function. Nerve cells, for example, are long and thin, which enables them to send information from one part of the body to another. Another type of cell, a red blood cell, carries oxygen throughout the body. This type of cell is in the form of a flattened disk that can flow through thin blood vessels. A sperm cell is a cell that travels to an egg cell. Each sperm cell has a strong tail that enables it to travel quickly.

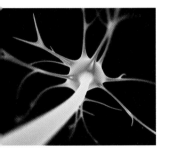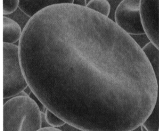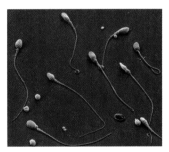

Figure 3.23 The shape of a cell is related to its function. A nerve cell (left) is long and thin. Red blood cells (center) are flattened and have a large area through which oxygen can pass. Sperm cells (right) have tails that propel them through fluids.

Levels of Organization

Your community could not function if many different people did not perform many necessary jobs. For example, letter carriers deliver the mail, trash collectors take away the garbage, and police officers help make sure that everyone is safe. Each person has a specific job to do to make life in the community possible. Groups of people with similar jobs often work together. Just as people perform different jobs within a community, cells perform specific tasks within an organism. Similarly, just as people can work together, so too can cells.

Tissues   The cell is the most basic unit of organization. In most multicellular organisms, groups of similar cells are organized into **tissues**. Each tissue has specific functions. Muscle tissues, for example, make the heart beat, the stomach digest food, and the body move.

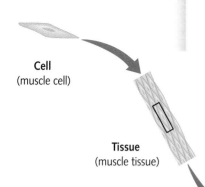

Cell
(muscle cell)

Tissue
(muscle tissue)

Organ
(stomach)

Organ system
(digestive system)

Organism
(Florida panther)

Figure 3.24 In multicellular organisms, such as this panther, cells are organized into tissues, organs, and organ systems. What is the most basic level of organization in the panther?

As You Read

Write a well-developed paragraph in your Science Notebook telling what this lesson is about. Describe how this lesson relates to the other lessons in the chapter.

Was your prediction about the content of the lesson correct, based on your preview? Explain.

Figure It Out

1. How do you find the surface area-to-volume ratio?

2. Calculate how doubling the length of the sides changes the volume of the cube. Then calculate the surface area.

Organs A group of related tissues is an **organ**. An organ performs a more complex function than do individual cells and tissues. The heart, brain, and stomach are examples of organs.

Organ Systems Organs that work together to perform a set of related tasks form an **organ system**. The stomach is one of several organs of the digestive system. Other organs of the digestive system include the esophagus and the intestines. Together, all of the organ systems in a multicellular organism carry out the processes needed to keep the entire organism alive.

Limits on Cell Size

You may be wondering why many small cells group together in an organism instead of simply growing larger. The small size of most cells is for a reason. As a cell grows larger, both its volume and its surface area increase. Recall that volume is the amount of space an object takes up. The surface area is a measure of the size of the outer surface of an object. For example, when you wrap a gift box, you cover the surface area of the box. The surface area of a cell determines the amount of material that can enter and leave the cell.

As shown in **Figure 3.25**, as a cell grows, the volume increases at a faster rate than does the surface area. If the cell grows too large, it will require more materials than can pass through its surface area. Therefore, there is a limit to how large a cell can grow.

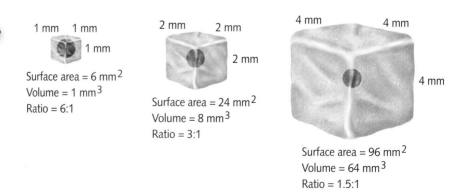

1 mm   1 mm
1 mm
Surface area = 6 mm^2
Volume = 1 mm^3
Ratio = 6:1

2 mm   2 mm
2 mm
Surface area = 24 mm^2
Volume = 8 mm^3
Ratio = 3:1

4 mm   4 mm
4 mm
Surface area = 96 mm^2
Volume = 64 mm^3
Ratio = 1.5:1

Figure 3.25 The ratio of a cell's surface area to its volume is one factor that limits cell size. When the sides of the cube double in length from 1 mm to 2 mm, the surface area-to-volume ratio changes from 6:1 to 3:1.

Cell Division

Instead of continuing to grow larger, cells form new cells that enable an organism to grow and develop tissues, organs, and organ systems. The process by which one cell (known as the parent cell) divides into two cells (called daughter cells) is known as **cell division**, or cell reproduction. The actual process of cell division depends on the type of cell involved.

Cell division is relatively simple in prokaryotic cells such as bacteria. Prokaryotes reproduce through **binary fission**. During this process, a cell's DNA is copied. The cell then splits into two parts. Each part receives one copy of the DNA.

Eukaryotic cells undergo a more complex process of cell division that is part of the cell cycle. The **cell cycle** is a sequence of events that lead to cell growth and division. The cell cycle is summarized in **Figure 3.26**.

CONNECTION: Math

For a regular object, such as a cube, you can find the surface area by first finding the area of each face and then adding the areas together. What is the surface area of a box that is 10 cm long, 6 cm wide, and 5 cm high?

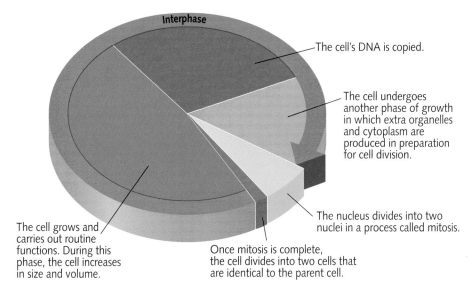

Interphase

The cell's DNA is copied.

The cell undergoes another phase of growth in which extra organelles and cytoplasm are produced in preparation for cell division.

The nucleus divides into two nuclei in a process called mitosis.

The cell grows and carries out routine functions. During this phase, the cell increases in size and volume.

Once mitosis is complete, the cell divides into two cells that are identical to the parent cell.

Figure 3.26 Existing cells divide as part of the cell cycle to produce new cells for growth or to replace old or damaged cells.

After You Read

1. How is cell shape related to cell function? Give an example.

2. What prevents most cells from becoming very large?

3. What is the purpose of cell division in multicellular organisms?

4. According to the preview and summary you wrote in your Science Notebook, how are cell organization and cell division related to each other?

KEY CONCEPTS

3.1 Cells: Basic Units of Life

- Robert Hooke and Anton van Leeuwenhoek were among the first to observe cells or cell parts under a microscope.

- Francesco Redi, John Needham, and Lazzaro Spallanzani conducted experiments to investigate spontaneous generation. It was Louis Pasteur who finally proved that living things do not arise spontaneously from nonliving matter.

- The cell theory states that all living things are made up of cells, cells are the basic units of life, and cells arise only from existing cells.

3.2 The Parts of a Cell

- Eukaryotic cells contain smaller parts and membrane bound organelles that carry out specific functions.

- Organelles include the nucleus, ribosomes, endoplasmic reticulum, Golgi apparatus, mitochondria, and lysosomes. All cells are bound by a cell membrane. Plant cells also contain chloroplasts, a cell wall, and a large vacuole.

- Prokaryotic cells are smaller than eukaryotic cells and lack a nucleus and membrane-bound organelles.

3.3 Cellular Transport

- Diffusion is the process by which particles move from an area of higher concentration to an area of lower concentration without the use of energy.

- Osmosis is the diffusion of water through a selectively permeable membrane.

- During active transport, the cell uses energy to move particles against a concentration gradient.

3.4 Energy and Cells

- During photosynthesis, energy from the Sun is converted into chemical energy stored in organic molecules.

- The energy stored in organic molecules is released in a form that cells can use during cellular respiration.

- When oxygen is not available, cells release energy using fermentation.

3.5 Cell Growth and Development

- The shape of a cell is often related to its function.

- The surface area-to-volume ratio of a cell limits how large the cell can grow and still obtain the materials it needs to survive.

- During cell division, a parent cell divides into two identical daughter cells.

VOCABULARY REVIEW

Write each term in a complete sentence, or write a paragraph relating several terms.

3.1
microorganism, p. 35
spontaneous generation, p. 36
cell theory, p. 38

3.2
organelle, p. 39
nucleus, p. 39
cytoplasm, p. 39
ribosome, p. 41
endoplasmic reticulum, p. 41
Golgi apparatus, p. 41
mitochondrion, p. 42
lysosome, p. 42
vacuole, p. 42
chloroplast, p. 43
cell membrane, p. 43
cell wall, p. 43
eukaryotic cell, p. 43
prokaryotic cell, p. 43

3.3
solution, p. 44
concentration, p. 44
diffusion, p. 45
equilibrium, p. 45
passive transport, p. 45
selectively permeable membrane, p. 46
osmosis, p. 46
facilitated diffusion, p. 46
active transport, p. 47

3.4
producer, p. 48
cellular respiration, p. 49
aerobic, p. 50
fermentation, p. 50
anaerobic, p. 50

3.5
tissue, p. 51
organ, p. 52
organ system, p. 52
cell division, p. 53
binary fission, p. 53
cell cycle, p. 53

MASTERING CONCEPTS

True or False

If the statement is true, write "true." If it is false, change the underlined word or words to make the statement true.

1. Using a microscope, <u>Theodor Schwann</u> saw "chambers" and called them cells.

2. Louis Pasteur conducted an experiment that disproved the theory of <u>spontaneous generation</u>.

3. An <u>endoplasmic reticulum</u> is an organelle on which proteins are assembled.

4. Worn-out organelles are digested in the <u>vacuoles</u> of a cell.

5. Water moves by <u>facilitated transport</u> across a selectively permeable membrane.

6. Endocytosis is a form of <u>active</u> transport.

7. During <u>cellular respiration</u>, organisms store energy in organic molecules.

8. Groups of similar cells are arranged into <u>organ systems</u> in multicellular organisms.

Short Answer

Answer each of the following in a sentence or brief paragraph.

9. Restate the cell theory in your own words. Be sure to identify scientists who made major contributions to its development.

10. You see a cell under a microscope. List features that indicate the cell came from a plant.

11. Describe the process by which an amoeba would ingest large food particles.

12. Discuss how producers are important to the survival of other organisms.

13. Through what process do multicellular organisms grow?

Critical Thinking

Use what you have learned in this chapter to answer each of the following.

14. **Infer** A liver cell requires a great deal of energy. Which organelle would you expect to find in large numbers in a liver cell? Explain your reasoning.

15. **Compare and Contrast** How are eukaryotic cells similar to and different from prokaryotic cells?

16. **Illustrate** Using diagrams, identify the conditions under which (a) water flows into a cell by osmosis and (b) water flows out of a cell by osmosis. In each diagram, indicate what the appearance of the cell will be as a result.

Standardized Test Question

Choose the letter of the response that correctly answers the question.

17. Which is the only structure responsible for conducting photosynthesis in plant cells?

A.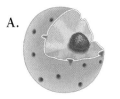

C.

B.

D.

Test-Taking Tip

Remember that qualifying words such as *only*, *always*, *all*, and *never* mean that the statement has **no** exceptions.

Chapter 4

Heredity and Genes

KEY CONCEPT Traits are passed along from one generation to the next through genes that are part of the DNA of chromosomes.

One of the most beautiful birds in the world is the Scarlet macaw. Living mostly in the rain forests of South America, the bird is known for its bright colors, loud calls, and powerful beak.

Macaws are often known to gather in large groups. The matching colors of the many birds make for a spectacular sight. What causes so many birds to share the same colors, sounds, and habits? Why does each generation of birds have the same traits? This chapter is about the traits of organisms and how they are passed down from parents to offspring.

Think About Making Observations

You can often tell that certain students in your school are related because they look alike. The same is true for other organisms, as well.

- Find a group of plants near your home or school. In your Science Notebook, draw the plants that you find and make a list of the features of the plants. For example, write whether they have large leaves or small leaves.

- Based on your observations, decide which plants are the same type and which are different.

SCI
LINKS
THE WORLD'S A CLICK AWAY

www.scilinks.org
Genes **Code:** WGB04A
Heredity **Code:** WGB04B

4.1 Basic Principles of Heredity

Before You Read

Preview the lesson by reading the headings and the Learning Goals. In your Science Notebook, write each heading as it appears here. Then turn each heading into a question. Leave at least five lines of space after each question. Use one or two of those lines to answer each question based on what you already know.

Learning Goals

- Examine the law of dominance and the experiments that led to it.
- Compare the genotype of an individual with its phenotype.
- Recognize that some traits can be understood as incomplete dominance or codominance.

New Vocabulary

trait
heredity
genetics
gamete
pollination
fertilization
hybrid
dominant
recessive
gene
allele
law of dominance
law of segregation
genotype
homozygous
heterozygous
phenotype
incomplete dominance
codominance

Have you ever heard people say that a certain person looks like her mother, her father, or maybe even her Aunt Inez? People tend to look like other members of their family because they share similar traits. The **traits** of an organism are its characteristics. The color of your eyes, the shape of your nose, and the texture of your hair are just some of your many traits.

Some traits are passed from parents to offspring. The passing of traits from one generation to the next is known as **heredity** (huh REH duh tee). The field of biology devoted to studying heredity is called **genetics** (juh NE tihks).

Gregor Mendel

In the mid-nineteenth century, an Austrian monk named Gregor Mendel became one of the first people to carefully study genetics. For his studies, Mendel observed pea plants from his garden. Pea plants reproduce when sex cells called **gametes** (GA meets) join together. Male gametes, or sperm, are produced in pollen grains formed in a male reproductive organ. Female gametes, or eggs, are produced in a separate female reproductive organ.

During **pollination** (pah luh NAY shun), pollen grains from a male reproductive organ are transferred to a female reproductive organ. There, the male gamete unites with the female gamete in a process called **fertilization** (fur tuh luh ZAY shun). The zygote, or the cell that results, then develops into a seed.

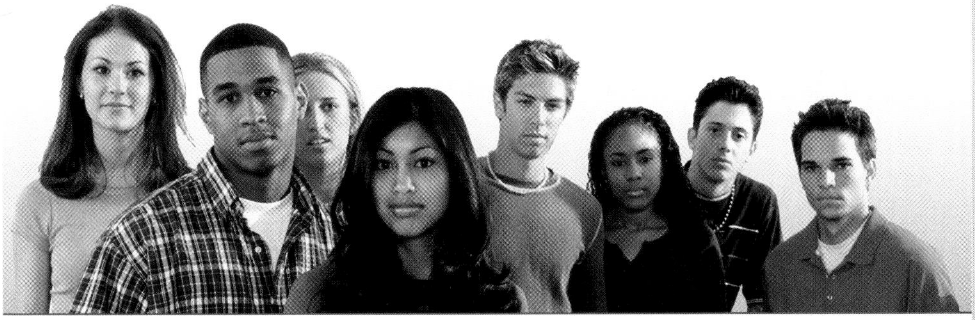

Figure 4.1 These students have many different traits. How would you describe your traits?

Monohybrid Cross

In pea plants, both male and female reproductive organs are located in the same flower. They are tightly protected by the flower petals. Pollen from other flowers is therefore unable to enter a flower on a different plant. Pollination that occurs within a single flower is known as self-pollination.

If Mendel wanted two different plants to reproduce, he could open the petals of a flower to remove the male reproductive organs. He could then brush pollen from a different plant onto the female reproductive organ. This process is known as cross-pollination and is illustrated in **Figure 4.3**.

What was so important about being able to cross different plants? By choosing which plants reproduced, Mendel could study how specific traits are inherited, or passed along. For example, he could observe the flower color in the plants produced when a plant with white flowers was crossed with a plant with purple flowers.

Mendel selected plants that were true-breeding for different versions of the same trait. A true-breeding plant is one that, when self-pollinated, always produces offspring with the same version of a trait. He then cross-pollinated them and observed their offspring.

The offspring of parents that have different forms of a trait are called **hybrids** (HI brudz). A cross between two parents that differ only in one trait is called a monohybrid cross. The prefix *mono-* means "one." An example of a monohybrid cross is one between two plants that are identical in every way except height.

Figure 4.2 The Austrian monk Gregor Mendel is sometimes known as the father of genetics. He made careful observations of pea plants over many generations to draw conclusions about how traits are inherited.

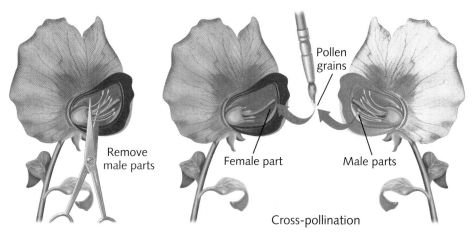

Pollen grains

Remove male parts

Female part

Male parts

Cross-pollination

Figure 4.3 Mendel would transfer pollen from one plant to another plant with different traits. This is known as crossing the plants, or making a cross.

Dominant and Recessive Traits

For one of his earliest experiments, Mendel crossed a tall pea plant with a short pea plant. The parent plants that are first crossed are known as the P_1 generation.

Mendel planted the seeds that resulted from the cross. The offspring plants are called the F_1 generation. Mendel discovered that all of the offspring in the F_1 generation grew to be tall. It was as if the shortness trait had disappeared.

Mendel then allowed the plants of the F_1 generation to self-pollinate. Again, he planted the seeds that were produced and observed the plants that grew. In this generation, known as the F_2 generation, some plants were tall and some were short. The shortness trait had reappeared.

Mendel conducted similar monohybrid crosses to test for other traits. He tested for such traits as seed shape, seed color, flower color, and flower position. In every case, Mendel discovered that one trait seemed to disappear in the F_1 generation and reappear in the F_2 generation.

Mendel concluded that each trait is controlled by two factors. He called the form of a trait that appeared in the F_1 generation the **dominant** (DAH muh nunt) trait. He named the form of the trait that disappeared until the F_2 generation the **recessive** (rih SEH sihv) trait.

As You Read

As you read the text under each heading, write a summary of the information in your Science Notebook. Then answer the questions you wrote about each heading.

What is a dominant trait?
What is a recessive trait?

Figure It Out

1. Which trait for height did the F_1 generation show?

2. What fraction of the plants of the F_2 generation was tall? What fraction was short?

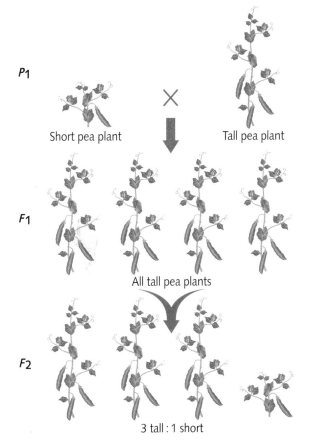

P₁

Short pea plant   Tall pea plant

F₁

All tall pea plants

F₂

3 tall : 1 short

Figure 4.4 Mendel crossed true-breeding tall pea plants with true-breeding short pea plants.

Explore It!

In domestic rabbits, black fur is dominant over white fur. A breeder crossed a true-breeding black rabbit with a true-breeding white rabbit. Draw a diagram predicting the color of the F_1 offspring. Then predict what would happen if a rabbit from the F_1 generation were crossed with an offspring from an identical cross.

	Seed Shape	Seed Color	Flower Color	Flower Position	Pod Color	Pod Shape	Plant Height
Dominant Trait	round	yellow	purple	axial (side)	green	inflated	tall
Recessive Trait	wrinkled	green	white	terminal (tips)	yellow	constricted	short

Figure 4.5 Mendel studied seven different traits in pea plants. Each trait had a dominant allele and a recessive allele.

Alleles

Scientists now call the factors Mendel described **genes**. Different forms of a gene are called **alleles** (uh LEELZ). Pea plants, for example, have one allele for tall height and one allele for short height. In pea plants, tall height is dominant over short height.

Each plant receives two alleles for a trait—one from each parent. The dominant trait will appear if the plant has at least one dominant allele for it. The recessive trait will appear only if the plant does not have a dominant allele and has two recessive alleles. This is known as the **law of dominance**.

To describe alleles, scientists use an uppercase letter for the dominant allele. The lowercase version of the same letter is used for the recessive allele. The allele for tall height, therefore, is written as T, and the allele for short height is written as t. If an individual has a dominant allele and a recessive allele, the dominant allele is usually written before the recessive allele (Tt).

The law of dominance explains what Mendel observed in his pea-plant crosses. Recall that he selected true-breeding plants. These plants had two copies of the same allele for a trait. The tall plants had two alleles for tall height (TT), and the short plants had two alleles for short height (tt).

The plants in the F_1 generation received one allele from each parent. Therefore, they received an allele for tall height from the tall parent and an allele for short height from the short parent. As a result, every plant in the F_1 generation had both types of alleles (Tt). Because tall height is dominant over short height, all of the plants were tall.

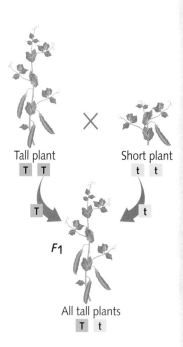

Tall plant
T T

Short plant
t t

T

t

F_1

All tall plants
T t

Figure 4.6 A plant receives one allele from each parent. The tall parent can give only the allele for tall height (T). The short parent can give only the allele for short height (t). Why is the offspring tall?

Law of Segregation

Each of the plants of the F_1 generation had both types of alleles—T and t. The gametes of these plants randomly receive only one allele. It can be either an allele for tall height (T) or an allele for short height (t). Mendel described this as the **law of segregation**. When two gametes combine, their alleles combine as well. **Figure 4.7** shows the possible combinations of alleles.

Three of the combinations include the dominant allele for tall height. As long as at least one of these alleles is present, the plant will be tall. If only the recessive alleles are present, as in the fourth combination, the plant will be short. This explains why three out of four of the offspring in the F_2 generation are tall and one out of four is short.

Genotypes and Phenotypes

Can you tell which alleles a pea plant has just by looking at it? Not necessarily. After all, a tall pea plant might have two alleles for tall height (TT) or one allele for tall height and one allele for short height (Tt). The combination of alleles is called an organism's **genotype** (JEE no tipe). An organism is said to be **homozygous** (hoe moe ZI gus) for a trait if its two alleles for a trait are the same. It is **heterozygous** (he tuh roe ZI gus) if it has two different alleles for a trait.

The trait that an organism displays is called its **phenotype** (FEE no type). When you see that a plant is tall or short, you observe its phenotype.

Figure It Out

1. How many different genotypes does the F_2 generation have?

2. How many different phenotypes does the F_2 generation have?

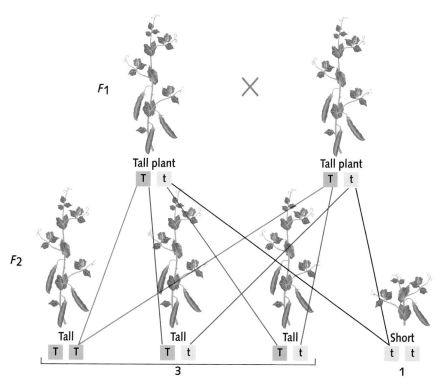

Figure 4.7 According to the law of segregation, a parent passes on at random only one allele for each trait to each offspring.

Incomplete Dominance

The traits that Mendel studied were either dominant or recessive. For some traits, however, no trait is completely dominant over another as shown in **Figure 4.8**. Had Mendel chosen a plant known as a snapdragon, he would have observed a very different result. The reason is that snapdragons display **incomplete dominance**, in which the offspring show a blend of the traits of each parent.

Snapdragon plants have alleles for red flowers and alleles for white flowers. Neither allele is dominant over the other. If a homozygous plant with red flowers is crossed with a homozygous plant with white flowers, all of the offspring will have pink flowers.

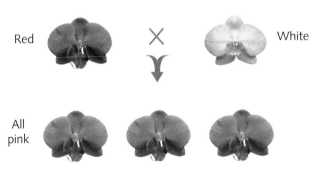

Red   X   White

All pink

Figure 4.8 Some organisms display incomplete dominance, for example, if neither allele for flower color is dominant over the other. How can you tell the genotype of the plant by looking at the phenotype of the flowers?

Codominance

When a certain variety of black chicken is crossed with a white chicken, the feathers of the offspring are black and white. In this type of inheritance, known as **codominance**, both traits are displayed in the offspring. As in incomplete dominance, neither allele is dominant over the other. However, in codominance, the alleles do not blend together.

Figure 4.9 This chicken is the offspring of a black chicken and a white chicken. What color would the chicken have been if its feather color were determined by incomplete dominance?

After You Read

1. In pea plants, round seeds are dominant over wrinkled seeds. What combination of alleles must a plant have to produce wrinkled seeds?

2. According to the summaries you wrote in your Science Notebook, how is the genotype of a homozygous organism different from the genotype of a heterozygous organism?

3. How is codominance similar to incomplete dominance? How are the two types of inheritance different?

4.2 Punnett Squares

Learning Goals

- Demonstrate the use of a Punnett square.

- Calculate the percent values that describe the offspring of a cross.

- Recognize the difference between probability and actual outcomes.

New Vocabulary

Punnett square
percent
probability

Before You Read

Create a K-W-L-S-H chart in your Science Notebook. In the *K* column, write a few notes describing what you already know about percents and probability. Include examples of events that are described using these two terms.

To learn about heredity, Mendel studied and recorded the details of his crosses over the course of many years. Is it possible, however, to predict the results of a cross in advance? In 1905, an English biologist named Reginald Punnett developed a chart that can be used to predict the possible genotypes of the offspring of a cross between two organisms. The chart he developed is called a **Punnett square**.

To use a Punnett square, a box with four squares in it is drawn. The genotype of one parent is written across the top, and the genotype of the other parent is written down the side of the chart. It does not matter which parent is written in which place. For example, look at the Punnett square in **Figure 4.10**. It shows the cross between two heterozygous pea plants. Remember that a heterozygous organism has two different alleles for a trait. In the box at the top left, the first allele from each parent is written. For this cross, the first allele from each parent is *T*, so *TT* is written in this box. In the box at the bottom left, the first allele from the parent shown at the top (*T*) and the second allele from the parent along the side (*t*) are written. The remaining two boxes are completed using the appropriate combinations.

Figure It Out

1. How many possible combinations are homozygous?

2. How would the results change if one of the parents were homozygous tall?

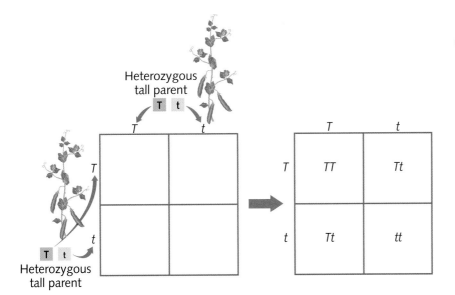

Figure 4.10 A Punnett square shows the alleles of the parents. It also shows the possible combinations of alleles of the offspring.

Genotypes
TT = 25%
Tt = 50%
tt = 25%

Phenotypes
tall = 75%
short = 25%

	T	t
T	TT	Tt
t	Tt	tt

Figure 4.11 Each possible outcome of a cross can be described by a percent (%).

As You Read

Complete the second column in your K-W-L-S-H chart. Fill in three things you want to know about percents and probability, and explain how they relate to heredity.

Explore It!

Flip a coin ten times, and record whether it lands heads-up or tails-up after each flip.

Flip the coin 20 more times, and again record your observations.

How do the actual results compare with the predicted probability of 50 percent? How does the comparison change as the number of flips increases?

Percents

Once the Punnett square is completed, you know the possible genotypes of the offspring. You can then describe them using percents. A **percent** describes a part of one hundred. Fifty percent, for example, means 50 out of 100.

In the Punnett square in **Figure 4.11**, one out of four boxes has the TT genotype. Another way to say "one out of four" is one-fourth, or $\frac{1}{4}$. A fraction can be written as a percent. The fraction $\frac{1}{4}$ is equal to 25 percent. So you can predict that 25 percent of the offspring will have the TT genotype. The same is true for the tt genotype. Half of the boxes, or two out of four, have the Tt genotype. The fraction $\frac{2}{4}$ is the same as the fraction $\frac{1}{2}$. This fraction is equal to 50 percent.

Now think about the phenotypes of the offspring. A plant will be tall as long as it inherits one dominant allele. Three of the four boxes, or $\frac{3}{4}$, include a dominant allele. This fraction is equal to 75 percent. Therefore, the Punnett square predicts that 75 percent of the offspring will be tall and that one out of four offspring, or 25 percent, will be short.

Probability

Do the percents you found mean that if two plants produce 100 offspring, exactly 25 will be short? No, they don't. Punnett squares help you predict the likelihood that combinations of alleles will be produced. The likelihood that a certain event will occur is known as **probability**. The actual outcome may be higher or lower than the predicted number.

As an example, think about flipping a coin. When you flip a coin, it might land heads-up or tails-up. There are two possible outcomes. The probability that the coin will land heads-up is one out of two, which is 50 percent. You can predict that if you flip a coin ten times, it will land heads-up 50 percent of the time, or five times. If you then go ahead and flip a coin ten times, it may land heads-up one time, four times, eight times, or even ten times. The result is a matter of chance.

It is the same way with living organisms. The Punnett square helps you make predictions about the offspring of a cross. Only careful observations tell the actual results of the cross.

After You Read

1. In some guinea pigs, black hair (*B*) is dominant over brown hair (*b*). A black guinea pig (*BB*) is crossed with a brown one (*bb*). In a Punnett square for this cross, how many boxes would have the genotype *Bb*?

2. What percent of the offspring of the cross in Question 1 might be black?

3. Complete the last three columns of your K-W-L-S-H chart. Based on your chart, explain how percents and probability are related to Punnett squares.

4.3 Chromosomes

Learning Goals

- Describe the structure and function of chromosomes.

- Contrast mitosis and meiosis.

- Explain how chromosomes determine gender and sex-linked traits.

New Vocabulary

chromosome
centromere
homologous chromosome
mitosis
cytokinesis
meiosis
sex chromosome
autosome
sex-linked trait

Before You Read
You can usually trace back every effect you observe to something that caused it. In your Science Notebook, describe an example of a cause and its effect from your everyday life. Then look for examples of cause and effect as you read this lesson.

Height in pea plants, color in snapdragons, and feather color in chickens are just a few of the many traits determined by genes. A single organism can have tens of thousands of different genes. Genes are not located randomly in cells. Instead, genes are lined up on structures called **chromosomes** (KROH muh sohmz). Each chromosome can contain more than one thousand genes.

Chromosomes are located within the nucleus of eukaryotic cells. Each chromosome is made up of a combination of DNA wrapped with proteins. Recall from Chapter 2 that DNA is an organic molecule that carries the information of heredity.

Much of the time, a chromosome is a structure that looks like a thin thread. Just before a cell divides, a chromosome shortens and takes on an X shape. Each side of an X-shaped chromosome is called a chromatid. The two chromatids, called sister chromatids, are held together at a point called the **centromere** (SEN truh mihr). Sister chromatids are exact copies of each other. Chromosomes can be seen clearly only when the cell is about to divide.

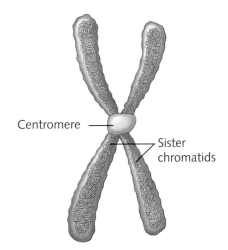

Centromere —

Sister chromatids

Figure 4.12 A chromosome is shaped like an X just before a cell divides. The sister chromatids are held together by a centromere.

Homologous Chromosomes

Chromosomes occur in pairs in the cells of organisms that have two parents. This type of reproduction is known as sexual reproduction. One chromosome in each pair came from the male parent. The other came from the female parent. The number of chromosomes is different from one organism to another. **Figure 4.13** lists the chromosome number of several species.

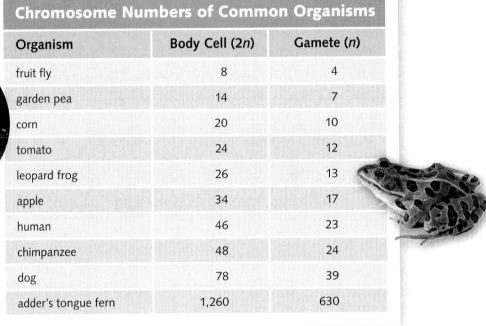

Chromosome Numbers of Common Organisms		
Organism	Body Cell (2*n*)	Gamete (*n*)
fruit fly	8	4
garden pea	14	7
corn	20	10
tomato	24	12
leopard frog	26	13
apple	34	17
human	46	23
chimpanzee	48	24
dog	78	39
adder's tongue fern	1,260	630

Figure 4.13 The body cells of each species have a characteristic number of chromosomes. This number is the diploid number, or 2*n*. The gametes of each species have a characteristic haploid number, or *n*.

The chromosomes in each pair are known as **homologous** (huh MAH luh gus) **chromosomes**. Both chromosomes have genes for the same trait arranged in the same order. However, the genes may be different alleles. Recall that alleles are genes that control different versions of the same trait. Therefore, the two chromosomes in a homologous pair are not necessarily genetically identical.

Extend It!

Sometimes, a new discovery disagrees with what scientists already believe. This can make it hard to accept the new finding. This is exactly what happened when a scientist named Barbara McClintock discovered jumping genes in 1950. Find out who she was and what jumping genes are.

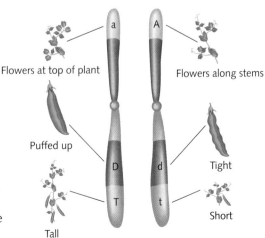

Figure 4.14 This diagram shows two homologous chromosomes from garden pea plants. These particular chromosomes contain genes that determine the position of the flowers, the shape of the pea pod, and the height of the plant. Each chromosome has a different allele for each trait.

Flowers at top of plant · Flowers along stems · Puffed up · Tight · Tall · Short

Mitosis

Recall from Chapter 3 that organisms grow when cells divide. Before they divide, cells make a complete copy of their chromosomes. The nucleus of the cell then divides in a process called **mitosis** (mi TOH sis).

Scientists often describe mitosis as having the four stages shown in **Figure 4.15**. During prophase, the first stage, the chromosomes group together and the nuclear envelope disappears. In metaphase, the second stage, the chromosomes line up across the center of the cell. During anaphase, the third stage, the centromeres split and the sister chromatids move to opposite ends of the cell.

In telophase, the final stage, the cell membrane pinches in at the center of the cell. A nuclear envelope reappears around each group of now single chromosomes. Afterwards, in a process called **cytokinesis** (si tuh kih NEE sus), the cell splits into two identical cells called daughter cells. Each cell has a complete set of chromosomes.

As You Read

In your Science Notebook, identify at least three of the cause-and-effect relationships described in this lesson.

What results are caused by mitosis and meiosis?

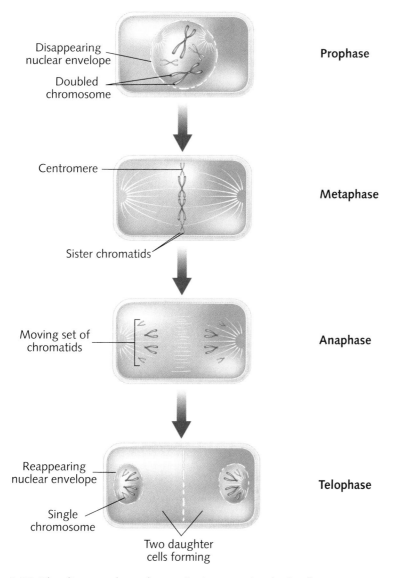

Prophase
- Disappearing nuclear envelope
- Doubled chromosome

Metaphase
- Centromere
- Sister chromatids

Anaphase
- Moving set of chromatids

Telophase
- Reappearing nuclear envelope
- Single chromosome
- Two daughter cells forming

Figure 4.15 The diagram shows how mitosis occurs in plant cells.

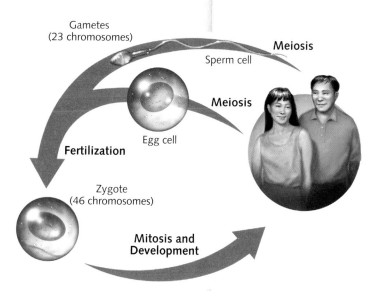

Figure 4.16 Gamete cells contain half the number of chromosomes that are in the body cells of an organism. When two gametes combine, their chromosomes join together in a nucleus. In this way, the zygote receives DNA from both parents.

Figure It Out

1. After which phase are gametes formed?

2. Which four stages most closely resemble mitosis?

Meiosis

During sexual reproduction, a male gamete combines with a female gamete. If two normal body cells combined, the resulting cell would have twice the number of chromosomes for an organism. A cell cannot function properly if it has too few or too many chromosomes. Sexual reproduction, then, must involve a form of cell division in which the nucleus divides in a way that is different from mitosis. The number of chromosomes must be reduced by half.

Gametes are produced during a type of cell division that involves **meiosis** (mi OH sus). Unlike mitosis, meiosis leads to cells with half as many chromosomes as there are in a body cell of an organism. Cell division that involves meiosis occurs only in the production of male gametes (sperm) and female gametes (eggs). When a sperm cell then combines with an egg cell, the zygote formed will have a nucleus with the correct number of chromosomes. The zygote then divides mitoticly to form the many cells of a multicellular organism.

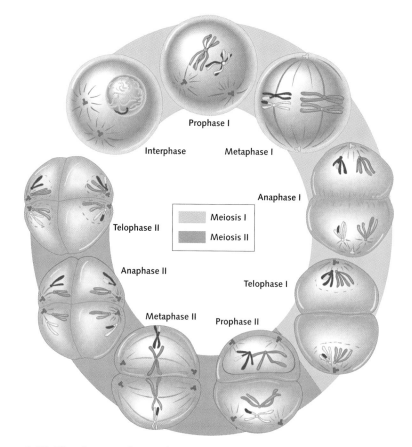

Figure 4.17 The diagram shows the phases of meiosis. Compare this diagram with the phases of mitosis shown in **Figure 4.15**.

Figure 4.17 shows that meiosis occurs in two parts. The first part, known as meiosis I, is similar to mitosis. Then in meiosis II, the nucleus splits again. In other words, the chromosomes are copied once, but the nucleus splits twice. The result is four cells, each with half the number of chromosomes that are in a body cell of an organism.

Determining Gender

Human body cells have 23 pairs of homologous chromosomes, for a total of 46 chromosomes. One pair is made up of **sex chromosomes** that determine the gender, or sex, of an individual. The other 22 pairs of chromosomes are called **autosomes**. They determine traits other than gender.

In females, sex chromosomes are described by the letters XX. Females have two X chromosomes and can pass only X chromosomes on to any gametes they produce. In males, sex chromosomes are described by the letters XY. Males have only one X chromosome and one Y chromosome. They can pass on either an X chromosome or a Y chromosome to their gametes.

It is the male gamete that determines the sex of the offspring. Look at the Punnett square in **Figure 4.18**. Every egg cell carries an X chromosome. If a sperm cell carrying an X chromosome fertilizes an egg cell, the resulting zygote will have two X chromosomes (XX). It will therefore develop into a female.

If a sperm cell carrying a Y chromosome fertilizes an egg cell, the resulting zygote will have an X chromosome and a Y chromosome (XY). It will develop into a male. The probability of producing a male is the same as that of producing a female—50 percent.

Explain It!

A couple has six children. Use probability to predict how many children should be girls and how many should be boys.

It turns out that all six children are boys. Explain how this can be possible. Write your answers in your Science Notebook.

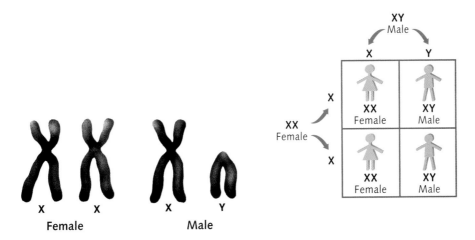

Figure 4.18 Human sex chromosomes are known as X and Y chromosomes. They are named for the letters they look like. If a couple has a child, what is the probability that it will be a girl?

Sex-Linked Traits

Like other chromosomes, sex chromosomes carry genes. The X chromosome carries many more genes than the Y chromosome does. Traits determined by genes on sex chromosomes are known as **sex-linked traits**.

One sex-linked trait is color blindness. People who are color blind cannot see the difference between certain colors, such as red and green. Red-green color blindness is caused by a recessive allele that is carried on the X chromosome. If a female inherits the allele for color blindness on one of her X chromosomes, she may have a normal allele on the other chromosome. If this is the case, the female will not be color blind. Because it is a recessive trait, a female will be color blind only if she receives the allele for color blindness from both parents. A female who carries an allele for a disorder but does not exhibit the disorder is considered to be a carrier.

	X	X^c
X	Normal XX	Carrier XX^c
Y	Normal XY	Colorblind X^{c}Y

Figure 4.19 In this Punnett square, the normal X chromosome is represented by X. The X chromosome carrying the gene for color blindness is represented by X^c. What is the probability that an offspring of these parents will be color blind?

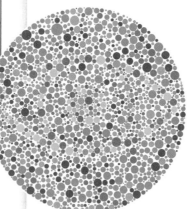

If a male inherits the allele for color blindness on the X chromosome he receives from his mother, he will be color blind. The reason is that a male has only one X chromosome. He does not have another X chromosome on which the dominant normal allele might appear. Like color blindness, many sex-linked traits are caused by recessive alleles on the X chromosome. As a result, sex-linked traits appear more often in males than they do in females.

⬤⬤ CONNECTION: History

An inherited disease called hemophilia was once known as the Royal Disease. Research the disease and the royal families of Europe to find out why. Write a summary of your findings to present to the class. If possible, include photos or diagrams in your presentation.

After You Read

1. How are DNA and genes related to chromosomes?

2. Give an example of a cell produced by mitosis and an example of a cell produced by meiosis. How are the cells different?

3. According to the notes in your Science Notebook, what causes a zygote to develop into a male?

4.4 The Role of DNA

Before You Read

Many events occur in specific orders. In your Science Notebook, describe something you do every day that takes place in a series of steps. Think about how the order of events is important. Imagine doing one of the steps out of order.

You have learned that chromosomes and the genes they carry are responsible for passing traits from parents to offspring. Now it is time to find out how they do this. As you read in Chapter 2, genes are made up of molecules of deoxyribonucleic acid, or DNA.

Structure of DNA

DNA is made up of smaller units known as nucleotides. A **nucleotide** (NEW klee uh tide) consists of a phosphate, a sugar, and a nitrogenous base. Each nucleotide contains one of four different nitrogenous bases—adenine (A), thymine (T), guanine (G), and cytosine (C). The nucleotides are arranged in a shape known as a double helix, which looks like a twisted ladder. The sugar and phosphate molecules of the nucleotides make up the sides of the ladder. Pairs of bases make up the steps and hold the two sides together. Only certain bases can pair together. Adenine pairs with thymine, and cytosine pairs with guanine.

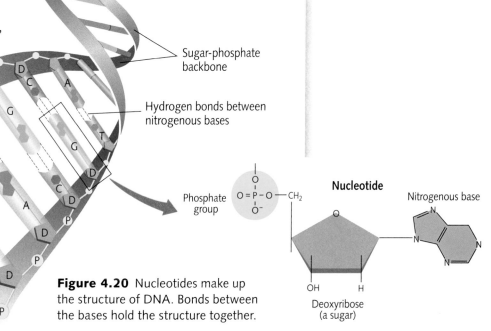

Key
D = Deoxyribose (a sugar)
P = Phosphate group

Figure 4.20 Nucleotides make up the structure of DNA. Bonds between the bases hold the structure together.

PEOPLE IN SCIENCE James Watson and Francis Crick

Details about the structure and function of DNA were not learned overnight. Nor were they discovered through the work of a single person. Instead, several scientists contributed to the development of a useful model of DNA. Among them were American biologist James Watson (top) and British physicist Francis Crick (bottom).

In the early 1950s, the pair was working toward making a three-dimensional model of DNA. No matter what they tried, nothing seemed to explain all of the observations other scientists had made about DNA. Then they learned about the research of a British scientist named Rosalind Franklin. She studied DNA by aiming a powerful X-ray beam at it. The X rays were spread out, or scattered, by the DNA. Franklin recorded the pattern of scattered X rays on a film. She found that the pattern was in the shape of an X.

Once they saw the X shape, Watson and Crick suddenly knew what they had been trying to figure out for so long. They developed a model in which the strands of DNA were twisted around each other. Their model immediately explained what was known about DNA and could be used to make predictions about it. On February 28, 1953, Watson and Crick celebrated because they knew that they had discovered the "secret of life." In 1962, they shared the Nobel Prize for their work.

DNA Replication

A copy of all of a cell's genetic material is passed along when cells divide. The process of copying a cell's DNA before cell division is called **DNA replication**. During DNA replication, the two strands of a DNA molecule separate from one another in a process that is similar to the unzipping of a zipper. New bases attach to each separated strand of DNA. As they do, the bases form pairs according to the same rule as in the original strand of DNA. For example, if the base on the old strand was adenine, the base thymine will attach to it. The process, shown in **Figure 4.21**, results in two exact copies of the original DNA molecule.

Figure It Out

1. How do the new DNA molecules compare with each other?

2. Describe the composition of each new molecule.

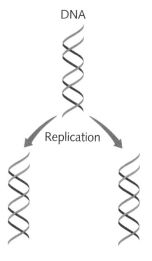

DNA

Replication

Figure 4.21 During DNA replication, two molecules of DNA are made from one. Each new molecule has one original strand and one new strand. In the diagram, the original strands are blue. The new strands are red.

Genetic Code

The order of the bases on a DNA molecule acts like a genetic code. This code directs how proteins will be put together. It is the proteins that then regulate the cell's activities. More specifically, proteins determine the traits of an organism. In this way, DNA is responsible for determining which traits are inherited from one generation to the next.

The Role of RNA

Proteins are assembled on ribosomes. Recall from Chapter 3 that ribosomes are organelles in the cytoplasm of a cell. DNA does not leave the nucleus of a cell. The cell therefore needs a way to get the information from the DNA in the nucleus to the ribosomes in the cytoplasm. This is where the nucleic acid RNA comes in. **RNA**, or ribonucleic acid, carries information from DNA to the ribosomes and then carries out the process by which proteins are made.

RNA is similar to DNA but has a few important differences. RNA is made up of a single strand of nucleotides rather than a double strand. In addition, RNA has a nitrogenous base called uracil (U) in place of thymine. The sugar in the nucleotides of RNA is ribose rather than deoxyribose, as is found in DNA nucleotides.

Protein Synthesis

The process by which proteins are made is known as **protein synthesis**. It begins when two strands of DNA separate. Bases then attach to the strands, much like they do during DNA replication. Now, however, the new strand is made of RNA instead of DNA. The RNA then leaves the nucleus and carries information into the cytoplasm. The genetic code is translated into a chain of amino acids. Recall from Chapter 2 that amino acids are the building blocks of proteins.

Although organisms have evolved many ways to protect their DNA from changes, changes in DNA sometimes occur. Any change in the DNA sequence is called a **mutation** (myew TAY shun). Mutations can be caused by errors in DNA replication or cell division, or by external agents such as nuclear radiation. Mutations can be harmful or have little or no effect.

As You Read

In your Science Notebook, create a sequence chart to describe the events that must occur for proteins to form.

How does information encoded in DNA get to ribosomes?

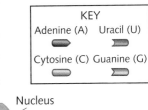

KEY
Adenine (A)   Uracil (U)
Cytosine (C) Guanine (G)

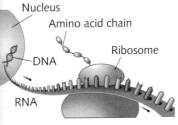

Nucleus
Amino acid chain
Ribosome
DNA
RNA

Figure 4.22 The information in RNA is used to direct the arrangement of amino acids to form proteins. On what organelle are proteins formed?

After You Read

1. Builders use blueprints that provide instructions about how to construct a building. How does DNA act as a blueprint for living things?

2. Why is RNA essential to living things?

3. Proteins determine the traits of organisms. Use the information from your chart to write a well-developed paragraph describing the sequence that explains how DNA is related to proteins.

Changing Heredity

Before You Read

Draw a T-chart in your Science Notebook. This chart can be used to compare and contrast processes. Label one side *Selective Breeding* and the other side *Genetic Engineering*. As you read the lesson, describe how the processes are alike and different.

Detailed knowledge of DNA, genes, and chromosomes is relatively new because of the complex technology scientists needed to study such small structures. That does not mean that the study of genetics is new, however. For thousands of years, people have been using their observations to breed plants and animals with specific traits.

Selective Breeding

In a process known as **selective breeding**, people choose plants and animals with preferred traits and allow them to reproduce. A farmer practicing selective breeding might allow only cows that give large quantities of milk or hens that lay large numbers of eggs to reproduce. In a similar way, a plant breeder might cross only plants that produce the largest vegetables or the sweetest fruits.

Increasing the frequency of a desired allele in a population is the goal of selective breeding. However, this often requires time and patience. For animals, it takes several generations of offspring before the desired trait becomes common within the population of the species.

Figure 4.23 Corn is one of many crops grown from seeds that are produced through selective breeding techniques. In the 1930s, midwestern farmers discovered that crossing different plants could result in a plant that was stronger and produced more corn than either plant alone.

Hybridization One selective breeding technique known as hybridization is used to take advantage of the most desirable traits of different organisms. In **hybridization**, breeders mate or cross organisms with different traits. A wheat plant that produces a large amount of wheat, for example, might be crossed with a wheat plant that is resistant to disease. The goal would be to create plants that produce a large amount of wheat and are resistant to disease.

Occasionally, organisms from different species are bred together. Female horses have been bred with male donkeys to produce mules, such as the one shown in **Figure 4.24**.

Figure 4.24 Although mules cannot reproduce, they are extremely useful. Mules are stronger, healthier, and can stand greater differences in temperature than either the donkeys or horses from which they are produced.

Figure 4.25 This thoroughbred racehorse comes from a line of horses that have been bred to run fast over long distances.

Inbreeding Another type of selective breeding known as **inbreeding** involves mating or crossing organisms with similar traits. The goal of this practice is to keep the traits in an organism the same over time and to eliminate undesired traits. A group of organisms within a species that has been bred to have specific traits is known as a breed. Breeders of dogs and horses often take advantage of inbreeding.

As You Read
On the correct side of the T-chart in your Science Notebook, add the terms *hybridization* and *inbreeding*. Make a list of the characteristics of each.

What is hybridization? What is inbreeding?

Genetic Engineering

Selective breeding is not a technique that occurs quickly. Changes occur over several generations and only after years of careful observation. A faster and more complex method of controlling the genetic makeup of an organism is **genetic engineering**. This term covers a variety of techniques that directly change the hereditary material of an organism.

Recombinant DNA   One method of genetic engineering involves cutting a portion of the DNA from one organism and inserting it into another organism. DNA formed by combining pieces from different sources is described as **recombinant** (ree KAHM buh nunt) **DNA**.

Scientists generally transfer DNA from a more complex organism to a simpler one. For example, they might transfer DNA from a human to a bacterial cell, as **Figure 4.26** shows. In a bacterial cell, some of the DNA is in a circle called a plasmid. Scientists take out the plasmid, cut it open, and insert the human DNA. The plasmid containing the human DNA is then returned to the bacterial cell.

The organism into which the recombinant DNA is inserted is known as the host organism. The host organism uses the foreign DNA as if it were its own. Recall that bacteria reproduce by binary fission. In this process, a copy of all the DNA in a cell is made. The cell then divides into two cells that are identical to the original cell. When a bacterial cell containing human DNA reproduces, the cells produced also contain human DNA.

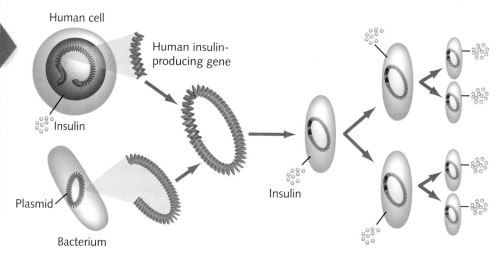

Human cell

Human insulin-producing gene

Insulin

Plasmid

Bacterium

Insulin

Figure 4.26 An enzyme is used to remove the human gene for insulin. The same enzyme is also used to cut open bacterial plasmids. The human gene is then inserted into the plasmid and returned to the bacterial cell.

Cloning   Another technique in which scientists alter the natural processes of heredity is cloning. **Cloning** involves using a single parent to produce genetically identical offspring.

To produce a clone of a mammal, scientists obtain an egg cell from a donor organism. They then remove the nucleus, and therefore the DNA, from the cell. The next step is to take a cell from the adult being cloned. The nucleus from this cell is fused into the egg cell. The result is a complete cell that can be made to divide. Once the single cell divides into a ball of cells, it is placed inside a host mother. There it develops normally into an offspring that is genetically identical to the original.

Benefits and Risks

Why would scientists want to insert human DNA into bacteria or clone animals? One reason is to produce substances that humans need. Insulin is a protein humans need to control the level of sugar in the blood. People with a condition known as diabetes must take injections of insulin every day. Genetically engineering bacterial cells that produce human insulin creates a plentiful supply of needed insulin. Researchers hope to use cloning to produce tissues and organs that people need.

Another benefit is the ability to alter the traits of organisms in a way similar to selective breeding. A number of crop plants, such as wheat, have been genetically engineered to grow larger, faster, and more resistant to disease. Cloning is widely used to produce large numbers of identical ornamental plants and fruit and nut trees.

Another goal is to use cloning to produce organisms that are on the verge of extinction.

Unfortunately, changing natural processes raises its own concerns. Fields of genetically identical wheat plants may be wiped out by a single disease. In addition, the risks that altered organisms pose to the environment may not be known for many years. When it comes to cloning animals, many individuals may be hurt or destroyed in the process of creating a single clone. Even when a mammal is cloned, it is often born with disorders that the original animal did not have. Scientists continue to learn more about these processes and their results so that people can make informed decisions about these technologies.

Figure 4.27 In 2005, this Afghan hound puppy, named Snuppy, became the first cloned dog. He was formed from an ear cell of an adult dog and developed in a yellow Labrador retriever.

CONNECTION:
Economics

Genetic engineering can have economic consequences. For example, plants that produce larger yields might bring greater profits to farmers. The hopes of higher profits, however, might lead people to ignore ethical concerns. Cloned pets are being sold for large sums of money. Although they are purchased by people hoping to replace a beloved pet, the cloned pets merely look like their originals. They may have different personalities or health problems.

After You Read
1. How is the goal of hybridization different from that of inbreeding?
2. How can bacteria be made to produce human insulin?
3. Examine the information in your T-chart. What is a benefit of genetic engineering? What is a possible risk, or problem, caused by genetic engineering?

Summary

KEY CONCEPTS

4.1 Basic Principles of Heredity

- A dominant allele is expressed over a recessive allele in an organism.

- The genotype of an individual is the alleles it has for a trait. Its phenotype is the form of the trait displayed.

- In incomplete dominance, two traits blend together. In codominance, two different traits are displayed in an offspring.

4.2 Punnett Squares

- A Punnett square can be used to predict the possible genotypes of the offspring of a cross between two individuals.

- The probability, or likelihood, of the outcome of a cross can be predicted. However, the actual outcome of a cross is a matter of chance.

4.3 Chromosomes

- A chromosome is a structure that contains genes.

- Mitosis occurs during cell division that results in two genetically identical cells.

- Meiosis occurs during cell division that results in four gametes with half the normal number of chromosomes.

- Sex chromosomes determine the gender of an individual. Autosomes determine traits other than gender.

4.4 The Role of DNA

- DNA carries the instructions that direct the formation of proteins through its arrangement of nucleotides.

- RNA carries the instructions from DNA to ribosomes and then organizes amino acids into proteins.

- During protein synthesis, the proteins that determine the traits of an organism are made.

- A mutation, or change in the DNA sequence, can result from an error in DNA replication or cell division, as well as from external agents.

4.5 Changing Heredity

- Hybridization involves crossing individuals with different traits, whereas inbreeding involves crossing individuals with similar traits.

- Recombinant DNA is formed when DNA from one organism is combined with DNA from another organism. Cloning occurs when a single parent is used to produce genetically identical offspring.

- Genetic engineering has both possible benefits and risks.

VOCABULARY REVIEW

Write each term in a complete sentence.

4.1
trait, p. 57
heredity, p. 57
genetics, p. 57
gamete, p. 57
pollination, p. 57
fertilization, p. 57
hybrid, p. 58
dominant, p. 59
recessive, p. 59
gene, p. 60
allele, p. 60
law of dominance, p. 60
law of segregation, p. 61
genotype, p. 61
homozygous, p. 61
heterozygous, p. 61
phenotype, p. 61
incomplete dominance, p. 62
codominance, p. 62

4.2
Punnett square, p. 63
percent, p. 64
probability, p. 64

4.3
chromosome, p. 65
centromere, p. 65
homologous chromosome, p. 66
mitosis, p. 67
cytokinesis, p. 67
meiosis, p. 68
sex chromosome, p. 69
autosome, p. 69
sex-linked trait, p. 70

4.4
nucleotide, p. 71
DNA replication, p. 72
RNA, p. 73
protein synthesis, p. 73
mutation, p. 73

4.5
selective breeding, p. 74
hybridization, p. 75
inbreeding, p. 75
genetic engineering, p. 76
recombinant DNA, p. 76
cloning, p. 77

True or False
If the statement is true, write "true." If it is false, change the underlined word or words to make the statement true.

1. A <u>dominant</u> trait will be expressed only if the organism has two alleles for the trait.

2. A fruit fly is <u>homozygous</u> if it has two different alleles for wing length.

3. A Punnett square is used to <u>predict</u> the genotypes of the offspring of a cross.

4. The sister chromatids of a chromosome are held together at the <u>clone</u>.

5. Gametes are formed through the process of <u>mitosis</u>.

6. DNA is in the form of a <u>single</u> helix.

7. The goal of <u>inbreeding</u> is to maintain the same traits in a line of individuals.

Short Answer
Answer each of the following in a sentence or brief paragraph.

8. In pea plants, green pods (*G*) are dominant over yellow pods (*g*). Compare the genotype and phenotype of a plant that is *Gg* for this trait with plants that are *GG* and *gg*.

9. In pea plants, yellow seeds are dominant over green seeds. Two homozygous parents, one with yellow seeds and one with green seeds, are crossed. What percent of the offspring will display the recessive trait? Use a Punnett square to find the answer.

10. Describe homologous chromosomes, and relate them to the parents of an organism.

11. Discuss how mitosis and meiosis are both essential to multicellular organisms that carry out sexual reproduction.

12. Explain how a single DNA molecule can form two identical copies.

13. Why is the nucleus removed from an egg cell at the start of the mammal cloning process?

Critical Thinking
Use what you have learned in this chapter to answer each of the following.

14. **Cause and Effect** Why is it difficult to determine the genotype of an organism in which one form of a trait is dominant over the other?

15. **Infer** Why are sex-linked traits observed more often in males than in females?

16. **Relate** How has selective breeding been used in farming?

Standardized Test Question
Choose the letter of the response that correctly answers the question.

	T	*t*
T	1	2
t	3	4

17. The diagram shows a cross between two organisms. Which of the following statements is true?

   A. Individual 1 is heterozygous.

   B. Individuals 1 and 4 are homozygous.

   C. Individuals 2 and 3 are homozygous.

   D. Individual 4 is heterozygous.

Test-Taking Tip

Make sure you understand what kind of answer you are being asked to provide. Pay attention to words such as *illustrate*, *list*, *define*, *compare*, *explain*, and *predict*. A graphic organizer might help you organize your thoughts if you see a word such as *list* or *compare*.

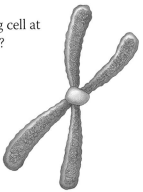

Creatures of the Dark

DEEP BENEATH the jagged mountains and towering trees of Sequoia National Park is a place unlike any other. It is a world of deep, marble caves full of strange creatures that live in total darkness. Recently, the caves gave up an amazing secret—the existence of at least 27 animal species never before seen on Earth.

You might wonder how it's possible to discover any new forms of life on Earth. Haven't we already found everything there is to find? Not at all! There are still places on Earth so hard to reach that they have yet to be thoroughly explored. These places include the ocean floor and the deepest parts of some caves.

Scientists have been exploring the caves in California's Sequoia and Kings Canyon national parks for years. Finding one or two new species of small invertebrates, or animals without backbones, wouldn't have been startling. But finding 27 new species shocked even the most experienced cave-exploring scientists.

The scientists collected several species of millipedes, centipedes, spiders, animals related to scorpions, and insects. Many of them are very strange-looking. One spider is bright orange and glows in the dark. Another is transparent, eyeless, and has extra-long legs. One spiderlike creature has jaws bigger than the rest of its body. What makes these creatures even more amazing is that they are found *only* in these caves—nowhere else. In fact, some are found only in one room of the caves!

Many of the animals are troglobites, or species adapted to spending their entire lives in caves. In fact, they cannot survive outside of this dark, moist environment. Each cave species is likely related to a species that lives on the surface. However, as a result of evolution, the cave dwellers are better adapted to life underground. Their adaptations are what make some of them so strange-looking. For

🔗 CAREER CONNECTION ZOOKEEPER

GIVE AN ELEPHANT A BATH. Make dinner for a rhino. Train a panda to stand still while getting a shot. It's all in a day's work for a zookeeper—someone who feeds, cares for, and sometimes trains zoo animals.

A zookeeper's most important job is daily animal care. First on the list is making sure the animals are fed. That might mean anything from chopping vegetables for rabbits to feeding meat to hungry lions. A keeper also cleans the animal enclosures. That may not be the best part of the job, but a zookeeper is glad to do it because it keeps the animals comfortable. Each zookeeper knows a lot about the animals he or she cares for. So if you're at the zoo and have a question about an animal, a zookeeper is a good person to ask.

Another important duty for a zookeeper is helping to keep the animals healthy. The keeper observes the animals constantly. If there is a change in the way an animal looks, acts, or even smells, it could mean an illness. So the keeper tells the zoo veterinarian right away. The keeper might also assist the veterinarian in examining the animal or giving it medicine.

Most people who become zookeepers like animals and are comfortable around them. They usually have a bachelor's degree in biology or animal science. Some have a two-year degree in zookeeping. Getting a job as a zookeeper can be difficult because there are few jobs available and many people who want them. So a good combination of education and experience working with animals is the best way to start.

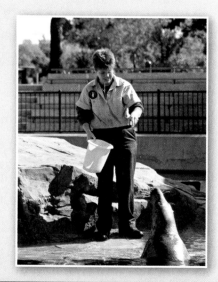

example, pigment isn't needed in the sunless world of a cave. Many animals are white or have clear bodies. Absence of eyes is also common. After all, if an animal lives where it's always black as night, there's no need for it to see. The sense of touch would be much more important. That's why cave species often have features such as elongated legs. The extra-long legs help the animals feel their way around and catch prey in the dark.

There isn't much food to eat deep in caves, either, so it's no surprise that many of the species also have small bodies that don't require a lot of food. Some of the newly discovered creatures are so tiny that even a small pair of tweezers was too big to pick them up. Scientists brushed the animals onto the tip of a paintbrush to carry them.

The scientists were eager to share their new discoveries with the world. So they preserved samples and shipped them to taxonomists, or scientists who classify living things. The taxonomists will determine exactly what these creatures are and give them names.

In the meantime, exploration of the caves will continue. There are more than 200 caves under the two national parks. Some are several kilometers long. Who knows what creatures hide in the dark, awaiting the light of curious scientists!

Research and Report

You have just learned about unusual cave creatures. Now do some research to find an animal or plant specially adapted to some other type of extreme environment. With a partner, report on that environment. Also report on the adaptations that allow the plant or animal to survive there. Share what you find with the rest of the class.

Baking with Fungus

EVEN IF you're more interested in eating bread than in baking it, you probably know that yeast makes bread rise. What you may *not* know is that yeast isn't a chemical. It's a microorganism—a one-celled organism so tiny that billions can fit in a small packet.

There are about 160 species of yeast, a type of fungus. But *Saccharomyces cerevisiae* (baker's yeast) is the one used to make bread and pizza dough. Mixed in with the dough, the yeast can produce enzymes that break down flour into sugars. Some of these sugars serve as a food supply for the yeast. As the hungry yeast feed on the sugar, two things are produced: ethanol, an alcohol that evaporates during baking, and carbon dioxide gas. The carbon dioxide gas forms little bubbles in the dough. This makes the dough rise. You see these bubbles as the little holes in baked bread or pizza crust.

Today, chemicals such as baking powder and baking soda do the same thing live yeast do. They produce bubbles of carbon dioxide gas that cause baked goods to rise. Baking powder is a mixture of baking soda and a dry acid. When added to a moist batter, the acid and baking soda react to produce carbon dioxide gas. Baking soda, or sodium bicarbonate, does not contain a dry acid. So the batter needs an acidic ingredient such as lemon juice, sour cream, buttermilk, or molasses to produce carbon dioxide gas. Baking powder and baking soda have one big advantage over yeast: they work almost instantly. Yeast takes several hours.

Even though there are other leavening agents, or substances that make dough rise, yeast remains an important one. So the next time you taste a piece of freshly baked bread, you can thank the baker . . . and the fungi that helped him or her make it!

Unit 2

Diversity of Living Things

Chapter 5

Change Over Time

How does fossil evidence support the theory that species evolve, or change over time, as the better adapted organisms survive to reproduce?

Chapter 6

Classification of Living Things

How can similarities and differences among organisms be used to classify living things?

Change Over Time

KEY CONCEPT A collection of assorted evidence shows that organisms evolve—or change over time.

Don't expect to see this ferocious cat anytime soon. The cat is a saber-toothed tiger, and it has not lived on Earth for at least 11,000 years. Yet scientists know a great deal about it: what it looked like, where it lived, what it ate, and why it disappeared.

How can modern scientists learn about organisms that lived so long ago? In this chapter, you will learn about the different types of evidence scientists used to piece together an ancient puzzle.

Think About Solving Mysteries

Do you like solving puzzles? Learning about a past event from bits of information is a lot like solving a puzzle.

- Think of an interesting event that might happen at home or in school. For example, your dog runs into the house from a muddy garden. He slides across the floor, jumps on the sofa, and heads out the back door.

- Imagine that you come in after the event has occurred. In your Science Notebook, write down the clues you see. Trade with a partner, and try to figure out what happened from the clues he or she provided.

www.scilinks.org
Evolution and Adaption
Code: WGB05

Before You Read

All people have traits that enable them to perform some tasks better than others. Some people can remember things easily, which helps them keep track of new information. Others can tell the difference between sounds, which helps them write or play music. In your Science Notebook, write how one of your traits helps you do something well.

Explain It!

The South African burrowing bullfrog inflates its body like a balloon when it senses danger. Explain what an adaptation is. Tell how this is an adaptation that helps the frog survive.

It's a hot afternoon on the Serengeti. A hungry pride of female lions is eyeing a group of zebras. As the zebras move, their stripes blend together. It is difficult for the lions to pick out a single zebra. They can't tell how many there are or exactly how far away they are. Unless one zebra strays from the crowd, the lions will go hungry—for now.

Stripes on a zebra are an example of an adaptation. An **adaptation** is any trait that helps an organism survive and reproduce itself in its environment. The mixture of stripes created when many zebras stand together makes it difficult for predators, such as lions, to attack. This helps each zebra survive.

Figure 5.1 These zebras cannot outrun or outfight the lions that hunt them, but they are not without protection. Their patterns of stripes confuse the lions. Because this helps the zebras survive, the stripes are considered an adaptation.

Prey Adaptations

Like a zebra's stripes, some adaptations protect an organism from predators. A **predator** (PRE duh tor) is an organism that kills and eats another organism for food. The organism that is eaten is known as the **prey** (PRAY).

Some adaptations help organisms hide from predators. Others, such as strong odors, cause predators to keep away. Still other adaptations, such as an antelope's fast speed, help prey escape from predators. **Figure 5.2** shows several more methods of defense against predators.

As You Read

In your Science Notebook, make a T-chart. On the left side, list several adaptations of living things. On the right, describe how each adaptation allows the organism to survive in its environment.

What is an adaptation?

Figure It Out

1. Why might a predator avoid eating a brightly colored grasshopper?

2. A harmless syrphid fly looks like a yellow jacket wasp. What is the name of this defense mechanism?

Mimicry Is this a highly poisonous coral snake? No, it's actually a harmless king snake. Mimicry is the ability to look like a different, often more dangerous, organism. Both king snakes and coral snakes have the same coloring in a slightly different arrangement. If this snake is lucky, predators will not try to find out which one it is.

Camouflage Do you see an insect? How about a stick? The color and slow movement of this stick insect make it look like one of the twigs on this tree. Camouflage is the ability to blend into the surroundings.

Protective Covering Come too close to this porcupine and you're in for a sharp surprise. The pointy quills that cover its body protect the animal from predators. Some plants have sharp needles for the same reason.

Warning Colors This poison arrow frog does not blend into its environment. In fact, its bright colors invite predators to see the frog. Like this frog, many brightly colored organisms are poisonous. The colors warn predators not to eat them.

Figure 5.2 Prey have many adaptations to protect themselves from predators. These photographs show several examples.

Figure 5.3 Predators have adaptations that help them catch prey. The python *(top)* uses heat and smell to find its prey. The owl *(bottom)* can see and hear its prey well.

Predator Adaptations

Like prey, predators have adaptations that help them survive. These include any traits that help them see and catch their prey. Female lions are very fast. This helps them catch their prey. Once they do, they have large, sharp teeth to hold and kill the prey.

Some other predators, such as owls, have excellent vision. This helps them spot a tiny mouse among the leaves on the forest floor. Most species of owls have outstanding night vision as well. In addition, their feathers direct sound toward their highly sensitive ears. This helps them find prey that might otherwise stay hidden.

Other animals, such as Burmese pythons, have heat sensors in their top lips. Along with a keen sense of smell, the sensors help them find prey. They also have jaws with a loose hinge. Once a snake has captured its prey, this loose hinge makes it possible for the snake to open its mouth wide enough to fit the prey inside.

Environmental Adaptations

Many adaptations help organisms survive in their environment. A polar bear, for example, has thick fur and a layer of fat to keep it warm in its cold surroundings. The only heat that a polar bear gives off comes from its breath. Another adaptation of the polar bear is its large feet. They help the bear paddle through water, and they spread out the bear's weight so that it can walk safely on ice.

The Saguaro cactus lives in the very dry environment of the desert. This plant can reach heights of up to 12 meters and can live to be 200 years old. The cactus has a thick, waxy stem with ribs that can expand to store water. The roots of the cactus do not grow deep into the soil. This makes it possible for the cactus to take in as much water as it can when it rains. The plant then saves some water for periods when there is no rain.

Figure 5.4 Organisms survive in the environments to which they are best adapted. A polar bear *(left)* would not survive in a hot climate because it is adapted to cold conditions. The Saguaro cactus *(right)* is adapted to a dry environment.

Variation

Remember the zebras you read about at the beginning of this lesson? As you may already know, all zebras have stripes. What you may not know is that no two zebras have the same pattern of stripes. A zebra's pattern of stripes is like a human fingerprint.

Any difference among individuals of the same species is known as a **variation** (ve ree AY shun). You can look around your classroom to see some of the variation among humans. Your classmates probably have a variety of heights, hair colors, eye colors, and skin colors.

Sources of Variation

Variation among members of the same species depends on genetics. Recall from Chapter 4 that genes are passed on to the offspring of sexual reproduction in a random way. This means that, with the exception of identical twins, individual offspring have a unique set of genes. Although the overall traits they have may be similar, there will be variation among them. This is known as **genetic diversity**. The word *genetic* refers to genes, and the word *diversity* means "difference" or "variety."

Another way that variation is introduced into a population is through mutation. A **mutation** (myew TAY shun) results when an error occurs during DNA replication, which is the process by which a copy of DNA is made. The error changes some of the proteins that are produced in an organism. If the mutation is in a sex cell, it can be passed on to offspring.

Some mutations are harmful and prevent an organism from surviving. Other mutations have little effect on an organism. Occasionally, a mutation is helpful and allows an organism to better survive in its environment.

Figure 5.6 A white tiger is a Bengal tiger that has a recessive mutation in the gene for color. The several hundred white tigers that exist today can all be traced back to the same tiger caught years ago in India.

Papilio ajax ajax

Papilio ajax ampliata

Papilio ajax curvitascia

Papilio ajax ehrmanni

Figure 5.5 These swallowtail butterflies all belong to the same species. However, they live in different parts of North America and have slight variations.

After You Read

1. Using the information you have recorded in your T-chart, hypothesize about how a wide, flat tail is an adaptation for a beaver, which lives mainly in water.

2. What does it mean to say that variation exists among the puppies in a litter?

3. How might a mutation that produces a white tiger affect the tiger's survival?

5.2 Evidence from the Past

Before You Read

It is often important to place items in order of age. In your Science Notebook, write the names of the students in your class. Work together to arrange the list in order of age from oldest to youngest. You will need to compare the months and days of all students' birthdays.

Scientists can learn a lot about plants and animals that live today just by looking at them. How can they find out about organisms that lived long ago? One way is to discover and study fossils. A **fossil** (FAH sul) is the preserved remains of an organism that was once living.

A fossil can be part of the organism itself, such as bones, teeth, or shells. It can also be a trace of an organism, such as an imprint of a leaf or a footprint. A fossil can even be something left behind by an organism, such as animal droppings.

After many years, a change in Earth's surface might cause a fossil to get near the surface. A scientist digging for fossils might find it. A scientist who uses fossils to study forms of life that existed in prehistoric times is known as a **paleontologist** (pay lee ahn TAH luh just).

Figure 5.7 Fossils, the preserved remains of once-living organisms, include marine animal shells (top), wasps preserved in amber (center), and shrimp imprints (bottom).

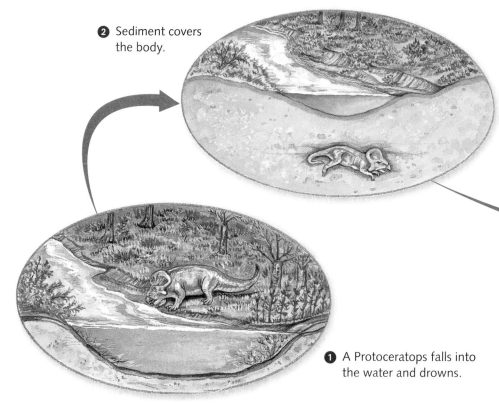

❷ Sediment covers the body.

❶ A Protoceratops falls into the water and drowns.

Figure 5.8 Few organisms become fossils. Sometimes, however, the conditions are just right for a fossil to form.

How Fossils Are Formed

When an organism dies, other organisms such as bacteria and fungi feed on it, breaking down its body. Sometimes, however, an organism is buried in mud, clay, or soil soon after it dies. Over many years, layers of sediment build up over the organism. The soft parts of the organism decay, but the harder parts are left behind.

As sediment builds up, the upper layers push down on lower layers. This pressure, along with chemical changes, causes the sediment to harden into rock. The type of rock formed in this way is known as sedimentary rock. The parts of organisms captured in the rock are fossils.

In some cases, the rock particles around an organism retain the shape of the organism even after the soft parts have decayed. This results in an imprint. In other cases, an organism is buried in ash from a volcano or in fine clay before it begins to decay. If this happens, the organism might be perfectly preserved.

Relative Dating

To understand the significance of a fossil and to recognize the conditions in which the organism lived, a scientist needs to know the age of the fossil. If layers of sedimentary rock are not disturbed, the oldest layers are at the bottom. The younger layers are at the top.

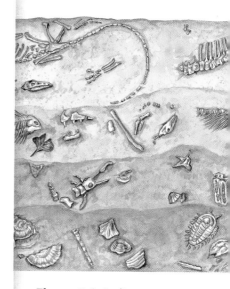

Figure 5.9 Sedimentary rocks usually form in horizontal layers. Which layer contains the oldest fossils?

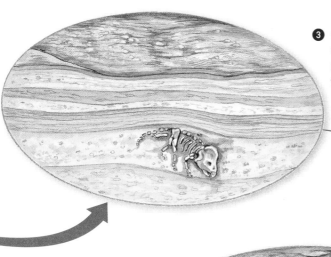

❸ Over time, additional layers of sediment compress the sediment around the body, forming rock.

❹ Earth movements or erosion may expose the fossil millions of years after it formed.

Figure It Out

1. What might cause a fossil to become exposed?

2. What might be the source of the layers of sediment that cover the body?

❺ Scientists carefully remove the fossil from the surrounding rock.

As You Read

In your Science Notebook, draw a sedimentary rock with four layers. Label the layers in order of age.

How is the relative age of a fossil different from its exact age?

Scientists can use the order of the layers of sedimentary rock to estimate the age of any fossils in the rock. Because lower layers of rock are older than upper layers, a fossil in a lower layer must be older than a fossil in an upper layer. This method of dating fossils, called **relative dating**, is used to determine which fossils are older or younger than others without giving an exact age.

Radioactive Dating

Sometimes scientists can find the exact age of a fossil. Organisms are made up of elements. Some elements are unstable. This means that they decay, or break down, into a different form. As they do so, they give off particles and energy called radiation.

An element that decays is called radioactive. Every radioactive element decays at a specific rate. By comparing the original form of the element to the form after decay, scientists can calculate the exact age of a fossil. The method of using radioactive elements to find a fossil's real age is known as **radioactive dating**.

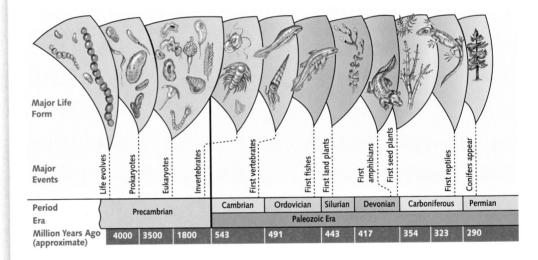

Major Life Form											
Major Events	Life evolves	Prokaryotes	Eukaryotes	Invertebrates	First vertebrates	First fishes	First land plants	First amphibians	First seed plants	First reptiles	Conifers appear
Period					Cambrian	Ordovician	Silurian	Devonian	Carboniferous		Permian
Era	Precambrian				Paleozoic Era						
Million Years Ago (approximate)	4000	3500	1800	543	491	443	417	354	323	290	

The Fossil Record

As paleontologists study fossils, they organize them into groups. They place similar organisms together and arrange them according to when the organisms lived. All of this information is combined to form the fossil record. The **fossil record** provides a snapshot of life on Earth over time.

The fossil record reveals that life on Earth has changed over time. In addition, certain fossils that appear in older rocks no longer appear in more recent rocks. This shows that some organisms have become **extinct** (ihk STINGT), which means the species died out. In fact, more than 99 percent of all the species that have ever lived on Earth are extinct. New species have also appeared at different points in time.

CONNECTION: Chemistry

An element often used in radioactive dating is carbon-14. This element is present in all living things. Even after an organism dies, carbon-14 continues to decay into nitrogen-14. Carbon-14 has a half-life of 5,730 years. By comparing the amounts of carbon-14 and nitrogen-14 in a fossil, scientists can determine the fossil's age.

Despite the tremendous number of fossils that have been found, the fossil record is far from complete. Only a small fraction of all the species that ever lived are preserved in fossils. One reason is that organisms with soft bodies or thin shells rarely form fossils. Another reason is that rocks containing fossils can be worn away or changed into other types of rocks over time. As rocks are altered, the fossils in them are usually destroyed.

The Geologic Time Scale

Scientists used the fossil record to develop the **geologic time scale**. A version of the scale is shown in **Figure 5.10**. The scale was originally developed by studying fossils and rock layers. Scientists then realized that major changes in the fossils had occurred in specific rock layers. Some of these major changes were mass extinctions. When a mass extinction occurs, many organisms disappear from the fossil record at the same time. These changes were used to divide the time scale into sections. The geologic time scale is divided into four large sections called eras. Each era is further divided into periods, and then into epochs.

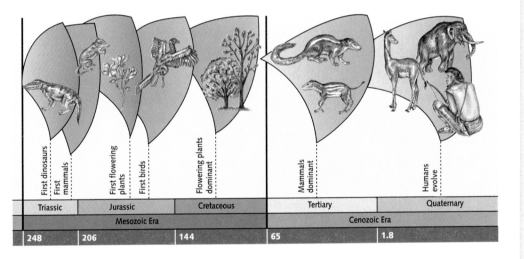

| First dinosaurs | First mammals | First flowering plants | First birds | Flowering plants dominant | Mammals dominant | Humans evolve |

| Triassic | Jurassic | Cretaceous | Tertiary | Quaternary |
| Mesozoic Era | | | Cenozoic Era | |

| 248 | 206 | 144 | 65 | 1.8 |

Figure 5.10 The geologic time scale is like a calendar of Earth's past. What evidence did scientists use to develop this scale?

Later on, when scientists were able to use radioactive dating, they identified the ages of specific rock layers. In doing so, they found that the divisions of the geologic time scale were not equal periods of time. The sections of the scale vary in length by millions of years.

After You Read

1. Why are fossils found in sedimentary rocks?

2. How is relative dating different from radioactive dating?

3. Review the sequences of ages you described in your Science Notebook. Consider how these ages are separated. What occurrences are used to separate the divisions of the geologic time scale?

Learning Goals

- Relate body structures to evolutionary change.
- Compare embryos to identify evolutionary relationships.
- Analyze how genetic material can serve as evidence of evolution.

New Vocabulary

evolution
homologous structure
analogous structure
vestigial structure
embryo

Before You Read

Preview the lesson. Read the headings and Learning Goals, and look at the pictures. Think about what you expect the lesson to be about. Write the headings in your Science Notebook.

The fossil record shows that organisms and environments have changed over time. The change in living things over time is **evolution** (eh vuh LEW shun). As an example, consider the modern camel. The camels that inhabit today's deserts look very different from the camels of long ago. **Figure 5.11** shows how scientists think camels have changed over millions of years.

Body Structure

The fossil record is not the only evidence that supports evolution. Scientists also consider the structures of the bodies of living things. For example, most animals with backbones have two pairs of limbs. Each kind of limb, such as forelimbs, has a different function. A whale uses its forelimbs to swim through the water. A crocodile uses them to swim and also to move on land. A bird uses its forelimbs to fly. Despite the different functions, the basic arrangement of bones is similar in each of these animals' forelimbs.

Figure It Out

1. How has the skull of a camel changed over time?

2. How might the size of the skull be related to the size of the camel's brain?

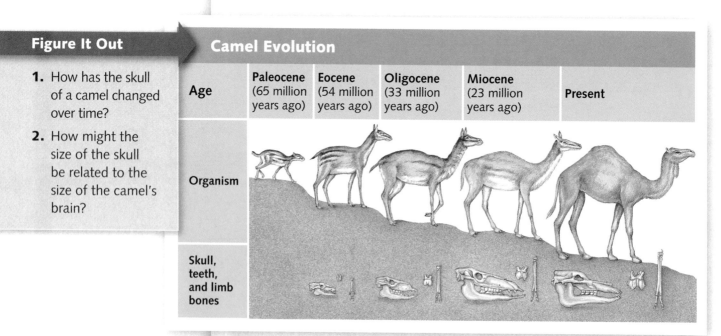

Age	Paleocene (65 million years ago)	Eocene (54 million years ago)	Oligocene (33 million years ago)	Miocene (23 million years ago)	Present
Organism					
Skull, teeth, and limb bones					

Figure 5.11 Scientists use fossils to try to determine how camels evolved.

Homologous Structures   The bones in the forelimbs of three vertebrates (animals with backbones) are shown in **Figure 5.12**. Many scientists suggest that it would be unlikely for different species to have such similar structures if each species arose separately. Instead, scientists conclude that these organisms evolved from a common ancestor. Structures that have the same evolutionary origin are called **homologous** (huh MAH luh gus) **structures**. These structures can have the same arrangement or function, or both.

As You Read

In your Science Notebook, paraphrase the information under each heading.

What are homologous structures?

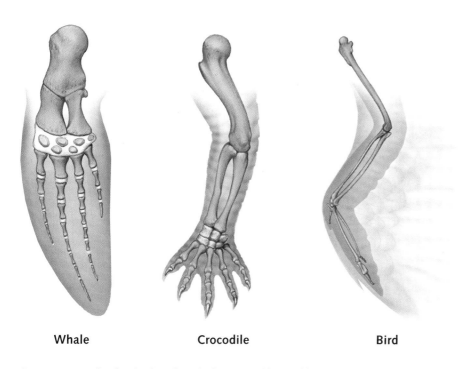

Whale   Crocodile   Bird

Figure 5.12 The forelimbs of a whale, crocodile, and bird are homologous structures. The similar bones are used to perform different functions. What are the functions?

Analogous Structures   Just because two organisms have a similar body feature does not necessarily mean they are related. Birds and butterflies, for example, both have wings. Both organisms use their wings to fly. However, the structure of a bird wing is quite different from that of a butterfly wing. These two types of wings evolved in different ancestors. Body parts that have similar functions but do not have a common evolutionary ancestor are called **analogous** (uh NAL uh gus) **structures**.

Vestigial Structures   Another type of body structure that provides evidence for evolution is a vestigial structure. An organism has a **vestigial** (veh STIH jee ul) **structure** if it has a body part that no longer serves its original purpose. The structure was probably useful to an ancestor.

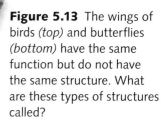

Figure 5.13 The wings of birds *(top)* and butterflies *(bottom)* have the same function but do not have the same structure. What are these types of structures called?

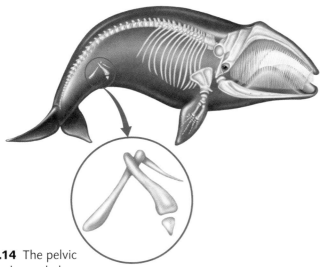

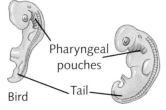

Figure 5.14 The pelvic bones of baleen whales are vestigal structures. The bones suggest that the whales arose from organisms that walked on land.

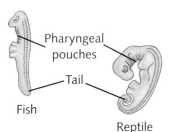

Figure 5.15 The embryos of these organisms reveal an evolutionary relationship. What structures do they have in common?

Human wisdom teeth, or third molars, are an example of a vestigial structure. At one time, humans had larger jaws with more teeth. They ate more plants and needed the teeth to grind the thick cellulose of plant tissues. As the human diet changed and included fewer plants, the size of the jaw decreased. Wisdom teeth still grow in, but they are often removed at some point because they are not needed.

Embryos

You would probably agree that an adult eagle looks very different from a rhinoceros. As embryos, however, they look very much alike. An **embryo** (EM bree oh) is the earliest stage of growth and development for plants and animals. If two embryos had the same features, scientists would conclude that they evolved from a common ancestor.

Figure 5.15 compares four embryos. The first is for a fish, such as a tuna. The second is for a reptile, such as a crocodile. The third is for a bird, such as an eagle. The last is for a mammal, such as a rhinoceros. If you look at the bottom of each embryo, you will see that it has a tail. Toward the top, each embryo has structures called pharyngeal (fuh RIN jee ul) pouches. In the fish, the pouches are related to gills. In the other organisms, they become parts of the ears, jaws, and throat.

Genetic Material

Scientists can compare the DNA of organisms to find out how they are related. Recall from Chapter 4 that DNA carries information that determines the traits of an organism. The code of bases on a DNA molecule is used to determine the order of amino acids in the proteins of an organism. The more similar the sequence of amino acids is between two organisms, the more closely the organisms are related.

After You Read

1. What does it mean to say that the fossil record suggests that evolution occurs among living things?

2. According to the notes you wrote in your Science Notebook, what are four types of evidence used to support the theory of evolution? Write a well-developed paragraph to answer this question.

3. What might a scientist conclude about the evolutionary relationship between two organisms with very similar DNA?

5.4 Humans and Evolution

Before You Read

Scientists often use time lines to place events in the sequence in which they occurred. In your Science Notebook, draw a time line as a horizontal line. Preview the lesson to look for important dates that are mentioned. Write any years you find in the correct order on your time line.

Monkeys, chimpanzees, and gorillas all have something in common. They are mammals known as **primates**. A primate can be as small as a mouse lemur with a mass of just 30 grams. It can be as large as the Eastern lowland gorilla, which has a mass of up to 250 kilograms. There are 235 different kinds of primates.

Primates

Despite how different they look, all primates share certain traits. For one thing, primates have rounded heads with faces that are generally flat. Compared with other mammals that live on land, primates have larger, more complex brains.

Most primates live in trees. Several adaptations help them survive there. These adaptations include shoulder and hip joints that have a wide range of motion. Primate hands and feet have fingers and toes with nails instead of claws. Most important, primates have a hand with an **opposable thumb**, which is a thumb that can bend across the palm of the hand. Together, these traits allow primates to grab and hold onto objects such as fruits and tree branches.

Have you ever looked through binoculars? These devices have two eyepieces to look through. Primates have binocular vision. Binocular vision enables primates to see an object through two eyes at the same time. Using two eyes at the same time, primates can figure out how far away objects are. In addition, primates are able to see in color.

Figure 5.16 These animals, the ring-tailed lemur *(left)* and the mountain gorilla *(right)*, are primates. There are several traits that describe a primate. What are some of these traits?

Extend It!

Like many other things in life, science can be affected by dishonest people. In 1912, Charles Dawson announced the discovery of a skull that "proved" that humans evolved from apes. In 1953, however, the skull was found to be a fake. Research this event, which came to be known as the Piltdown hoax. Find out what happened and how it affected people's views of evolution and science. If possible, find out who is believed to have been responsible for the hoax.

Evolution of Primates

Although scientists do not know for certain how humans developed from early primates, several theories exist. One leading theory suggests that ancient primates split into different evolutionary lines. One line evolved into modern lemurs, aye-ayes, and similar animals. Another line evolved into monkeys and hominoids. Apes and humans are known as **hominoids**. The evolutionary line that led to humans is known as the **hominid** line.

In 1924, a South African named Raymond Dart discovered the skull of an early hominoid child. It was dated to be between 2.5 million and 2.8 million years old. The shapes of the face and the region around the brain were similar to the shapes of those parts on an ape. However, the structure of the brain suggested that it belonged to an organism that walked upright.

Dart concluded that it must be a species of primate that had not been discovered before. He named it *Australopithecus africanus*, which means "southern ape from Africa." When the name of a species is written, the first word is often shown as a single letter (*A. africanus*). Specimens such as these are described as australopithecines. They are hominids that have some features of apes and some of humans.

Figure It Out

1. What are two ways to describe the groups into which gibbons are placed?
2. Which two groups of organisms are anthropoids but not hominoids?

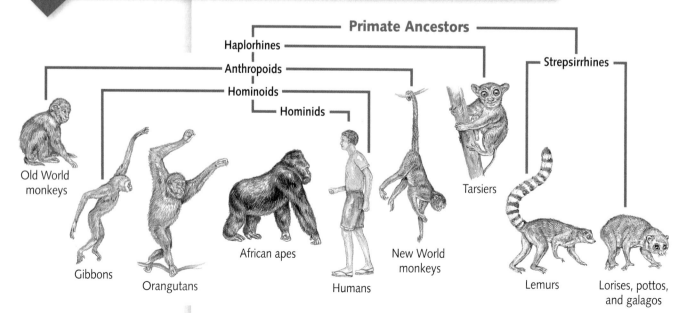

Figure 5.17 This diagram shows how ancient primates split into different groups. Follow the lines to see the organisms that evolved.

Lucy

After Dart's discovery, many similar sets of fossils were found. An American named Donald Johanson discovered one of the most complete skeletons in 1974. The skeleton, which he called "Lucy," was determined to be about 3.2 million years old. Johanson concluded that Lucy was a new species, which he named *Australopithecus afarensis*.

Individuals described as *A. afarensis* are more like humans, but their brains are still much smaller. **Figure 5.18** shows the skulls and pelvic bones of a chimpanzee, *A. afarensis*, and a modern human.

As You Read

Write a note summarizing the events described in the lesson under the related years on the time line you drew in your Science Notebook.

Why was Raymond Dart's discovery significant?

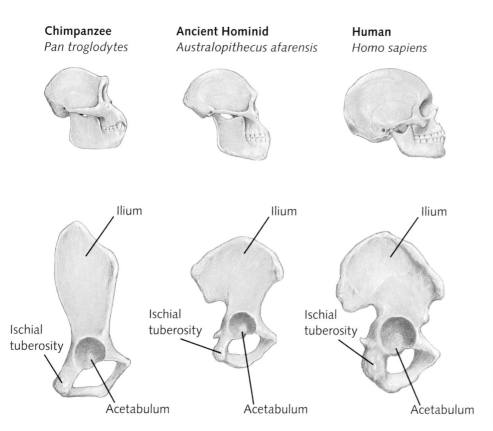

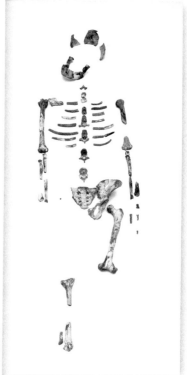

Chimpanzee
Pan troglodytes

Ancient Hominid
Australopithecus afarensis

Human
Homo sapiens

Ilium — Ilium — Ilium

Ischial tuberosity — Ischial tuberosity — Ischial tuberosity

Acetabulum — Acetabulum — Acetabulum

Figure 5.18 Some of the features of an australopithecine, such as Lucy, are intermediate between modern apes and humans. The reconstructed skeleton of Lucy is shown at the far right.

🔊 CONNECTION: Music

Donald Johanson named the australopithecine skeleton he found after a song, "Lucy in the Sky with Diamonds." This song was written by John Lennon and recorded by the Beatles in 1967.

More Humanlike Primates

Scientists wondered when hominids developed the large brain that humans have today. The first clue came in 1964, when anthropologists Louis and Mary Leakey discovered a different kind of skull. An **anthropologist** (an thruh PAH luh just) is a scientist who studies the development of humans. This skull, found in Tanzania, Africa, was more like a human's than any other skull that had been discovered. The region that surrounded the brain was larger and the jaw was smaller.

The Leakeys placed this skull into a new grouping known as *Homo*. They named it *Homo habilis*, which means "handy human," because stone tools were found near the skull. The skull was found to be between 1.5 million and 2.5 million years old.

Modern Humans

No one knows for certain exactly how modern humans, *Homo sapiens*, evolved. The fossil record indicates that *H. sapiens* appeared about 100,000 to 500,000 years ago.

PEOPLE IN SCIENCE The Leakey Family

It's not often that an entire family can be credited with making important scientific discoveries. This is exactly the case, however, with the Leakey family. The main members of this family are Louis and Mary (top) and their son, Richard (bottom).

Louis Leakey was born in 1903 in Kenya. Growing up in Africa, he came to learn about the land and the culture. As a child, he eagerly collected and studied stone tools and other materials he found. Louis left Africa to attend college, but he returned to search for fossils in order to learn about the evolution of humans.

During his work, Louis met and married Mary Nicol. Born in 1913 in England, Mary traveled quite a bit with her family. She took a strong interest in fossils. She entered the field using her talent for drawing pictures of fossils and other discoveries. Eventually, Mary learned enough to discover and analyze fossils by herself.

One of Louis and Mary's three sons, Richard, took up his parents' interest in studying fossils. Born in 1944, Richard originally had planned to pursue a different career, but he eventually found himself studying fossils, just as his parents had.

Together, the Leakeys made many discoveries of fossils. They found that the human evolutionary line went farther back than anyone had imagined. They forced other scientists to rethink their understanding of human development, and they provided insight into early cultures and peoples.

In the process of investigating, anthropologists gathered information about other *Homo* species. One well-known species is the Neandertals (nee AN dur tawlz). Fossils show that Neandertals had thick bones. They had large faces with large noses. Their brains were about the same size as those of modern humans.

The discovery of Neandertal bones in the 1820s raised many questions. Scientists wondered if they were the ancestors of modern humans. However, studies of DNA taken from a Neandertal bone showed that Neandertals did not evolve into modern humans. Instead, Neandertals eventually became extinct.

Fossil evidence also shows that another group lived at the same time as Neandertals. This group, sometimes known as Cro-Magnons (kroh MAG nunz), was an early form of humans. They were about the same height as modern humans. They also had the same skull structure, tooth structure, and brain size. Cro-Magnons were most likely the earliest humans.

Figure 5.19 Neandertals *(left)* and Cro-Magnons *(right)* showed many traits of modern humans.

After You Read

1. Describe how an opposable thumb helped primates survive.
2. According to your time line, were fossils of human ancestors discovered in the order in which they were formed? Explain.
3. How are Neandertals and Cro-Magnons believed to be related to modern humans?

Before You Read

An important way to explain or understand a new idea is through examples. In your Science Notebook, draw three boxes in which you can write a few sentences. As you read the lesson, look for examples that are presented.

Paleontologists and archaeologists had discovered a tremendous amount of evidence suggesting that living things evolve. The big question that had yet to be answered, however, was "Why?" What causes the adaptations of organisms to change over time?

The basis of the modern theory of why things evolved was established by the naturalist Charles Darwin. In 1831, Darwin set sail on an English ship called the HMS *Beagle*. The ship sailed around the world on a five-year voyage. During that time, Darwin's job was to collect and study fossils and other specimens along the way.

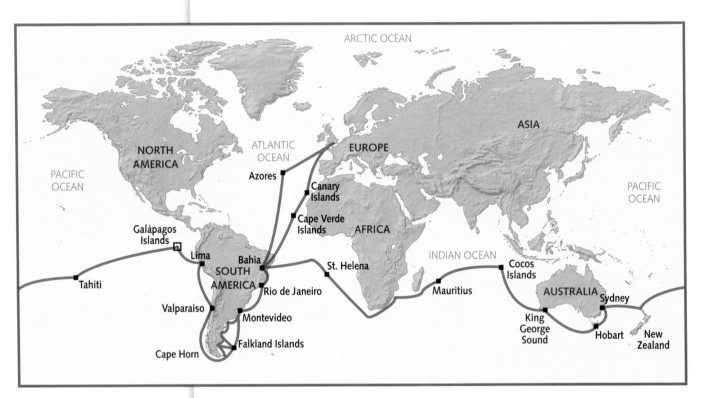

Figure 5.20 Darwin's five-year voyage took him around the world. Along the way, he collected fossils and made detailed notes about his observations.

The Galápagos Islands

Darwin's travels took him to a group of small islands off the west coast of South America known as the Galápagos Islands. The islands were close to one another, yet they had different climates and plants. Some islands were hot and dry with few plants. Others received more rainfall and had a larger variety of plants.

Darwin collected information about many species of organisms. In particular, he made extensive observations of giant tortoises that lived on the islands. Darwin noticed that the shells of the tortoises were different on each island. For example, one type of tortoise, the saddleback tortoise, had a long neck and a curved shell. A tortoise on a different island, the domed tortoise, had a shorter neck and a shell that looked like a dome.

In addition, Darwin observed 13 different species of a bird called a finch. Each type of finch had a beak with a different shape. The shape of the beak was adapted to the type of food present on the island where the bird lived.

As You Read

Fill in the boxes in your Science Notebook with examples of natural selection. Include information about the conditions that caused an organism to evolve.

What caused the finches on the Galápagos Islands to evolve with different beaks?

Figure 5.21 The domed tortoise *(left)* and saddleback tortoise *(right)* are two of the types of tortoises Darwin observed on the Galápagos Islands. Which tortoise has a longer neck?

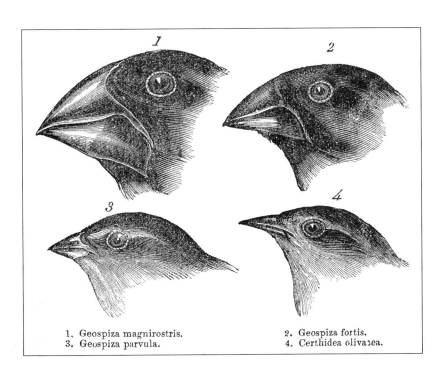

1. Geospiza magnirostris.
2. Geospiza fortis.
3. Geospiza parvula.
4. Certhidea olivaiea.

Figure 5.22 These are some of the finches Darwin observed on the Galápagos Islands. Notice how their beaks are different. The shape of the beak is related to the food available where each finch lives.

Figure It Out

1. Why is fast speed a useful variation for a fish?

2. How might a fish with a bold color be less fit than a fish that blends into the ocean floor?

Figure 5.23 According to Darwin's theory, natural selection can explain how species change over time.

Darwin's Theory

After returning to his home in England, Darwin spent many years trying to explain the observations he had made. Then, in 1859, Darwin presented his theory in a book titled *On the Origin of Species by Means of Natural Selection.* Darwin's theory is known as **natural selection**.

Darwin's theory was based on the idea that most organisms produce more offspring than are able to survive. The offspring must compete with each other for the things they need, such as food, water, and space. Some offspring have variations that will make them better able to survive. These are known as favorable variations. The organisms are said to be more "fit" than others.

Those organisms that are fit are more likely to live to reproduce. This is sometimes known as **survival of the fittest**. When organisms survive to reproduce, they pass the favorable variations on to their offspring. Over time, the favorable variations are found in more and more members of the species. In this way, the traits of a species evolve. Sometimes, the changes are so great that a new species is formed.

Natural selection can be used to explain changes in most species over time. **Figure 5.23** shows how natural selection might change a species of fish over time.

❶ In nature, organisms produce more offspring than can survive. Fishes, for example, can sometimes lay millions of eggs.

❷ In any population, individuals have variations. Fishes, for example, may differ in color, size, and speed.

❸ Individuals with certain useful variations, such as speed, survive in their environment, passing those variations to the next generation.

❹ Over time, offspring with certain variations make up most of the population and may look entirely different from their ancestors.

Understanding Darwin's Tortoises and Finches

The theory of natural selection explains what Darwin observed about tortoises and finches. Darwin suggested, for example, that all of the tortoises evolved from a common ancestor. The domed tortoise, which generally has a short neck and a shell that does not allow it to stretch its neck easily, was found on an island with plenty of plants. It didn't need to stretch its neck to reach food, so it thrived on this island.

The saddleback tortoise lived in a dry habitat with fewer plants. It needed to stretch to reach higher plants. Tortoises with long necks and shells that were open around the legs and neck were favored by natural selection. As a result, they were more likely to survive and reproduce. Over many generations, the adaptations—a saddleback shell and a long neck—increased within this species. Eventually, only saddleback tortoises could be found on the dry, sparsely vegetated islands.

The finches could be explained in a similar way. A single species of finch, which came from nearby South America, originally inhabited the islands. Each island provided a different source of food. Some birds needed a strong beak to crack open seeds. Others needed a narrow beak to reach into plants for food. Natural selection favored those birds with the variation that made them better able to obtain food on their island. Over many generations, the populations of birds on each island became different from one another.

CONNECTION: Economics

Sometimes people apply Darwin's theory of natural selection to business. Though the process is not exactly the same as that which occurs in nature, the idea is related. Often, too many businesses of the same kind compete for customers. For example, there might be more computer companies than available customers in a particular area. As a result, those companies that best serve the customer are more likely to stay in business. Those that do not serve the customer as well usually do not survive.

After You Read

1. How does Darwin's theory of natural selection relate to variations among the individuals of a species?

2. How does producing a large number of offspring affect natural selection?

3. According to the notes you wrote in your Science Notebook, how could the giant tortoises of the Galápagos Islands be explained through natural selection?

Summary

KEY CONCEPTS

5.1 Adaptations and Variations

- An adaptation is a trait that helps an organism survive and reproduce itself in its environment.

- Some adaptations help predators or prey. Others help organisms survive in a particular habitat.

- Although the members of a species are similar, there are variations among them. Variations are the result of sexual reproduction and mutations.

5.2 Evidence from the Past

- Fossils are the remains of organisms that once lived.

- Relative dating compares the age of a fossil with other fossils and rock layers. Radioactive dating gives an exact age of a fossil according to amounts of radioactive elements.

- The fossil record indicates that living things change over time.

5.3 Modern Evidence of Evolution

- Homologous structures have the same evolutionary origin, whereas analogous structures do not.

- Organisms that are related have similar embryos even though they look different as adults.

- Organisms with similar genetic material are likely to share a common ancestor.

5.4 Humans and Evolution

- Primates have feet and hands with toes and fingers that have nails. They have flexible joints and opposable thumbs.

- Hominid fossils show that there were several types of pre-human primates.

- Neandertals were a humanlike species that became extinct. Cro-Magnons were ancestors of modern humans.

5.5 Natural Selection

- Darwin observed many plants and animals on the Galápagos Islands and elsewhere. He noticed that related species had different adaptations on different islands.

- Darwin's theory of natural selection suggests that when more offspring are produced than can survive, they must compete for resources. Only the fittest organisms survive to reproduce. Over time, this changes the traits of a species.

VOCABULARY REVIEW

Write each term in a complete sentence, or write a paragraph relating several terms.

5.1
adaptation, p. 84
predator, p. 85
prey, p. 85
variation, p. 87
genetic diversity, p. 87
mutation, p. 87

5.2
fossil, p. 88
paleontologist, p. 88
relative dating, p. 90
radioactive dating, p. 90
fossil record, p. 90
extinct, p. 90
geologic time scale, p. 91

5.3
evolution, p. 92
homologous structure, p. 93
analogous structure, p. 93
vestigial structure, p. 93
embryo, p. 94

5.4
primate, p. 95
opposable thumb, p. 95
hominoid, p. 96
hominid, p. 96
anthropologist, p. 98

5.5
natural selection, p. 102
survival of the fittest, p. 102

PREPARE FOR CHAPTER TEST

To prepare for the chapter test, create a question from each Learning Goal. Use the information in your Science Notebook to answer each question. Then use these answers to write a well-developed essay about the chapter. Use the Key Concept on the first page of this chapter as your topic sentence.

True or False
If the statement is true, write "true." If it is false, change the underlined word or words to make the statement true.

1. A <u>predator</u> is a trait that helps an organism survive and reproduce itself in its environment.

2. A <u>variation</u> is an error that occurs when DNA is copied.

3. A <u>fossil</u> is the preserved remains of a once-living organism.

4. A species becomes <u>preserved</u> when it no longer lives on Earth.

5. Two structures are <u>homologous</u> if they have the same function but did not evolve in the same way.

6. The <u>Neandertals</u> were an early form of modern humans that had the same height, size, and tooth structure.

7. <u>Natural selection</u> is sometimes described as survival of the fittest.

Short Answer
Answer each of the following in a sentence or brief paragraph.

8. How might an adaptation help an organism survive in a cold environment?

9. Identify the cause(s) of variations within a population of organisms.

10. Explain how a vestigial structure can be used to support evolution.

11. Many scientific theories attempt to explain human evolution. Explain why some of these theories may need to be revised at a later point in time.

12. An antibiotic is a drug that is used to kill bacterial cells. Construct an argument to explain why some bacterial diseases are no longer controlled by an antibiotic.

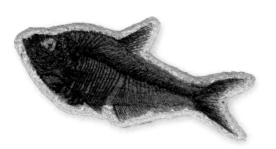

Critical Thinking
Use what you have learned in this chapter to answer each of the following.

13. **Apply** Plants that grow on the floor of a rain forest often have large, flat leaves. How is this structure an adaptation that helps them survive?

14. **Compare** How do the skull and brain sizes of modern humans compare to the skulls of early primates, such as australopithecines?

15. **Relate** What natural selection process might have led to a smaller jaw with fewer teeth in humans?

16. **Infer** About the time of the Industrial Revolution in England, light gray trees turned black from soot. Peppered moths that are white or black lived on these trees. Birds preyed upon the moths. Suggest how the change in tree color affected the population of these moths.

Standardized Test Question
Choose the letter of the response that correctly answers the question.

Decay Rate of a Radioactive Element

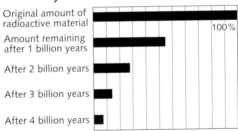

17. The bar graph shows how much of a radioactive element remains after different periods of time. What is the most likely half-life (how long it takes for half of the element to decay) of this element?

    A. 1 billion years

    B. 2 billion years

    C. 3 billion years

    D. 4 billion years

Test-Taking Tip

Remember that words such as *most likely* and *not* may change the meaning of a question. Be sure to read the question carefully before examining all the answer choices.

Classification of Living Things

KEY CONCEPT The classification of living things is based on similarities and differences among organisms.

Imagine going into a music store to find a CD. As you look around, you discover that the CDs are not arranged in any particular order. The only way to find what you want is to look through the thousands of CDs one by one. It could take you all day!

Your best bet is to go to a store that organizes its CDs in an easy-to-use system. Organizing items into groups is useful and important for many things people do every day. Life scientists organize living things into groups in order to describe and study them.

Think About Classifying

Describe one way in which you organize into groups objects such as books, clothing, or coins. What do you use as the basis for your organizational system? Now describe a way in which you might organize living things. What is the basis for this system?

- Compare your two systems. How are they alike? How are they different?
- Write one or two sentences in your Science Notebook summarizing how things can be organized into groups.

www.scilinks.org
Classification **Code:** WGB06

6.1 What Is Classification?

Before You Read

Create a K-W-L-S-H chart in your Science Notebook. Think about the title of this lesson. In the column labeled *K* on the chart, write what you already know about classifying things. In the column labeled *W*, write what you want to learn about how living things are classified.

Do you think that the more than 2.5 million types of organisms already discovered include all of the organisms on Earth? If so, think again. Scientists predict that several million more organisms are yet to be discovered!

Recently, damp caves in Sequoia National Park were found to be home to 27 types of spiders, centipedes, scorpions, and other creatures never seen before. Some of the organisms exist only in one section of a single cave! When scientists discover a new organism, they work to describe it, name it, and relate it to other organisms based on how they are alike and different. In other words, they classify it.

Early Classification

Arranging organisms into groups according to how they are alike is known as **classification**. The branch of biology that involves identifying, classifying, and naming organisms is **taxonomy** (tak SAH nuh mee). Scientists who study taxonomy are called taxonomists.

The science of taxonomy is not new. The Greek philosopher Aristotle (384–322 B.C.) developed the first widely accepted system of classification. Aristotle classified organisms into two basic groups: plants and animals. He further divided each group according to characteristics he could observe. For example, plants could be described as herbs, shrubs, or trees. Animals were classified according to whether they swam in water, flew through the air, or walked on land.

Figure 6.1 Aristotle classified more than 500 organisms as either plants or animals.

As You Read

In the column labeled *L* in your K-W-L-S-H chart, write three or four important things that you have learned about classification.

Identify the scientist who developed the classification system used today.

A New Approach

Aristotle's system of classification lasted for centuries. Over time, however, many organisms were discovered that did not fit into the system. In the late eighteenth century, the Swedish botanist Carolus Linnaeus (luh NAY us) (1707–1778) developed a new method of classification.

Linnaeus's system was based on physical and structural similarities of organisms. Using observations he made, he placed the organisms into groups and subgroups. You will learn more about Linnaeus's levels of classification in the next lesson. You will also discover how he used his system to name organisms.

Linnaeus's system of classifying and naming organisms survives to this day. However, the method of grouping organisms has changed. Modern classification is based on the idea that organisms that share a common ancestor are related and also share a similar evolutionary history.

CONNECTION: Chemistry

Classification is not limited to life science. Scientists in all fields of study benefit from organizing information. Chemists, for example, classify chemical elements. To date, there are more than 110 known chemical elements.

In 1863, when only 63 elements had been discovered, the Russian scientist Dmitri Mendeleev recognized the need to organize them. He compiled a list of the known physical and chemical properties of the elements and wrote the information for each element on a separate card. He arranged and rearranged the cards until he recognized a pattern. The result became known as the periodic table of the elements because the properties of the elements repeated in a periodic, or regular, fashion.

Mendeleev's table was first published in 1869. The organization of the table made it possible to predict the existence and properties of elements that had not yet been discovered.

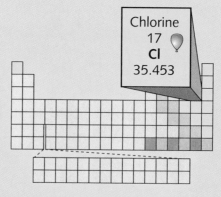

Modern version of the periodic table

Basis for Classification

When scientists try to determine the evolutionary relationships among organisms, they gather a variety of information. This information includes the similarities between the structures of organisms, the structures of chromosomes, and the DNA sequences.

Structural Similarities   If two organisms have similar structures, they likely evolved from a common ancestor. Lynxes and bobcats, for example, are more similar to each other than to members of any other group. This suggests that they share a common ancestor. Taxonomists identify the characteristics of organisms and compare them to those of known organisms. They use this information to make inferences about the evolutionary history of the organisms.

Chromosomes   Scientists can study the number and structure of chromosomes to learn about relationships. If the structure of the chromosomes of two organisms is similar, the organisms may have shared a common ancestor. Taxonomists classify cabbage, cauliflower, and kale together because their chromosomes are almost identical in structure.

DNA   Organisms that are closely related have similar DNA sequences. Recall that DNA is made up of sequences of nucleotides. In general, the more similar the nucleotide sequences of two organisms are, the more closely related they are. Consider the giant panda and red panda in **Figure 6.3**. Although they are both called pandas, the giant panda's DNA sequences are more similar to a bear's than to a red panda's.

Figure 6.2 Why do you think taxonomists classify the lynx (top) and the bobcat (bottom) in the same group?

Figure 6.3 DNA evidence shows that giant pandas (left) are more closely related to bears than to red pandas (right). Red pandas are related to raccoons.

Dichotomous Key

Taxonomists often use the information they collect to develop **dichotomous keys** for identifying an unknown organism. The word *dichotomous* means "two parts." A dichotomous key is made up of several pairs of opposing statements. A person trying to identify an organism must choose the one of the two statements that describes the organism.

At the end of each statement is either the name of an organism or a directive to go to another pair of statements. If the person selects the statement with the name of an organism, he or she has identified the unknown organism. No further statements need to be considered. If the person selects the statement with a directive, he or she must move on to another pair of statements.

A person must work through the pairs of statements one at a time in order to identify an unknown organism. A dichotomous key for identifying common beans is shown in **Figure 6.4**.

Explain It!

Write a paragraph explaining how a dichotomous key is useful. Then create a dichotomous key to classify items in your home, such as books, utensils, or sports equipment.

Figure It Out

1. Which is the only type of bean that is round?

2. Suppose you discover a bean that is reddish-brown, oblong, and has pigments that are spread out evenly. How would you classify it?

Dichotomous Key

1a.	The bean is round.	**Garbanzo bean**
1b.	The bean is elliptical or oblong.	Go to 2
2a.	The bean is white.	**White northern bean**
2b.	The bean has dark pigments.	Go to 3
3a.	The pigments are spread out evenly.	Go to 4
3b.	The pigments are mottled.	**Pinto bean**
4a.	The bean is black.	**Black bean**
4b.	The bean is reddish-brown.	**Kidney bean**

Figure 6.4 This dichotomous key contains four sets of statements to identify common beans.

After You Read

1. What is classification?

2. What characteristics do taxonomists use to classify an organism?

3. Using the completed K-W-L columns in your chart, describe how a dichotomous key is used to identify an organism. Complete your chart by indicating what you would still like to know about classification in the *S* column and how you can find this information in the *H* column.

6.2 Levels of Classification

Learning Goals

- Describe how scientists classify organisms.

- Identify the seven levels of classification.

- Explain the system used to give organisms scientific names.

New Vocabulary

kingdom
phylum
class
order
family
genus
species
binomial nomenclature

Before You Read

Create a lesson outline. Use the lesson title as the outline title. Label the headings with the Roman numerals *I* through *III*. Use the letters *A, B, C*, etc., under each heading to record information you want to remember.

If you were asked to identify yourself as part of a group, which group would you choose? You could choose your family, your class, or your school. Perhaps you would select a club or sports team. There are many groups of varying sizes to which you belong. The same is true of the more than 2.5 million living things that exist today. How are these groups organized? Read on to learn about the levels of classification scientists use to bring order to the study of living things.

What Is It?

Whenever a new organism is discovered, scientists carefully study its characteristics. They determine how many cells make up the organism, how the organism gets its food, and how it reproduces. For many years, scientists used this information to classify newly discovered organisms as either plants or animals.

With advances in technology, scientists learned even more about the great variety of living things. For example, they discovered tiny organisms, such as the slime mold shown in **Figure 6.5**. These organisms, now known as protists, have characteristics of plants as well as of animals. They are classified in their own group.

Figure 6.5 Many microscopic organisms have characteristics of both plants and animals, yet they are different enough to need their own group. What type of organism is the slime mold (center)?

As You Read

Look at your outline. Make sure that you have recorded additional information. Share your outline with a partner, and add any missing information.

Explain what organisms within the same species have in common.

A System of Levels

As you just learned, plants, animals, and protists are classified into different groups. Not all of the organisms within each group are identical. For example, dogs, whales, and worms are all considered animals. They are identified as animals because, among other things, they are multicellular (many-celled) and must consume other organisms as food. They have little else in common, however.

For this reason, scientists divide each major group into smaller groups based on how organisms are alike or different. The system of classifying organisms that Linnaeus devised is hierarchical (hi uhr AR kih kul), which means that it has levels. Each successive level is more specific and contains fewer types of organisms. The organisms in each successive level share more characteristics with one another than they do with the organisms in the level above.

As you can see in **Figure 6.7**, the largest and most general group into which an organism can be classified is known as a **kingdom**. Animals, plants, and protists make up three kingdoms of organisms. You will learn more about kingdoms in the following lesson.

Each type of organism within a kingdom is then organized into a smaller and more specific group called a **phylum** (FI lum, plural: phyla).

Members of each phylum are further divided into different **classes**. Classes are more specific than phyla and kingdoms. A class is made up of different **orders**.

An order is separated into **families**. Each family consists of at least one **genus** (JEE nus, plural: genera).

Each genus is then divided into **species**. A species is the smallest and most specific group into which organisms are classified. Organisms within a species have similar characteristics and can mate and produce offspring of the same type.

Terrier   Chihuahua   Spaniel

Figure 6.6 These dogs are members of the same species. No other type of organism, such as house cats or killer whales, can belong to this species.

Figure It Out

1. How many different types of organisms would you expect to find at the species level?

2. Which animal or animals disappear at each successive level? Why?

Classification of the Domestic Dog

Kingdom Animalia

Dogs are in the kingdom Animalia. All but one of the animal phyla contain only animals without a backbone. Dogs belong to only phylum that contains animals with backbones.

Phylum Chordata

Dogs are classified in the phylum Chordata. Animals in this phylum have a spinal cord and most have a backbone. Bears, foxes, turtles, wolves, and whales are also in this phylum.

Class Mammalia

Dogs belong to the class Mammalia. Mammals are warm-blooded animals with backbones whose females nourish young with milk. Turtles belong to a different class.

Order Carnivora

Dogs are members of the order Carnivora. The ancestors of these mammals had special teeth for tearing meat. Wolves, foxes, and bears also belong to this order.

Family Canidae

Dogs, as well as wolves and foxes, belong to the family Canidae. Members of this family are generally hunters that can run long distances.

Genus *Canis*

Dogs and wolves both belong to the genus *Canis*. Of all the animals shown in this table, dogs and wolves are the most closely related.

Species *Canis familiaris*

Only domestic dogs make up the species known as *Canis familiaris*. A species is the smallest and most specific classification group.

Figure 6.7 Many types of organisms can belong to the same kingdom. A kingdom is the largest and most general group. Fewer organisms belong in each successive level.

Grevy's Zebra
Equus grevyi

Onager
Equus hemionus

Domestic Horse
Equus caballus

Figure 6.8 All three of these animals are in the same genus. Each one belongs to a different species. How can you tell?

Naming Organisms

When you describe an animal as a cat or a dog, you are using its common name. Common names vary among people in different places. In fact, a single organism may have as many as 50 common names. The panther is an endangered species of wildcat. In Florida, this animal is called the Florida panther. In other parts of the country, it is called a cougar, puma, mountain lion, or catamount.

Linnaeus recognized the need to use a unique name for each type of organism. He gave each one a two-part scientific name made up of the genus name and a species descriptor. He used Greek and Latin words because those languages were understood by most scientists of his day. Linnaeus's system is known as **binomial nomenclature**. The word *binomial* means "two names," and the word *nomenclature* means "naming."

Look at the name of the domestic dog in **Figure 6.7** on page 113, *Canis familiaris*. This name is also the species of the domestic dog. The first part of the name identifies the genus: *Canis*. Remember, a genus is a group of closely related species. The second part of the name, *familiaris*, specifies which species. Since only one type of organism makes up a species, only domestic dogs belong to the species *Canis familiaris*. All breeds of domestic dogs belong to this species and have this scientific name.

Notice that in a scientific name, the genus is capitalized and the species descriptor is not. Also notice that the scientific name is written in italic letters. Sometimes the genus is abbreviated as a letter. For example, the scientific name for humans is *Homo sapiens*. This name can also be written as *H. sapiens*.

CONNECTION: Art

You are a scientist who has discovered a new species of animal. It is closely related to the wolf and the domestic dog. Make a detailed drawing of this new species in your Science Notebook. Describe the animal in detail, and name it using binomial nomenclature.

After You Read

1. Describe a hierarchical classification system.

2. Explain what you can learn from the scientific name of an organism.

3. Use the information in your outline. Think about how the classification of living things is similar to the organization of time using calendars and clocks. In your Science Notebook, write a well-developed paragraph summarizing your thoughts.

6.3 Domains and Kingdoms

Learning Goals

- Relate domains and kingdoms.
- Identify and compare the six kingdoms.

New Vocabulary

domain
Eubacteria
Archaebacteria
Protista
Fungi
Plantae
Animalia

Before You Read

Read the lesson title, the Learning Goals, and the headings, and look at the photos and table. Predict what you think you will learn in this lesson. Write two or three sentences in your Science Notebook.

As you learned in Lesson 6.2, all organisms are grouped into kingdoms. Although Aristotle used two kingdoms and Linnaeus used three, most modern biologists classify organisms into six kingdoms: Eubacteria, Archaebacteria, Protista, Fungi, Plantae, and Animalia.

Three Domains

Since the time of Linnaeus, biologists have developed a new taxonomic category called the **domain**. The domain is larger than the kingdom. There are three domains. The six kingdoms are grouped into these three domains. The domain Bacteria includes the kingdom Eubacteria. The domain Archaea includes the kingdom Archaebacteria. The domain Eukarya includes protists, fungi, plants, and animals.

Four basic characteristics can be used to compare and contrast the kingdoms. These are cell type, cell structures, number of cells, and method of obtaining nutrition.

Cell Type Recall that there are different types of cells. Cells with a nucleus and organelles bound by membranes are called eukaryotic cells. Organisms with these types of cells are called eukaryotes. In prokaryotes, the cells do not have a membrane-bound nucleus or organelles. In addition, the DNA is not organized into X-shaped chromosomes. These types of cells are prokaryotic.

Cell Structures Many cell structures help the cell function properly. One such structure is a cell wall, which surrounds and protects the cell membrane in some cells. Another structure is the chloroplast, a membrane-bound organelle that contains pigments that capture the energy of sunlight in some eukaryotic cells. Identifying cell structures helps scientists classify organisms.

Number of Cells Some organisms consist of a single cell. In these unicellular organisms, all life functions are carried out by one cell. Multicellular organisms are made up of more than one cell. Cells or groups of cells are usually specialized to perform a specific function.

Extend It!

Research the development of the six-kingdom system. Find out why some scientists still use a five-kingdom system. Based on your research, predict whether additional kingdoms will be proposed in the future.

As You Read

Refine your predictions using any information you have learned in this lesson. What four characteristics are used to compare kingdoms?

Domains and Kingdoms

1. What are four characteristics that a protist can have in common with a plant?

2. Which kingdoms do not contain unicellular organisms?

Domain	Kingdom	Cell Type	Cell Structures	Number of Cells	Nutrition
Bacteria	Eubacteria	prokaryotic	cell walls with peptidoglycan	unicellular	autotrophic or heterotrophic
Archaea	Archaebacteria	prokaryotic	cell walls without peptidoglycan	unicellular	autotrophic or heterotrophic
Eukarya	Protista	eukaryotic	cell walls of cellulose in some; chloroplasts in some	most are unicellular; some are multicellular	autotrophic or heterotrophic
	Fungi	eukaryotic	cell walls of chitin	most are multicellular; some are unicellular	heterotrophic
	Plantae	eukaryotic	cell walls of cellulose; chloroplasts	multicellular	autotrophic
	Animalia	eukaryotic	no cell walls or chloroplasts	multicellular	heterotrophic

Figure 6.9 Organisms can be classified into six different kingdoms. How are the kingdoms organized?

Nutrition All organisms need nutrients in order to survive. Autotrophic organisms use inorganic nutrients and make organic nutrients through photosynthesis. Heterotrophic organisms must obtain nutrients by eating other organisms as food.

Domain Bacteria

This domain contains the kingdom **Eubacteria**. Bacteria are unicellular. Their cells are prokaryotic and have a thick cell wall around a cell membrane. Some bacteria are autotrophic, and others are heterotrophic.

Domain Archaea

The members of this domain look much like bacteria. They are unicellular prokaryotes that can be autotrophic or heterotrophic. They belong to the kingdom **Archaebacteria**.

Archaea are found in some of the most extreme environments on Earth. They exist in volcanic hot springs, very salty water, and thick swamps. Some live near hot vents in the ocean. Archaea have even been found in the digestive tracts of cows and termites.

Figure 6.10 Archaebacteria such as *Halococcus* are found in harsh environments, such as these seawater evaporating ponds (top). Eubacteria such as cyanobacteria (bottom) live in freshwater.

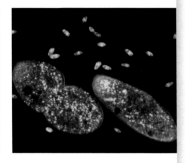

Figure 6.11 Slime molds (left) are fungus-like protists that live in damp forests. Kelps (center) are multicellular plant-like protists. Paramecia (right) are animal-like protists that move through water.

Domain Eukarya

All of the organisms in this domain are eukaryotes. This domain contains four kingdoms.

The kingdom **Protista** shows the greatest variety. Most protists are unicellular, but some are not. Some protists conduct photosynthesis, while others do not. Some protists, such as slime molds, have features in common with fungi. Other protists, such as algae, resemble plants. Still others, such as protozoa, share characteristics with animals.

Members of the kingdom **Fungi** are heterotrophs. Most fungi (singular: fungus) obtain nutrition by feeding on dead or decaying organic matter. Some fungi, such as mushrooms, are multicellular. Others, such as yeasts, are unicellular.

The kingdom **Plantae** contains multicellular organisms that are photosynthetic autotrophs. This means they can use inorganic nutrients to produce organic nutrients that heterotrophs can use as food.

Members of the kingdom **Animalia** are multicellular and heterotrophic. Animals obtain nutrition by eating plants or other animals that have eaten plants. Almost all animals are motile, which means they are able to move from one place to another.

Figure 6.12 The cheetah (left) is one species of animal. Tropical ferns (right) are species of plants. What are some ways in which these organisms are similar? How are they different?

After You Read

1. Which two domains include only organisms that are prokaryotes?

2. Explain to which kingdom you belong.

3. According to your predictions, if the cells of a multicellular organism have cell walls, to which kingdoms could the organism belong?

Summary

KEY CONCEPTS

6.1 What Is Classification?

- Classification is the practice of using similarities and differences to organize living things into groups.

- Taxonomy is the branch of biology that identifies, classifies, and names living things.

- The Swedish botanist Carolus Linnaeus developed a system of classification in which organisms are placed into groups and subgroups according to their physical and structural similarities.

- A series of opposing statements known as a dichotomous key can be used to identify unknown organisms.

6.2 Levels of Classification

- Scientists classify organisms by comparing their characteristics and grouping organisms with similar characteristics together.

- An organism is classified into a kingdom, phylum, class, order, family, genus, and species.

- A kingdom is a large and general group. A species is the smallest and most specific group.

- The system of assigning a two-part scientific name to an organism is known as binomial nomenclature.

- The scientific name of an organism is its species name, which consists of its genus and a species descriptor.

6.3 Domains and Kingdoms

- The domain is a level of classification into which kingdoms can be grouped.

- The three domains are Bacteria, Archaea, and Eukarya.

- The organisms in the domain Bacteria and the domain Archaea are all prokaryotes.

- All of the organisms in the domain Eukarya are eukaryotes. This domain contains four kingdoms: Protista, Fungi, Plantae, and Animalia.

VOCABULARY REVIEW

Write each term in a complete sentence, or write a paragraph relating several terms.

6.1
classification, p. 107
taxonomy, p. 107
dichotomous key, p. 110

6.2
kingdom, p. 112
phylum, p. 112
class, p. 112
order, p. 112
family, p. 112
genus, p. 112
species, p. 112
binomial nomenclature, p. 114

6.3
domain, p. 115
Eubacteria, p. 116
Archaebacteria, p. 116
Protista, p. 117
Fungi, p. 117
Plantae, p. 117
Animalia, p. 117

PREPARE FOR CHAPTER TEST

To prepare for the chapter test, create a question from each Learning Goal. Use the information in your Science Notebook to answer each question. Then use these answers to write a well-developed essay about the chapter. Use the Key Concept on the first page of this chapter as your topic sentence.

MASTERING CONCEPTS

True or False
If the statement is true, write "true." If it is false, change the underlined word or words to make the statement true.

1. A scientist who works to identify, classify, and name organisms practices <u>taxonomy</u>.

2. New organisms can often be classified by selecting statements from a <u>domain</u> key.

3. A <u>genus</u> is a group of related classes.

4. <u>Binomial nomenclature</u> is a two-word system of naming organisms.

5. A multicellular eukaryotic organism belongs in the kingdom <u>Protista</u> if it is heterotrophic and its cells do not have cell walls.

6. Many members of the kingdom <u>Eubacteria</u> are heterotrophic eukaryotes that feed on dead and decaying organic matter.

Short Answer
Answer each of the following in a sentence or brief paragraph.

7. Explain why taxonomy is important to biologists.

8. Why is Linnaeus often called the Father of Taxonomy?

9. Describe the levels of classification in Linnaeus's system.

10. A bird known as the white ibis has the scientific name *Eudocimus albus.* Identify the information provided by its name.

11. How do members of the domain Eukarya differ from members of the other two domains?

12. Explain why classification systems are updated as scientists learn new information about organisms.

Critical Thinking
Use what you have learned in this chapter to answer each of the following.

13. **Compare and Contrast** How is modern taxonomy different from early taxonomy? How is it similar?

14. **Infer** The structure of the chromosomes of two organisms is very similar. What is probably true about the relationship between the organisms?

15. **Apply Concepts** Why might a classification system be described as organization by elimination?

Standardized Test Question
Choose the letter of the response that correctly answers the question.

	Northern Red Oak	American Chestnut	Southern Red Oak
Kingdom	Plantae	Plantae	Plantae
Phylum	Anthophyta	Anthophyta	Anthophyta
Class	Dicotyledones	Dicotyledones	Dicotyledones
Order	Fagales	Fagales	Fagales
Family	Fagaceae	Fagaceae	Fagaceae
Genus	*Quercus*	*Castanea*	*Quercus*
Species	*Q. rubra*	*C. dentata*	*Q. falcata*

16. According to the table, which trees are most closely related?
   A. Northern Red Oak and American Chestnut
   B. Northern Red Oak and Southern Red Oak
   C. American Chestnut and Southern Red Oak
   D. all of these

Test-Taking Tip
If "all of these" is one of the choices in a multiple-choice question, be sure that none of the choices is false.

Where Is Linnaeus When We Need Him?

IN THE 1700S, the scientist Carolus Linnaeus created a system to organize the many new species being discovered. His system is still used today. Linnaeus believed that there were fewer than 15,000 species of plants and animals on Earth. If he were alive today, he would be amazed that more than 2.5 million species have been identified!

The Problem: Keeping Track of Millions of New Species Names

Believe it or not, scientists have not yet named all of the living things on Earth. Better technology and increased funding for research have led to an increase in the number of species being discovered each year. However, there is a problem—scientists do not have a central place to record the name of every new species that is discovered. Without a central place to record the names and information about new species, important information may be lost.

Solutions to the Naming Problem

Recently, a group of scientists suggested a solution to the naming problem. They created a Web-based tool called ZooBank. One goal of ZooBank is to create a complete record of the scientific names of animals. This is something that does not yet exist. It will be a long, difficult project. There is also no such record of scientific plant names, fungi names, or protist names. Although scientists agree that a naming system is needed, they disagree about who should control the system. They also disagree about how it should work.

There are many ideas about how to best track and name new species. They include BioCode, uBio, the ALL Species Foundation, Wikispecies, Species 2000, the Electronic Catalogue of Names of Known Organisms, the Taxome Project, and many more. Each approaches the listing and naming of species in a different way. It will be a challenge to develop a system that works for all scientists.

Research and Report

With a partner, research ZooBank and a few of the other projects mentioned in the article. Then come up with your own system to help scientists keep track of new species names and other important species information. Describe how your system works. What are its advantages and disadvantages? Create a poster with your information.

CAREER CONNECTION GENETIC ENGINEER

DID YOU EVER WONDER if there is a way to make food healthier, tastier, and easier to grow? Genetic engineers do this for a living. They create plants that produce more food. They also make fruits and vegetables that ripen faster and stay fresh longer. They are even researching ways to make vegetables that contain vaccines and pigs that lower our cholesterol!

Charles Darwin observed that plants and animals change over time through variation and adaptation. But genetic engineers don't have to wait thousands of years for organisms to change. They can alter the genes of organisms to create new plants and animals with desired traits. This process is called genetic engineering.

Farmers and scientists have used a type of genetic engineering called hybridization since the 1800s. To make a hybrid plant, one type of plant is fertilized using pollen from another type of plant. The new plant that grows

New Newt Species
Raises Difficult Questions

IT IS NOT every day that scientists find a new species that people want as a pet. When they do, it can lead to trouble.

In 1999, a researcher in the Southeast Asian country of Laos made an exciting discovery. Living in certain wet places were strange, colorful new animals that no one had seen before. They were newts, or salamanders, but they looked very different from any other type of newt. Scientists have named them Laos warty newts, but there is still a lot to learn about the species.

This new newt looks like a black lizard and can fit in your hand. Like other newts in its genus, it has stumpy legs, a long tail, a head shaped like a triangle, and warts on its skin. The Laos warty newt has something special, though, that helps it survive. It has bright yellow stripes down its dark back, and the skin of its belly has a pattern almost like a leopard's coat.

The bold colors and patterns are not just for decoration. These warty newts live in bright, shallow streams. They can swim around during the day and at night because they blend in with their habitat. This makes them hard for predators to see.

Now that people know they exist, the Laos warty newts have new predators who hunt them for their colors—people. In 2002, scientists published a paper about the new species, which is known as *Paramesotriton laoensis*. Just a few years later, the bright newts were being sold as pets!

Scientists are worried that the pet trade could be a big problem for the Laos warty newt. If pet dealers take away too many newts or disrupt their habitat, scientists might not be able to study them. Not enough is known yet about these animals to add them to the endangered species list. Scientists need to learn more about the newts before they can be legally protected.

The case of the Laos warty newt raises a difficult question. When scientists discover a species that people might want for a pet, should they publish their findings as they normally would? Or should they keep their discoveries private until the species can be protected? Perhaps governments could create laws to protect new species as soon as they are announced to the world.

The Laos warty newts are so rare and have such an unusual look that they have been sold for as much as $170 each.

from the resulting seed is a hybrid. This means it has some DNA from each type of plant. Most of the corn and apples we eat today come from hybrid plants.

Since the 1950s, genetic engineers have been altering genes using a more exact method. They select a single gene that is the "code" for a desired trait. They then insert that gene into an organism's DNA. This kind of genetic engineering has been used to make rice plants that produce more grain and corn plants that repel damaging caterpillars. It has also been used to grow tomatoes with thick skins that make them easier to store and transport. Genetic engineers today also work with the genes of animals.

In recent years, some people have worried that foods from plants and animals whose genes were altered might not be safe to eat. However, the U.S. Food and Drug Administration (FDA) requires that all of these foods be tested for safety before they are sold.

Because of discoveries about DNA, it is an exciting time to work in a genetic engineering lab. Genetic engineers are creative thinkers who are excited about science. They usually have advanced experience and doctoral degrees. Lab assistants, who usually have a bachelor's degree, help genetic engineers with their research.

Microorganisms and Fungi

Chapter 7 **Viruses and Bacteria**

What are viruses and bacteria, and how do they affect humans and the environment?

Chapter 8 **Protists and Fungi**

What are the characteristics of these diverse groups of organisms, and what is their role in the world?

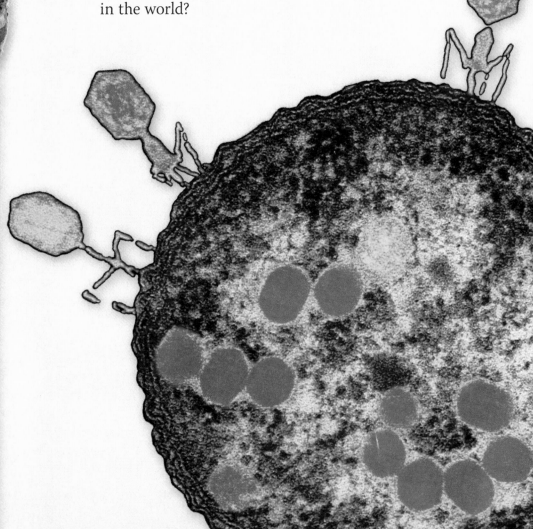

Viruses and Bacteria

KEY CONCEPT Viruses and bacteria share many common characteristics and impact humans and the environment.

Imagine spending the day at a nature preserve. Some animals and plants would be more familiar to you than others. Now imagine that you have the ability to see microscopic objects. You would surely be surprised by all the viruses and bacteria that inhabit the nature preserve! There are viruses and bacteria in the air, water, and soil. You would even see them on and in your body.

Think About Viruses and Bacteria

You know that viruses and bacteria are often responsible for making people sick. But viruses and bacteria do much more than that!

- In your Science Notebook, make a T-chart. Label one column *Disease* and the other *Cause*. Think about diseases that you have had or that you know about and list them in the first column. In the second column, predict the cause of each disease. Is the disease caused by a virus or by a bacterium?

- Can you think of ways in which viruses and bacteria affect you or the environment positively and negatively? Make a second T-chart with columns labeled *Positive Impact* and *Negative Impact*. Write your ideas in the appropriate columns.

NSTA
SCiLINKS
THE WORLD'S A CLICK AWAY

www.scilinks.org
Bacteria **Code: WGB07A**
Viruses **Code: WGB07B**

Learning Goals

- Describe a virus.
- Explain how viruses reproduce.
- Identify ways in which viruses affect the world.

New Vocabulary

virus
host
parasite
lytic cycle
lysogenic cycle

7.1 Viruses

Before You Read

In your Science Notebook, write the word *virus* vertically down the left side of the page. Create a "biopoem" using words or phrases that you predict will describe a virus. Use the letters in the word *virus* to begin each of your descriptive words or phrases.

You could probably write a movie script in which the bad guys sneak into an airport, take control of it, and then run it for their own benefit. Fortunately, the hero in your movie would probably find a way to regain control of the airport and save the day. A similar scenario occurs thousands of times a day as viruses invade living cells and hijack their operations.

What Is a Virus?

Sitting on a desk, viruses are no more "alive" than your pencil or textbook. But just like the bad guys in your movie script, when a virus connects with the right type of cell, it sneaks in, takes control, and runs the cell for its own benefit. In the photos of viruses in **Figure 7.1**, each virus has found the right type of cell and has taken control of it. A **virus** is a nonliving particle that invades and uses parts of a cell to reproduce itself and distribute more viruses.

Living or Nonliving? A virus does not exhibit the characteristics of living things because it is not made of cells and it does not have the cell structures that carry out the basic functions of an organism. The only way in which viruses resemble living things is in their ability to reproduce. However, they can only reproduce by using a cell. The living cells used by viruses to reproduce are called **hosts**. Hosts provide a home and energy for a parasite. A **parasite** (PER uh site) is an organism that lives on or in a host and does harm to it. Viruses act like parasites because they harm the host cell, often destroying it.

Figure 7.1 Viruses can invade the cells of all types of living things. These photographs show: **a)** rabies virus (animals), **b)** tobacco mosaic virus (plants), **c)** bacteriophages (bacteria), and **d)** measles virus (humans). Why do you think viruses invade living cells?

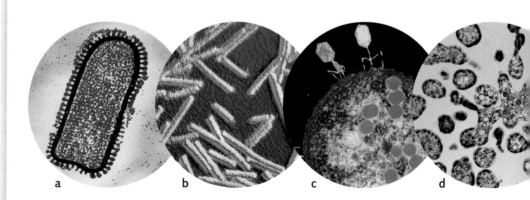

a   b   c   d

Naming Viruses   Because viruses are not living things, they are not classified and named like other organisms. A virus may be named for the disease it causes. The rabies virus infects nerve cells in animals and causes rabies. Some viruses, such as the tobacco mosaic virus, are named for the first host they infect and the type of damage they cause. Others are named for the place where the virus was first found or the scientist who first identified it. The deadly Ebola virus was named for the Ebola River in the Democratic Republic of Congo in Africa, where it was first identified.

Structure of Viruses   Although viruses vary in size and shape, they all share the same basic structure. A virus contains a core of genetic material surrounded by a protective protein coat. The genetic material in a virus contains instructions for making copies of the virus after it invades the host. The genetic material can be the nucleic acid DNA or RNA.

The arrangement of proteins in a virus's protective coat determines the shape of the virus. It also plays a role in determining which type of cell the virus can invade. The arrangement of proteins in the protein coat creates a unique shape that allows the virus to recognize and attach to a matching spot on its host cell. Like puzzle pieces that fit together perfectly, the protein coat of the virus snaps into place on the host cell, and the invasion of the host cell begins.

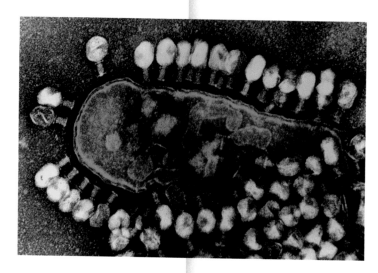

Figure 7.2 This bacterial cell is being attacked by a number of viruses called bacteriophages—viruses that attack only bacteria cells. What is the host for a bacteriophage?

As You Read

Use what you have learned in this lesson so far to add to or correct the information in your "biopoem."

How do these characteristics compare with your predictions?

Figure 7.3 The protein coat of each of these viruses has a specific shape that determines the type of cell to which the virus can attach. Once attached, the virus can invade the cell with its genetic core.

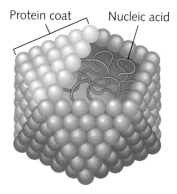

Polyhedral viruses, such as the papilloma virus that causes warts, resemble small crystals.

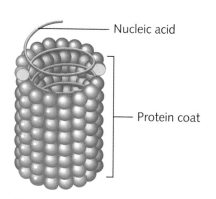

The tobacco mosaic virus has a long, narrow helical shape.

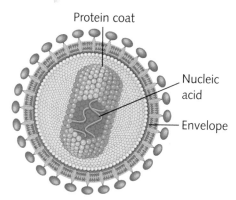

An envelope studded with projections covers some viruses, including the AIDS-causing virus.

Reproduction of Viruses

The structure of viruses allows them to succeed in their main mission—reproduction.

Lytic Cycle   Once attached to a host cell, a virus injects its nucleic acid into the cell. The nucleic acid takes over the normal operation of the host cell and produces multiple copies of the virus's protein coat and nucleic acid. Once produced, the protein coats and the nucleic acids are assembled into new viruses. As the host cell fills with newly assembled viruses, it bursts, just like a balloon with too much air. The host cell then dies, and the released viruses begin searching for the next host cell. This type of viral reproduction is called a **lytic** (LIT ihk) **cycle**. The steps of a lytic cycle for a bacteriophage are illustrated in **Figure 7.4**.

Lysogenic Cycle   Some viruses, such as herpes and HIV, enter the host cell but remain hidden for years. Even though the viral nucleic acid becomes part of the host cell's chromosome, it does not seem to affect the functions of the cell. At some point, however, the viral nucleic acid becomes active. It separates itself from the host cell's genetic material, takes over the functions of the cell to produce new viruses, and destroys the host cell as the new viruses are released. This type of viral reproduction is called a **lysogenic** (li suh JE nihk) **cycle**. The steps of a lysogenic cycle are also shown in **Figure 7.4**.

Explain It!

Once a computer virus infects a machine, the virus can spread to other disks, programs, and even other computers. In your Science Notebook, explain why the term *virus* is a good description of these computer invaders.

Figure It Out

1. What are the steps in a lytic cycle?

2. Compare and contrast a lytic cycle and a lysogenic cycle.

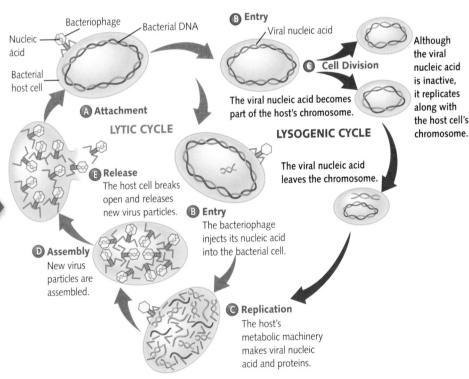

Figure 7.4 In a lytic cycle (red arrows), the virus uses the host cell's organelles to make new viruses. In a lysogenic cycle (blue arrows), the virus "hides" in the host cell's chromosome until it becomes active and uses the host cell's organelles to reproduce.

Viruses in the World

If you have ever had a cold or the flu or a childhood illness such as measles or chicken pox, you know that viruses can make you sick. But did you know that viruses can also protect you from diseases and save your life? As a young child, you received vaccinations (vak suh NAY shunz) made of dead or weakened viruses to protect you against viruses that try to invade your body.

Scientists are also using the ability of viruses to invade cells as a "delivery service" for genetic information that a person needs to treat or cure a disease. In gene therapy, viruses deliver normal genes that replace missing or faulty genes. Although gene therapy is a relatively new medical treatment, doctors are hopeful that it will be used widely in the future.

Viruses also affect plants and other animals. Your pets are vaccinated against diseases such as rabies and distemper to prevent infection by these deadly viruses. Other viruses that infect plants and animals can have an important economic impact. In the 1990s, cattle in Europe were infected with the virus that causes hoof-and-mouth disease. Thousands of animals had to be destroyed to prevent the spread of the disease. This caused a loss of millions of dollars.

Viruses can also be used to control pests that destroy crops used as food. The advantage of treating crops in this way is that it destroys harmful insects without polluting the environment.

Figure 7.5 These young people are receiving vaccinations for polio. Vaccines protect and save millions of lives. What vaccines have you received?

Figure 7.6 The stripes on this tulip were produced by the tulip mosaic virus.

CONNECTION: History

European explorers were able to conquer the Americas with small armies because they unknowingly brought with them a powerful weapon—an invisible army of viruses. Native Americans had never been exposed to these viruses, and the deaths caused by mumps, measles, diphtheria, and smallpox allowed the Europeans to defeat even the mighty Aztec empire.

Extend It!

Most schools require students to receive specific vaccinations at specific times before they can enroll or remain in school. Ask your school nurse or other administrator for a list of the vaccinations required by your school. Review the requirements and prepare a chart that shows the vaccination schedule for students in your grade. Why do you think it is important to have this requirement?

After You Read

1. Review the words and phrases you used in your "biopoem." Correct any of your predictions now that you know more about viruses. Select a title for your poem that you think best describes viruses. Are all viruses harmful? Explain your answer in a well-developed paragraph.

2. Describe the structure of viruses.

3. Compare the two ways in which viruses reproduce.

4. Is a virus a host or a parasite? Explain your answer.

7.2 Bacteria

Before You Read

In your Science Notebook, create a K-W-L-S-H chart. Think about the title of this lesson and read the Learning Goals. In the column labeled *K*, write what you already know about bacteria. In the column labeled *W*, write what you want to learn about bacteria.

Bacteria are everywhere. They can live in the saltiest waters on Earth, in hot springs and volcanic vents, in the freezing ice of the arctic, and even on your body. You cannot see or feel them, but bacteria live in your hair, lungs, mouth, stomach, and intestines. They practically cover your skin. They especially like your warm, moist armpits and your sweaty feet.

What Are Bacteria?

Microscopic prokaryotic cells are called **bacteria** (bak TIHR ee uh, singular: bacterium). As discussed in Chapter 3, prokaryotes do not have a membrane-bound nucleus or membrane-bound organelles, and their DNA is not organized into chromosomes.

Living or Nonliving? Even though a bacterium is microscopic and composed of only one cell, it is considered a living thing. Unlike the viruses you studied in Lesson 7.1, bacteria can sense and respond to stimuli, adapt to their environment, reproduce, and use energy to grow and develop. This is similar to the behavior of more complex organisms.

Classifying Bacteria Because bacteria are among the most numerous organisms on Earth, an amazing variety of them exists. Biologists group bacteria into two kingdoms, Archaebacteria and Eubacteria. *Archaebacteria* means "ancient bacteria." Scientists believe that these bacteria resemble Earth's first forms of life. Archaebacteria are often found in very harsh environments. Eubacteria is a much larger group of bacteria, and members of this kingdom are found everywhere that members of Archaebacteria are not. These are the bacteria that are found on and in your body, as well as in the soil, the air, and the water.

Types of Bacteria

There may be thousands of different types of bacteria, but they all have one of the three basic shapes shown in **Figure 7.7**. These basic shapes provide biologists with one way to identify different bacteria.

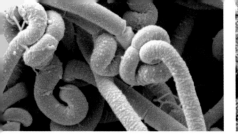
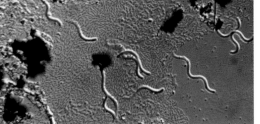
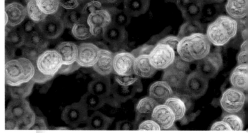

Bacilli   Spirilla   Cocci

Figure 7.7 Bacteria can be identified by their shapes.

Bacteria that are shaped like sticks or rods are called **bacilli** (buh SIH li, singular: bacillus). Bacteria that are shaped like globes or spheres are called **cocci** (KAH ki, singular: coccus). Bacteria that are shaped like corkscrews or spirals are called **spirilla** (spi RIH luh, singular: spirillum).

Structure of Bacteria   The cells of bacteria are prokaryotic cells. Because bacteria are cells, they have some of the same structures that were discussed in Chapter 3. As you read this section, locate each structure in **Figure 7.8**.

The outer wall of most bacteria is the cell wall. The cell wall is rigid and tough, and it protects the bacterial cell and determines its shape. Inside the cell wall is the cell membrane. The cell membrane controls what substances enter and leave the bacterial cell. Inside the cell membrane is the jelly-like cytoplasm that contains all the other structures found in a bacterial cell.

DNA, the bacterial cell's genetic material, is the rope-like tangle in the cytoplasm. Because bacteria are prokaryotes, their DNA is not enclosed in a nucleus. However, the DNA still controls the activities of the cells. The production of proteins is carried out by the ribosomes found throughout the cytoplasm.

Another structure found on some bacteria is the flagellum. **Flagella** (fluh JEH luh, singular: flagellum) are whiplike structures that extend outward from the cell membrane into the bacterial cell's environment and move the cell through that environment. Bacteria without flagella must depend on air or water currents or other living organisms to move from one place to another.

As You Read

In the column labeled *L* in your K-W-L-S-H chart, write three or four things you have learned about bacteria.

Are bacteria prokaryotes or eukaryotes? How do you know?

Cell wall A cell wall surrounds and protects the cell and gives the bacterium its shape.

Cell membrane A cell membrane controls what enters and leaves the bacterium.

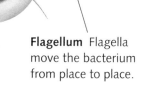

DNA A DNA molecule controls all the activities of the bacterium.

Flagellum Flagella move the bacterium from place to place.

Cytoplasm Cytoplasm is found inside the cell membrane and contains the bacterium's ribosomes.

Figure 7.8 A typical gram-negative bacterial cell would have the structures shown in this diagram.

Bacteria can reproduce very quickly, especially if conditions are suitable. If a bacterium reproduces by binary fission every 30 minutes, how many bacteria would there be after four hours? Construct a chart to show the number of bacteria present at each hour of the four-hour period. Then create a graph using this information.

Explore It!

Working in a small group, observe the prepared slides of the different types of bacteria your teacher provides. Design and complete a chart that contains the following information about each bacteria sample: name, diagram, and type.

Reproduction of Bacteria

When conditions are favorable, bacteria can grow and reproduce at amazing rates. Some bacteria can reproduce every 20 minutes! Bacteria reproduce by both asexual and sexual processes.

Asexual Reproduction When a bacterium has doubled in size, it makes a copy of its single chromosome and divides in half, producing two identical bacterial cells. This type of reproduction is called binary fission. **Binary fission** (BI nuh ree • FIH zhun) is a type of asexual reproduction because it requires only one parent and the daughter cells are identical to the parent cell.

Sexual Reproduction Some bacteria are able to exchange part of their genetic information in a form of sexual reproduction called **conjugation** (kahn juh GAY shun). During conjugation, a hollow tube connects the two bacterial cells, and one bacterium transfers part of its DNA along the tube to the other bacterium. Conjugation is a type of sexual reproduction because two parent cells are involved and the new bacterial cells are genetically different from the parent bacteria. Once the transfer of DNA is complete, the bacteria separate and then reproduce by binary fission.

Figure It Out

1. How do two bacteria transfer genetic information during conjugation?

2. Explain how the offspring produced by the process of binary fission are different from those produced by conjugation.

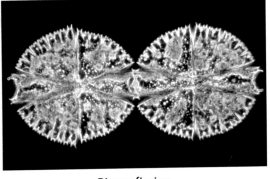
Binary fission

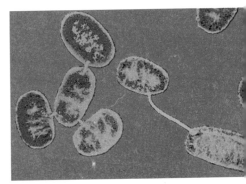
Conjugation

Figure 7.9 Most bacteria reproduce asexually by binary fission. Some bacteria reproduce sexually through conjugation followed by cell division.

Bacteria in the World

Figure 7.10 Bacteria play many important roles in the world.

Nitrogen-fixing bacteria in the nodules of legumes replenish nitrogen in the soil.

Bacteria can change the harmful chemicals in an oil spill into substances safe for the environment.

Some bacteria are used to produce antibiotics, which kill disease-causing bacteria.

As bacteria carry out normal life processes, they help produce foods such as cheese.

Bacteria break down nutrients in dead plants and animals and return them to the environment.

PEOPLE IN SCIENCE On the Trail of a Killer

On July 21, 1976, the American Legion began its annual convention in Philadelphia. Several weeks later, 34 of those who had attended the convention would be dead from a mysterious form of pneumonia.

On August 2, a team of epidemiologists—scientists who investigate disease—arrived in Philadelphia. They came from the Centers for Disease Control and Prevention (CDC) to track the killer, which had been named Legionnaires' disease. The team pinpointed the location of the outbreak as the Bellevue Stratford Hotel. Then they began testing tissue samples from those who had died, looking for bacteria or viruses.

In January 1977, the "disease detectives" announced that they had found the bacteria that had infected 221 people and killed 34. The bacteria were named *Legionella pneumophila*.

Eight years after the initial outbreak, the CDC team found the source of the bacteria—the hotel air-conditioning system. Today, air conditioners and other systems that hold water are regularly disinfected to prevent Legionnaires' disease.

After You Read

1. Draw a bacterial cell. Label the parts and describe their functions.

2. Compare the two types of bacterial reproduction.

3. Using the completed K-W-L columns in your chart, explain how bacteria are both helpful and harmful to other living things. Then complete the chart by indicating in the *S* column what you would still like to know about bacteria. Write in the *H* column how you would find this information.

Summary

KEY CONCEPTS

7.1 Viruses

- A virus is a nonliving particle that invades a cell and uses the cell's parts to reproduce and distribute more viruses.

- Viruses contain a core of genetic material surrounded by a protective protein coat. The genetic material contains the instructions by which viruses make copies of themselves inside the host cell.

- In a lytic cycle, a virus enters the host cell, makes copies of itself, and then destroys the cell to release the new viruses.

- In a lysogenic cycle, a virus stays hidden within the host cell's chromosome until the virus becomes active. Once active, it makes copies of itself and then destroys the cell to release the new viruses.

- Viruses can cause diseases in plants, humans, and animals. However, they can also protect humans from disease. Viruses are used in gene therapy and in crop-pest control.

7.2 Bacteria

- Bacteria are microscopic prokaryotic cells whose genetic material is found in the cytoplasm. Bacterial cells do not have nuclei.

- There are three different shapes of bacteria. Bacilli are rod-shaped bacteria, cocci are sphere-shaped bacteria, and spirilla are spiral-shaped bacteria.

- Bacteria reproduce asexually by binary fission. Two identical daughter cells are produced during binary fission. Some bacteria reproduce sexually by conjugation. Conjugation produces bacteria with new combinations of genetic information.

- Bacteria play many important roles in the world. Bacteria produce food, medicines, and fuel. They help clean up the environment and recycle nutrients for use by other living things.

VOCABULARY REVIEW

Write each term in a complete sentence, or write a paragraph relating several terms.

7.1
virus, p. 124
host, p. 124
parasite, p. 124
lytic cycle, p. 126
lysogenic cycle, p. 126

7.2
bacterium, p. 128
bacillus, p. 129
coccus, p. 129
spirillum, p. 129
flagellum, p. 129
binary fission, p. 130
conjugation, p. 130

PREPARE FOR CHAPTER TEST

To prepare for the chapter test, create a question from each Learning Goal. Use the information in your Science Notebook to answer each question. Then use these answers to write a well-developed essay about the chapter. Use the Key Concept on the first page of this chapter as your topic sentence.

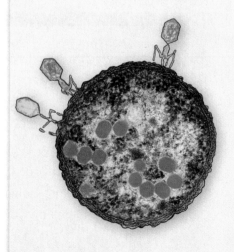

True or False

If the statement is true, write "true." If it is false, change the underlined word or words to make the statement true.

1. Particles that are made up of a protein coat and a core of genetic material are <u>bacteria</u>.

2. Viruses are <u>hosts</u> that cause harm to the cells that they invade.

3. In order to reproduce, a <u>virus</u> must invade a host cell.

4. Bacteria reproduce asexually by <u>conjugation</u>.

5. A flagellum is a whiplike structure that helps a <u>virus</u> move.

6. A bacterial cell does not have a <u>nucleus</u>.

Short Answer

Answer each of the following in a sentence or brief paragraph.

7. How are viruses different from living cells?

8. What is the difference between a lytic cycle and a lysogenic cycle?

9. What characteristic of bacteria determines their type?

10. Can viruses be grown in the laboratory using a nonliving nutrient substance? Explain your answer.

11. Why are bacteria considered prokaryotes?

Critical Thinking

Use what you have learned in this chapter to answer each of the following.

12. **Compare and Contrast** Identify ways in which reproduction in viruses is similar to and different from reproduction in bacteria.

13. **Apply** Farmers often rotate their crops with legumes such as soybeans. When the soybeans are mature, farmers plow the soybeans into the ground and then plant another crop, such as cotton. Explain the role that bacteria play in this process.

14. **Make Inferences** Scientists hypothesize that viruses could not have existed before bacteria or other organisms. What do you know about viruses that would support this hypothesis?

Standardized Test Question

Choose the letter of the response that correctly answers the question.

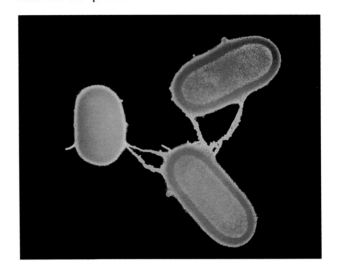

15. Which reproductive process is shown in the photo?
    A. lytic cycle
    B. lysogenic cycle
    C. conjugation
    D. binary fission

Test-Taking Tip

After you read a multiple-choice question, answer it in your head before reading the choices provided. This way the choices will not confuse or trick you.

Protists and Fungi

KEY CONCEPT Protists and fungi are diverse groups of organisms that play many important roles in the world.

Imagine that it is a hot, humid summer day and you are ready for a swim in your favorite pond. The pond, however does not look very inviting—it is covered with bright green scum!

While you might describe the pond scum as disgusting, what you are really looking at are millions of microorganisms that are very important to you. They are responsible for much of the oxygen you breathe. Beneath that pond scum are other microorganisms. You probably used some of them this morning to brush your teeth. Protists and fungi are found everywhere, and you could not live without them.

Think About Protists and Fungi

Have you ever seen bread that did not look quite right? Bread should be soft and spongy, not covered with green, powdery dots that are probably mold.

- What caused the green dots to appear? Why should you throw away the bread and not eat it? What other types of food spoilage are you familiar with?

- Write one or two sentences in your Science Notebook to describe the appearance of bread or other foods that have spoiled.

NSTA

SCI LINKS
THE WORLD'S A CLICK AWAY

www.scilinks.org
Protists **Code: WGB08A**
Fungi **Code: WGB08B**

8.1 Protists

Before You Read

Create a concept map by writing the lesson title *Protists* in the middle of a sheet of paper. As you read the lesson, scatter words or phrases that describe these organisms around the title and connect them to the title with lines.

Learning Goals

- Describe the characteristics of protists.
- Explain on what basis biologists divide protists into groups.

New Vocabulary

protist

Is there a closet in your home that contains a variety of different things? It might be full because it holds everything that does not fit anywhere else in the house. You can think about the kingdom Protista as the "hold-everything" kingdom of living things because organisms in this kingdom do not fit anywhere else. Even though these organisms are incredibly diverse, they do have some characteristics in common.

What Is a Protist?

A **protist** (PROH tihst) is any organism that is a eukaryote but is not a plant, an animal, or a fungus. Remember that a eukaryote is a living thing made of cells with a membrane-bound nucleus that directs the activities of the cell. Most protists live in moist environments such as salt water, freshwater, or very moist soil.

As You Read

Use the headings and the vocabulary term in this lesson to add descriptive words to your concept map.

What is a eukaryote?

Types of Protists

Because of the diversity among protists, one way that biologists group them is based on how they obtain their food. The three categories of protists are animal-like protists, plant-like protists, and fungus-like protists.

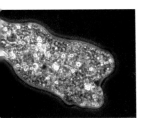  

Figure 8.1 Kingdom Protista includes animal-like, plant-like, and fungus-like protists. A microscopic amoeba *(left)* lives in freshwater environments. Giant kelp *(center)* can reach heights of more than 50 m and grows in saltwater kelp forests. A slime mold *(right)* is a fungus-like protist.

Figure It Out

1. Which photograph depicts a plant-like protist?

2. What characteristics might scientists use to identify a protist as an animal-like protist?

After You Read

1. What characteristic do all protists share?
2. Name the three groups of protists.
3. Review your concept map. Which word or phrase best describes all protists?

- Compare and contrast the different types of animal-like protists.
- Identify the roles played by animal-like protists in the living world.

New Vocabulary

heterotroph
protozoan
pseudopod
contractile vacuole
cilia
spore

Recall Vocabulary

binary fission (p. 130)
conjugation (p. 130)
flagellum (p. 129)
parasite (p. 124)

Before You Read

Create a lesson outline. Use the lesson title as the outline title. Label the subheadings with the Roman numerals *I* through *IV*. Use the letters *A, B, C,* etc., under each subheading to record important information.

Animal-like protists get their name because they share two characteristics with animals—they are able to move and they are heterotrophs. **Heterotrophs** (HE tuh roh trohfs) are organisms that cannot convert inorganic materials into organic molecules and must consume other organisms for food. Animal-like protists are called **protozoans** (proh tuh ZOH unz). Unlike an animal, however, a protozoan is unicellular.

Types of Protozoans

Protozoans are usually divided into groups based on how they move. Some protozoans use "false feet," others use structures similar to oars or whips, and still others rely on their hosts.

Protozoans with Pseudopods   This group includes protozoans that have "false feet" called pseudopods. **Pseudopods** (SOO duh pahdz) are extensions of the cell membrane that fill with cytoplasm. As the cell membrane bulges outward, the cytoplasm streams into the extension and the rest of the cell follows.

Figure 8.2 shows an amoeba using its pseudopods to capture prey. **Figure 8.3** illustrates how an amoeba encloses the prey in a food vacuole. The amoeba's food vacuole functions like a digestive system. Protozoans, such as amoebas, have a **contractile** (kun TRAK tul) **vacuole** (VAK yuh wohl) that collects extra water that enters the cell and pumps it out.

Most amoebas reproduce by asexual reproduction, which requires only one parent. The offspring are genetically identical to the parent. When the amoeba's cytoplasm divides in half, two identical amoebas are produced.

Figure 8.2 *Protozoa* means "first animals." Protozoans are the main hunters in the microscopic world. Notice how this amoeba has moved to capture its food. What type of organisms must consume other organisms for food?

Figure 8.3 These diagrams show the structure of a typical amoeba, such as *Amoeba proteus*.

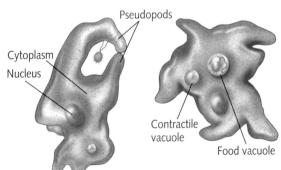

❶ The food becomes trapped by the pseudopods. When the pseudopods meet, they join to form a food vacuole.

Pseudopods

Cytoplasm

Nucleus

Contractile vacuole

Food vacuole

❷ Food is broken down inside the food vacuole by digestive enzymes, and the nutrients then move out into the cytoplasm.

Protozoans with Cilia Another type of animal-like protist is the ciliate, or protozoan with cilia. **Cilia** (SIH lee uh) are tiny hairlike projections that move with wavelike rhythm. The cilia act like tiny oars that propel the protozoan through the water.

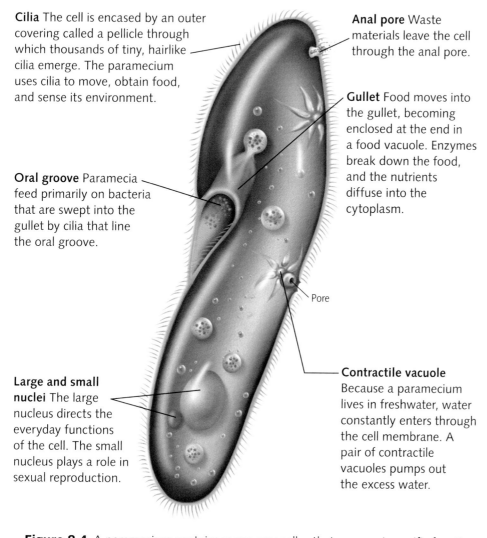

Cilia The cell is encased by an outer covering called a pellicle through which thousands of tiny, hairlike cilia emerge. The paramecium uses cilia to move, obtain food, and sense its environment.

Oral groove Paramecia feed primarily on bacteria that are swept into the gullet by cilia that line the oral groove.

Large and small nuclei The large nucleus directs the everyday functions of the cell. The small nucleus plays a role in sexual reproduction.

Anal pore Waste materials leave the cell through the anal pore.

Gullet Food moves into the gullet, becoming enclosed at the end in a food vacuole. Enzymes break down the food, and the nutrients diffuse into the cytoplasm.

Pore

Contractile vacuole Because a paramecium lives in freshwater, water constantly enters through the cell membrane. A pair of contractile vacuoles pumps out the excess water.

Figure 8.4 A paramecium contains many organelles that carry out specific functions.

Ciliates usually reproduce asexually by binary fission, and the daughter cells are genetically identical. Ciliates can, however, reproduce sexually by conjugation. Two ciliates join together temporarily to exchange genetic material. Once the exchange is complete, they separate and divide by binary fission. Both types of reproduction are shown in **Figure 8.5**.

Protozoans with Flagella Animal-like protists called zooflagellates move using flagella. Zooflagellates also use their flagella to sweep bacteria and other protists into their mouthlike openings. Many species of zooflagellates live inside other organisms.

Some zooflagellates are disease-causing parasites. African sleeping sickness is a serious disease caused by a zooflagellate that spreads to humans when they are bitten by the tsetse fly.

Figure It Out

1. What organelles are used for catching and digesting food?

2. Compare the amoeba in **Figure 8.3** with the paramecium in **Figure 8.4**. How are these organisms similar? How are they different?

As You Read

Review your outline. Then write a brief summary in your Science Notebook about protozoans with cilia and protozoans with flagella.

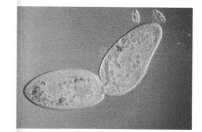

Figure 8.5 Ciliates usually reproduce asexually *(top)*, but when environmental conditions change, they can also reproduce sexually *(bottom)*.

Extend It!

In a small group, research the disease amoebic dysentery. Identify the protozoan that causes the disease, how the protozoan moves, how the disease is transmitted, and its symptoms and treatment. You may want to include additional information, such as where the disease occurs, how many people it affects yearly, and how outbreaks of the disease can be controlled. Report your findings in a short paragraph or a chart in your Science Notebook.

Protozoans That Produce Spores   Another type of animal-like protist is a parasite that produces spores. This protozoan depends on its host for movement and a supply of food. Called sporozoans, these organisms produce reproductive cells called **spores**. Spores develop into new organisms when environmental conditions are favorable. Spores pass easily from host to host when food that contains spores is eaten, or when a mammal, bird, or fish is bitten by an insect that carries spores in its body. Sporozoans often have complex life cycles that can require more than one host organism.

Perhaps the best-known sporozoan is *Plasmodium* (plaz MOH dee um), the protist that causes malaria. Malaria spreads when a healthy mosquito bites a person with the disease and becomes infected with *Plasmodium* spores. When the infected mosquito then bites a healthy person, the spores pass into that person's body, infect red blood cells, and cause the cells to rupture. Symptoms of malaria can last for weeks at a time and include high fevers that alternate with severe chills.

Protozoans in the World

Protozoans play many important roles in the world. Parasitic protozoans cause diseases such as malaria and African sleeping sickness. These diseases affect millions of people all over the world and cause many deaths. Two major goals of health organizations are to control the protozoans that cause these diseases and to develop new treatments.

Not all protozoans are parasitic. Termites would starve to death without the protozoan *Trichonympha* in their intestine to digest the wood they eat. Other protozoans are decomposers that recycle nutrients in dead plants and animals. Many protozoans are important links in freshwater and saltwater food chains. Protozoans eat bacteria and other protists and then become a meal for small animals that are, in turn, eaten by larger animals.

After You Read

1. Describe the characteristics of an animal-like protist.

2. Use your outline to produce a chart that identifies the four groups of protozoans, describes how they move, and includes interesting facts about each group.

3. Use the information in your outline to write a well-developed paragraph summarizing the roles protozoans play in the living world.

8.3 Plant-Like Protists: Algae

Before You Read

Create a table to organize the information presented in this lesson. Use the lesson title as the table title and create columns titled *Structure, Habitat, Nutrition,* and *Important Information.* Each subheading that identifies a type of plant-like protist should have a row in your table.

What Are Algae?

If you have seen brown seaweed washed up on a beach, or green scum floating on top of a pond, you are familiar with plant-like protists. Plant-like protists are autotrophs called **algae** (AL jee, singular: alga). **Autotrophs** (AW tuh trohfs) are organisms that capture energy and matter from their surroundings. Autotrophs are food for many heterotrophs. Algae live in several different environments and come in a variety of shapes, sizes, and colors. Algae have different colors because they contain different **pigments** (PIG munts), or chemicals that produce colors. These pigments allow algae to capture and use energy from the Sun to produce organic molecules through the process of photosynthesis.

Types of Algae

Algae are divided into several groups based on the pigments they contain.

Dinoflagellates   These unicellular algae, whose flagella spin them through the water, look like tiny toy tops. Like euglenoids, they can be both autotrophic and heterotrophic. Although beautiful, some of these algae produce harmful toxins.

A population explosion of toxin-producing dinoflagellates has deadly consequences. When there is an increase in nutrients in the water, the dinoflagellate population can explode, or "bloom," to produce a red tide. Shellfish, such as clams and oysters, that eat large quantities of these algae store toxins in their cells. When these shellfish are eaten by fish or by humans, the toxins are passed through the food chain.

Figure 8.6 Algae range in size from microscopic diatoms that look like pieces of jewelry to giant kelp found in underwater forests.

As You Read

Review the table you developed at the beginning of this lesson.

What characteristic do euglenoids, dinoflagellates, and diatoms have in common?

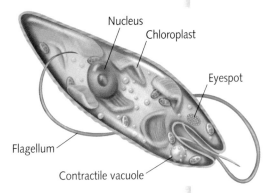

Figure 8.7 The eyespot of *Euglena gracilis* is not really an eye, but it contains pigments that are sensitive to light. The flagella of these organisms help them move toward light or food sources.

Explain It!

Compare the euglena in **Figure 8.7** with the paramecium in **Figure 8.4**. Explain how these organisms are similar and how they are different.

Euglenoids This type of alga is closely related to the protozoans. Although euglenoids are green, unicellular algae usually found in freshwater, they can be heterotrophs. When light is not available for photosynthesis, they can find and capture food from their environment.

Diatoms Another type of plant-like protist is the diatom. These unicellular algae are enclosed in a two-part cell wall made of the element silicon, the main component of glass. Diatoms are often called "golden algae" because they contain pigments that give them a golden yellow color. These algae store the organic molecules they produce as oils, not starches. Because oil is less dense than water, it helps diatoms float in the upper part of the water so that they can absorb light from the Sun.

Figure 8.8 Diatom cell walls are covered with patterns that give each species a unique design.

CONNECTION: Earth Science

When diatoms die, their hard silicon cases fall to the floor of an ocean or lake and form deposits called diatomaceous earth, or diatomite. These deposits are important tools for studying events in Earth's history.

Diatoms are sensitive to changes in water conditions such as temperature, salinity (the amount of salt in water), and nutrient level. This sensitivity to the environment allows Earth scientists to use the fossilized remains of diatoms to reconstruct events such as ice ages and plate tectonic activity. By studying the diatom populations in coastal waters, Earth scientists can document the changes in water quality produced by the clear-cutting of forests, the damming of rivers and streams, and the use of fertilizers.

Diatomaceous earth, or diatomite, is used in fine china, swimming pool filters, acoustic tile, potting soils, and many household powders for cleaning or polishing. Because diatomite is also highly absorbent and does not react with other chemicals, it is also used in industrial absorbents that clean up oil and chemical spills and in pet litter to absorb odors. It can also be added to soil, where it acts as an insecticide. The sharp edges of the diatoms' silicon cell walls puncture the bodies of insects as they crawl through the soil.

Red, Brown, and Green Algae These three types of algae are the most plant-like. Like plants, most of these algae are multicellular. They also have cell walls, pigments that are identical to those of plants, and reproductive cycles that are very similar to those of plants. The major difference among these three groups of autotrophic algae is the type of photosynthetic pigments they use to absorb light.

Figure It Out

1. What differences do you see among red, brown, and green algae?

2. Which type of algae is thought to be an ancestor of modern plants?

Figure 8.9 Red algae include seaweeds that can grow at depths of 100 m or more. They range in color from red to green to purple to black. They might look like flat sheets, rock formations, or clumps of hair swaying in the water. Red algae provide food for the animals that form coral reefs.

Figure 8.10 Brown algae contain yellowish-brown pigments that give them their brown color.

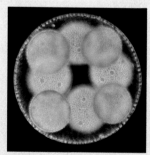

a) This diver is swimming in a forest of giant kelp, which can grow up to 50 m long. Kelp forests are ecosystems that provide many marine organisms with homes and food.

b) Some brown algae float as huge mats of seaweed at or near the ocean's surface, where there is plenty of sunlight. This carpet of brown algae is home to many different animals.

Figure 8.11 Green algae have cell walls made of cellulose, contain chlorophyll *a* and *b* pigments, and store their food as starch. Many scientists hypothesize that green algae and plants may have a common ancestor.

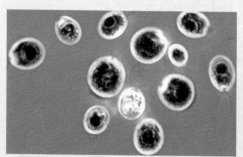

a) *Chlamydomonas* is a unicellular green algae found in freshwater and moist soil.

b) Some green algae, such as *Volvox*, live in multicellular colonies.

c) *Ulva*, commonly called sea lettuce, are multicellular green algae that live along rocky coasts.

Algae in the World

Perhaps the most important role of algae is that of producers. Algae make up a large part of phytoplankton. **Phytoplankton** (FI toh PLANK tun) are photosynthetic organisms that live floating near the surface of the ocean and carry out much of the photosynthesis that occurs on Earth. Phytoplankton are the main producers in many aquatic food chains and food webs. Through the process of photosynthesis, algae produce about half of the oxygen found in Earth's atmosphere!

Algae are also used as a food supply for humans. In some parts of the world, algae are an important source of protein and are actually grown on large farms. In the United States, algae are used in ice cream, pudding, and candy to thicken them and to keep them from separating.

Algae are also used to study microorganisms such as bacteria. Agar, which is made from seaweed, is used to thicken the nutrient mixtures used to grow bacteria so that the bacteria can be studied.

Algal blooms also occur in freshwater. When nutrients from the land run off into a pond or lake, algae populations explode, or "bloom." The rapid growth of algae makes it impossible for sunlight to reach other algae and plants beneath the surface. These organisms die and sink to the bottom, where bacteria break down their bodies. As the bacteria use the oxygen in the water, fish begin to die. The only organisms that survive these population explosions are the algae themselves. Freshwater blooms of algae are easier to control than red tides are because ponds and lakes have definite boundaries. Wastes from human sources, such as crop fertilizers or pipes that leak sewage, are often responsible for the increased nutrients. If the source of the nutrients can be eliminated, the lake can eventually return to normal.

Figure 8.12 The seaweed used to wrap the fish, rice, and vegetables in Japanese sushi is a type of algae.

 Explore It!

Go on a scavenger hunt in your home, and identify as many products as you can that contain algae. Record your observations in your Science Notebook.

After You Read

1. Describe the characteristics of plant-like protists.

2. How are algae important to living things on Earth?

3. Use the information in your table to compare dinoflagellates, euglenoids, and diatoms with red, brown, and green algae. Write a well-developed paragraph describing the main differences among these groups of algae.

8.4 Fungus-Like Protists

Learning Goals

• Compare and contrast the different types of fungus-like protists.

• Identify the roles played by fungus-like protists in the living world.

New Vocabulary

plasmodium
hypha

Before You Read

Record each heading in this lesson in the form of a question. Write the questions in your Science Notebook. As you read, write answers to the questions.

Imagine stepping into your backyard to find a huge, reddish, jelly-like mass on the grass. Many questions would probably come to mind. Is it alive? Is it an animal, a plant, or a new life-form? Is it dangerous? In 1973, a Dallas resident had exactly this experience. Biologists called to the scene were able to assure neighbors that a menacing creature was not about to ooze its way throughout the area, eating everything in its path. The biologists identified the mass as a slime mold. Slime molds, water molds, and mildews are fungus-like protists.

What Are Fungus-Like Protists?

Fungus-like protists resemble fungi. You will learn more about fungi in Lessons 8.5 and 8.6. To understand what fungus-like protists are, you need to know that fungi are organisms that are similar to both animals and plants. Like animals, fungi are heterotrophs that obtain food by consuming organic molecules from the environment. Like plants, fungi have cells that are surrounded by cell walls.

Fungus-like protists are heterotrophs that obtain food by decomposing dead plants and animals and then absorbing the nutrients. These protists have cells with cell walls, and they reproduce by spore formation.

Figure 8.13 Slime molds grow in moist, shaded environments where there are plenty of dead organisms to feed on.

Figure 8.14 Fungus-like protists such as the water mold on this fish are similar to both the plant-like protists and animal-like protists discussed in earlier lessons.

Types of Fungus-Like Protists

Fungus-like protists include organisms known as slime molds, water molds, and mildews.

Slime Molds One of the more interesting types of fungus-like protists is the slime mold. A slime mold is made up of a large number of individual cells that come together to form a jelly-like mass called a **plasmodium** (plaz MOH dee um). The plasmodium of most slime molds is visible to the unaided eye and can be more than a meter in diameter. Slime molds ooze along like giant amoebas and eat decaying plants, bacteria, fungi, and even other slime molds.

As long as there is enough food and moisture in the environment, the cells of a slime mold work together. But when food and moisture become scarce, the slime mold begins the reproductive process shown in **Figure 8.15**. The plasmodium grows stalks topped by spore-producing structures called fruiting bodies. Spores released into the environment are carried by rain or wind to new locations, where they can germinate and produce new cells that allow the slime mold to start a new life cycle.

As You Read

According to the questions and answers in your Science Notebook, what are the characteristics of fungus-like protists?

Figure It Out

1. What do spores released by a slime mold become?

2. At what point in this cycle would you infer that food and water have become scarce in the environment?

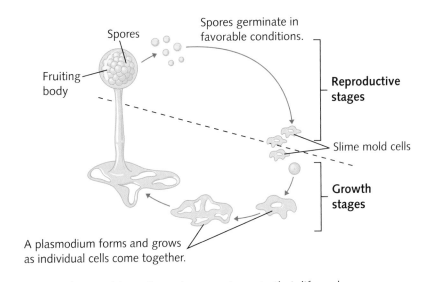

Figure 8.15 Slime molds go through many stages in their life cycles.

Figure 8.16 Water molds and mildews live in moist environments and cause plant and animal diseases. What function do hyphae of fungus-like protists have?

Water Molds and Mildews These types of fungus-like protists live in wet environments such as freshwater or the upper layers of moist soil. These organisms absorb their food from the surrounding water or soil, or they may invade the body of another organism to feed. The fish in Figure 8.14 on page 143 has been attacked by a water mold. Water molds and mildews that attack fish and plants are disease-causing parasites.

The thin, fuzzy threads growing on the fish in Figure 8.14 are called **hyphae** (HI fee, singular: hypha). Digestive enzymes ooze out of the hyphae onto the fish and break down the nutrients in its body. Once digested, the nutrients are absorbed by the water mold through the hyphae.

Fungus-Like Protists in the World

Slime molds, water molds, and mildews play an important role in the decomposition and recycling of organisms that have died. When decomposers break down dead organisms, they return nutrients to the environment. Rich, dark topsoil that provides nutrients to plants is produced by the decomposition efforts of fungus-like protists.

Parasitic species of fungus-like protists that destroy important food crops have actually had an impact on history. The Great Potato Famine in Ireland, caused by the parasitic water mold *Phytophthora infestans,* is an example of how a protist changed history. The potato was introduced to Europe by Spanish explorers, and it quickly became an important food crop. *Phytophthora infestans* infects the stem and leaf tissue of the potato and kills the plant. It can also infect the underground stem, or tuber, which is the part of the potato plant that stores starch and is eaten by humans. The disease caused by this parasite is called potato blight. Potatoes that have blight look normal when they are harvested, but within a few weeks, they become sacs of spores and dust. The cool, damp conditions that allow this water mold to spread rapidly are common in western Europe.

The summer of 1845 was unusually wet and cool in Ireland, which meant that conditions were perfect for the spread of potato blight. By the end of the 1845 harvest, potato blight had destroyed almost 60 percent of the Irish potato crop. During a single week in 1846, almost the entire Irish potato crop was destroyed by this parasite. Because potatoes were the main food of the working class, as the potato crop died, so did nearly a million of Ireland's working class. More than a million Irish people and their families immigrated to other countries, particularly the United States. These new immigrants brought their religion, culture, and customs with them to the American cities in which they settled.

Extend It!

Research the parasite *Plasmopara viticola,* also known as the downy mildew of grapes. Find out how this parasite almost destroyed the economically important French wine industry.

Figure 8.17 If your ancestors were Irish, it was probably *Phytophthora infestans* that brought them to the United States when it destroyed the Irish potato crop in 1845 and 1846. Many immigrants arrived in New York in the late 1840s, bringing all of their possessions with them.

After You Read

1. Describe the characteristics of fungus-like protists.
2. How do hyphae function?
3. What important role do fungus-like protists play on Earth?
4. Use the information in your Science Notebook to describe the impact that potato blight had on history.

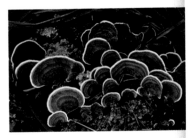

Figure 8.18 Fungi come in a variety of different shapes, sizes, and colors. What determines these varieties?

Before You Read

Read the Learning Goals and headings in this lesson. In your Science Notebook, write three facts that you already know about fungi. Leave six lines of space below each. As you read the lesson, add at least two more facts to each fact you already know.

A tiny piece of dust lands on the back of an insect. This piece of dust is alive—and it is a killer. The dust particle sprouts tiny threads that begin to grow into the insect's body. As the threads grow, they dissolve the insect's cells and use them as nutrients. Once the insect has been "eaten," a thin stalk grows out of the threads and a lump forms at the end. The lump explodes and releases thousands of other pieces of dust looking for an insect meal.

What Are Fungi?

The insect-killing dust just described is actually a fungus (FUN gus, plural: fungi, FUN ji). Fungi can be found on food in your refrigerator, growing like shelves from the side of a tree, and even between your toes or under your nails. Most **fungi** share three important characteristics: They are eukaryotes, they are heterotrophs that feed on other organisms, and they use spores to reproduce.

Structure of Fungi

A few types of fungi, such as baker's yeast, are unicellular, but most fungi are multicellular. Multicellular fungi are composed of threadlike tubes called hyphae, which are one cell thick and grow and branch in various patterns to produce fungi with many different shapes. Multicellular fungi have a thick mat of tangled hyphae called a **mycelium** (mi SEE lee um). There are different types of hyphae within the mycelium. Some function to obtain food or anchor the fungus, while others may play a role in reproduction.

Fungi eat whatever they are growing on. Some fungi are parasites that feed on living organisms. Athlete's foot is a parasitic fungus that feeds on chemicals in a person's skin. Wheat rust is a parasitic fungus that attacks wheat plants. Other fungi feed on dead organisms. Through their hyphae, fungi release enzymes that digest the food source. The mycelium of a fungus can be very large and thus provide a great deal of surface area through which the hyphae can absorb nutrients.

Reproduction in Fungi

Depending on the type of fungus and the environmental conditions, most fungi reproduce both asexually and sexually.

Asexual Reproduction   Fungi are able to reproduce asexually by fragmentation, by budding, or by the production of spores. During fragmentation, pieces of hyphae that are separated from the mycelium grow into a new mycelium. Fragmentation often occurs when the ground is disturbed in preparation for planting. Plows and shovels used to prepare the soil cut through the underground mycelia of fungi. Each fragment grows into a new mycelium.

Most fungi use spores to reproduce. A spore is a reproductive cell that can grow into a new organism. Some fungi produce spores in reproductive hyphae that have grown up from the mycelium. The reproductive hyphae that produce spores are called **sporangia** (spuh RAN jee uh, singular: sporangium). Above ground structures such as mushrooms are fruiting bodies that contain many closely packed sporangia. When a fungal spore is transported to a favorable environment, hyphae sprout from the spore and a new fungus begins to grow. If the spore lands in an unfavorable environment, it protects the new fungus until conditions improve or the spore is carried to a new location.

Fungi have developed a number of ways to distribute spores in the environment. Some fungi produce a fluid that smells like rotting meat and attracts flies. As the flies eat the fluid, they also eat the spores and then deposit them as they fly around. Some fungi actually throw their spores away from the mycelium. Many spores are carried by wind and animals.

Sexual Reproduction   Fungi can also reproduce sexually, primarily when environmental conditions are unfavorable. During sexual reproduction, the hyphae of two fungi grow together and exchange genetic material. A new spore-producing structure grows from these joined hyphae. Spores produced by this process are genetically different from either parent fungus.

Figure It Out

1. What is contained in the cloud above the puffballs?

2. Do you think most fungi reproduce sexually or asexually?

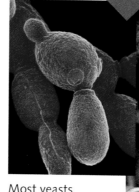

Most yeasts reproduce asexually by budding.

Puffballs are fruiting bodies that can produce up to a trillion spores to guarantee species survival.

Figure 8.19 Fungi reproduce asexually by fragmentation, budding, or the production of spores.

After You Read

1. What characteristics do all fungi have in common?

2. How do fungi reproduce?

3. Review the facts you recorded in your Science Notebook. Which fungi fact is the most interesting to you? Justify your selection.

Learning Goals

- Describe how scientists classify fungi.
- Name and describe the four groups of fungi.
- Identify the roles played by fungi in the living world.

New Vocabulary

mold
sac fungus
club fungus
imperfect fungus
lichen

As You Read

Reread the definitions of the vocabulary terms. Add to or correct your definitions.

Which is the largest group of fungi? Which is the most well-known group?

Figure It Out

1. Which group of fungi would be unwelcome in your refrigerator?

2. What would you need to know in order to classify a fungus you found?

Before You Read

Select three of the terms from the vocabulary list. Write them in your Science Notebook. Under each, write what you think the term means. As you encounter each term while you read, correct or add to your definition.

Fungi are classified into four groups based on the shape of their spore-producing structures and how they reproduce. **Figure 8.20** shows examples of these four groups.

Types of Fungi

Threadlike fungi, or **molds**, are easily identified as the fuzzy growths on rotten food. Most molds reproduce asexually by producing spores in sporangia. Molds function in nature as decomposers or parasites.

The largest group of fungi are the **sac fungi**. They are called sac fungi because most of them produce spores during sexual reproduction in structures that look like small sacs. When the sacs burst, spores are released into the environment. Some sac fungi are multicellular and are visible as they grow. Other sac fungi, such as yeast, are unicellular and reproduce by budding.

Mushrooms are probably the most well-known fungi. Mushrooms are **club fungi** that produce spores sexually in structures shaped like clubs. The clublike structures are found on the gills of the mushroom.

Fungi that are not known to reproduce sexually are called **imperfect fungi**. The best known of the imperfect fungi is *Penicillium*, which is a mold that grows on fruit and is the source of the antibiotic penicillin.

Figure 8.20 The four groups of fungi—*(left to right)* molds, sac fungi, club fungi, and imperfect fungi—exhibit a wide range of shapes, sizes, and colors.

Fungi in the World

Fungi are adapted to playing many important roles in Earth's ecosystems.

Recycling Nutrients   Fungi are decomposers. Decomposers break down the remains of dead plants and animals and return nutrients to the environment. The most important role played by fungi in Earth's ecosystems is that of decomposer.

Food   The smell of fresh-baked bread, the sight of blue streaks in bleu cheese, and the taste of mushrooms in a salad would not be possible without fungi. Fungi are sources of food, and they are used in the production of some foods. Yeast is needed to make bread. Molds are important in the production of some cheeses. The blue streaks in bleu cheese are patches of spores from the mold *Penicillium roqueforti.*

Medicines   In 1928, Alexander Fleming observed that mold in a petri dish not only prevented bacterial growth but also killed the bacteria near it. This led to the development of the first antibiotic—penicillin. The mold *Penicillium* provided an effective treatment against bacteria. Many other antibiotics have been developed from other fungi.

Relationships with Other Living Things   Fungi have different relationships with other organisms in ecosystems. Fungi in parasitic relationships cause diseases in plants and animals. Corn smut and wheat rust are club fungi that attack crops as they grow. Some fungi destroy crops already harvested and stored, such as wheat, rice, and rye.

Parasitic fungi can also attack humans. The itch and irritation of athlete's foot is produced as the mycelium of the fungus grows into the outer layers of the skin. The red sores that form release the spores of the fungus, which spread easily to other human feet that come in contact with the spores. Fortunately, athlete's foot and other irritations caused by fungi are easily treated with antifungal medications.

Some fungi form mutualistic relationships with other organisms in which both organisms benefit. One example is a lichen. A **lichen** (LI kun) is the partnership between a fungus and an alga. The fungus provides a stable, moist environment for the alga, and the alga provides food for the fungus. This relationship allows lichens to live in places where algae and fungi could not live alone.

Figure 8.21 The mushroom *Amanita muscaria (top)* is extremely poisonous. The mushroom *Amanita caesarea (center)* is edible. The multicolored spots on the rocks *(bottom)* are lichens.

After You Read

1. How do scientists classify fungi?
2. Describe roles that fungi play in the world. Identify the most important.
3. Review the vocabulary terms in your Science Notebook. Use them to develop a concept map to illustrate groups into which fungi are classified.

Extend It!

Research the discovery and history of penicillin. Summarize your research in your Science Notebook.

KEY CONCEPTS

8.1 Protists

- Protists are eukaryotic organisms that cannot be classified as animals, plants, or fungi.
- Protists are divided into three general groups based on how they obtain nutrition: animal-like protists, plant-like protists, and fungus-like protists.

8.2 Animal-Like Protists: Protozoans

- Protozoans are unicellular heterotrophs that can move.
- Heterotrophs are organisms that must consume other organisms for food.
- Protozoans are divided into four groups based on how they move.
- Amoebas move by extending pseudopods, ciliates use cilia, zooflagellates use flagella, and sporozoans depend on their host organism.
- Some protozoans are parasites, and some are decomposers.

8.3 Plant-Like Protists: Algae

- Algae are unicellular and multicellular autotrophs.
- Algae are a source of food and oxygen for many organisms.

8.4 Fungus-Like Protists

- Fungus-like protists are heterotrophic decomposers.
- Slime molds, water molds, and mildews make up this group of protists.
- Some fungus-like protists cause diseases in plants and animals.

8.5 Fungi

- Fungi are heterotrophs that obtain food by digesting the substances on which they are growing.
- Fungi are made of hyphae, which grow in a tangled mass to form a mycelium.
- Fungi reproduce using spores, which can be produced through both asexual and sexual reproduction.

8.6 Diversity of Fungi

- The four groups of fungi are molds, sac fungi, club fungi, and imperfect fungi.
- Most fungi are decomposers, but some are involved in parasitic or mutualistic relationships.
- A lichen forms when an alga and a fungus form a mutualistic partnership.
- Some fungi are disease-causing parasites that attack both plants and animals.

VOCABULARY REVIEW

Write each term in a complete sentence, or write a paragraph relating several terms.

8.1
protist, p. 135

8.2
heterotroph, p. 136
protozoan, p. 136
pseudopod, p. 136
contractile vacuole, p. 136
cilia, p. 137
spore, p. 138

8.3
alga, p. 139
autotroph, p. 139
pigment, p. 139
phytoplankton, p. 142

8.4
plasmodium, p. 144
hypha, p. 144

8.5
fungus, p. 146
mycelium, p. 146
sporangium, p. 147

8.6
mold, p. 148
sac fungus, p. 148
club fungus, p. 148
imperfect fungus, p. 148
lichen, p. 149

PREPARE FOR CHAPTER TEST

To prepare for the chapter test, create a question from each Learning Goal. Use the information in your Science Notebook to answer each question. Then use these answers to write a well-developed essay about the chapter. Use the Key Concept on the first page of this chapter as your topic sentence.

True or False

If the statement is true, write "true." If it is false, change the underlined word or words to make the statement true.

1. A population explosion of dinoflagellates is called a <u>lichen</u>.

2. Phytoplankton in aquatic food chains are <u>protozoans</u>.

3. All protists are <u>eukaryotes</u>.

4. In protists, <u>flagella</u> eliminate extra water that might cause the cell to burst.

5. Hyphae are threadlike tubes that form the <u>mycelium</u> of fungi.

6. <u>Yeasts</u> are fungi that reproduce by budding rather than by spore formation.

7. <u>Cilia</u> are structures in fungi that produce spores.

8. <u>Fungi</u> can be unicellular or multicellular, but they are all heterotrophs.

Short Answer

Answer each of the following in a sentence or brief paragraph.

9. Describe how a paramecium obtains and digests food.

10. Compare how animal-like, plant-like, and fungus-like protists obtain nutrients. Use the words *autotroph* and *heterotroph* and their definitions in your answer.

11. Describe how sexual reproduction in fungi occurs.

12. Explain why lichens are considered to be a mutualistic relationship.

Critical Thinking

Use what you have learned in this chapter to answer each of the following.

13. **Hypothesize** A neighboring farmer recently fertilized his fields. After a heavy rain, you notice that the surface of a nearby pond is covered with a thick layer of green algae. Develop a hypothesis to explain your observation.

14. **Predict** What would happen to life on Earth if protists suddenly disappeared? Explain your answer.

15. **Apply Concepts** Design a poster for the locker room to educate athletes on how to prevent athlete's foot.

Standardized Test Question

Choose the letter of the response that correctly answers the question.

16. Malaria is most common in tropical regions of the world, such as Africa, Southeast Asia, and Central and South America. Why is malaria **not** common in North America and Europe?

    A. The climate in tropical regions of the world provides a warm, moist environment in which the mosquito host for the malaria parasite can live and reproduce.

    B. The fungus that causes malaria is not found outside tropical regions of the world.

    C. The climate in North America and Europe is not as warm and moist as that in tropical regions of the world.

    D. People in North America and Europe are resistant to the malaria parasite.

Test-Taking Tip

If more than one choice for a multiple-choice question seems correct, ask yourself if each choice completely and directly answers the question. If a choice is only partially true or does not directly answer the question, it is probably not the correct answer.

Pollution-Fighting Microbes

MOST INDUSTRIAL PROCESSES create pollution. Pollution is the presence of harmful or unwanted substances in the environment. Foul-smelling clouds rise from factories. Mines leak waste into soil and streams. Coastal waters become coated with oil as a result of oil spills. Although there are ways to clean up this pollution, they're not simple. Most require a lot of equipment and a lot of time. In cases where pollution reaches deep into soil or groundwater, cleanup may be impossible.

Fortunately, nature has provided a way to clean up pollution. Many microbes actually eat some of the pollutants that dirty our soil and water. Today, scientists are putting these microbes to work in a process called bioremediation. In this process, bacteria and fungi help clean up harmful substances in water and soil by eating and digesting the pollutants. Then they turn substances that were once toxic, or poisonous, into harmless ones, such as carbon dioxide and water.

Scientists have been testing bioremediation for several years. In some cases, the pollution-eating microbes are already in the soil, but there are too few present for them to be useful. Researchers have learned that adding certain chemicals to the soil increases the microbe populations. The more microbes there are, the more pollution gets eaten. For example, the microbe *Geobacter sulfurreducens* can help take uranium out of groundwater. By adding vinegar to the soil, researchers can increase the number of the hungry bacteria living there. As the microbes eat the uranium in the groundwater, they cause reactions that change it from a soluble form to an insoluble form. The metal precipitates out of the groundwater and can be collected and removed. Seaweed does a similar job. Scientists mix small amounts of seaweed into soil contaminated with the pesticide DDT. The seaweed causes the growth of microbes that destroy DDT and clean the soil.

Researchers have evidence that bioremediation works. The goal now is to get the processes to work on larger and larger scales. With help from hungry microbes, cleaning up pollution could one day be an easy task.

Research and Report

You've just read about some of the uses of bacteria and fungi—from cleaning up pollution to baking bread. With a partner, do some detective work of your own. Find another use that people have for bacteria or fungi. Report to the class on the type of bacterium or fungus you have researched and what it is used to do.

CAREER CONNECTION MYCOLOGIST

WHAT TYPE OF PERSON becomes a mycologist, you might wonder? Someone who just *loves* fungi. That's right, a mycologist is a scientist who studies fungi—organisms such as yeasts, molds, and mushrooms.

Mycologists do an amazing variety of things. Some are disease detectives. Several years ago, dogwood trees started to die mysteriously. Mycologists were called in to track down the killer. Medical mycologists study fungi that cause human diseases. They also look for ways to use fungi to make medicines. Some fungi are already the source of important drugs. A current hepatitis B vaccine is based on an antigen that yeasts produce. The work of medical mycologists has also led to the development of antibiotics such as penicillin and streptomycin.

Mycologists work with border and port security officers, too. They guard our borders and ports against unwanted visitors— of the fungal variety.

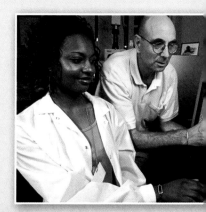

Fungus-Washed Jeans

AS YOU PROBABLY KNOW from experience, jeans are always more comfortable after you've worn them for a while. They feel soft and lived-in. Companies that make jeans figured this out long ago and came up with a process called stonewashing.

Originally, stonewashing was just what it sounds like. Brand-new jeans were thrown into huge washing machines with pumice stones. During the wash cycle, the stones pounded on the jeans, wearing out the cotton fabric. The process made the jeans look worn even before they left the factory.

Stonewashing was a big success. People loved the soft, faded look. However, there were problems with the process, which was hard to control. Sometimes, the jeans became too beaten-up. Seams were torn and the shiny rivets and buttons were often scratched. At other times, the jeans didn't look or feel worn enough. In addition, the stones clanking around inside the washers did plenty of damage.

As an alternative to stonewashing, acid-washing was tried for a while. The acid bath faded the jeans. But this process, too, had its problems. Luckily, at this point jeans manufacturers turned to a fungus for help.

A new process called biostoning became the choice for making worn-looking jeans. In fact, there's a good chance your jeans were treated using this process. In biostoning, an enzyme wears out jeans by eating away at the cotton fibers. The enzyme, called cellulase, was first found in the fungus *Trichoderma reesei*. This enzyme digests cellulose, a major component of the fibers of plants such as cotton. Scientists have isolated the gene that makes the fungus produce cellulase. By putting that gene into bacteria, large amounts of the enzyme can be produced quickly. Cellulase attaches to the surface of the cotton fibers in jeans. Then, it loosens dye particles and starts to break down the cotton fibers.

Of course, the process has to be controlled so that the enzyme does not destroy the jeans. But the benefits of this process are many. It creates less pollution than other methods. It also does less damage to clothing and machinery. Drains do not clog up with pumice grit, and there is no acid-disposal problem. Also, the material on jeans that is not cotton—polyester trim or plastic buttons—is unharmed by the enzyme. It works only on cotton. So the next time you wear your favorite faded jeans, you can thank a fungus!

Foods, plants, and products made from plants enter the United States all the time. Some carry harmful fungal pests. If security officers find something suspicious, they can call on government mycologists to check it out. The mycologists would identify the fungus. Then they would determine if it is harmless, or if the shipment containing it should be destroyed or kept out of the country.

Lots of foods are fungi (like mushrooms) or are made using fungi (like cheese, bread, soy sauce, beer, and wine). So mycologists are at work at mushroom farms, helping farmers grow the best mushroom crops possible. They also work with huge food manufacturing companies that use fungi in their production processes.

People who work as mycologists usually have a background in biology and a doctorate degree in mycology. Mycologists work in a variety of places, including government agencies such as the United States Department of Agriculture, drug companies, and wineries. Mycologists also work at universities, where they teach and do research. Mycologists agree that there is still a lot to learn about fungi and that many new species are yet to be discovered. For mycologists, the learning and discovery process is often the best part of the job.

Plants

Introduction to Plants

KEY CONCEPT Plants are able to capture energy and matter from their surroundings and are found on most of Earth's land.

Most years, Death Valley, California, earns its name. It is a difficult place in which to live. Temperatures there commonly reach 54°C (120°F) during a hot summer day. At night, temperatures can fall below freezing.

In an average year, Death Valley gets only about 5 cm (2 in.) of rain. Yet, Death Valley is full of life. When the heavy spring rains fall, plants quickly grow and bloom in large numbers, delighting those who visit this unusual place. How do plants grow in these and other conditions? Why are plants important to all life on Earth?

Think About Plants

If you have ever planted a garden or taken care of a houseplant, you know something about plants. Where did your plants grow? What did you do to take care of them? How well did they grow?

- Think about what plants need to grow. Do plants need the same things that animals need in order to survive?

- In your Science Notebook, summarize the things that you did to help your plants grow well. What things are essential to a plant's survival?

www.scilinks.org

Plant Characteristics **Code: WGB09**

What Is a Plant?

Before You Read

Create a K-W-L-S-H chart in your Science Notebook. Think about the title of this lesson. In the column labeled *K* on the chart, make a list of things you already know about plants. In the column labeled *W*, write what you want to learn about what plants are and how they live.

It would be difficult to define the word *plant* by looking at the plants that grow around you. Plants vary widely in appearance. They can be as small as the soft green mosses that grow only a few centimeters above a forest floor. They can be as tall as giant redwoods that can grow more than 100 m high.

Characteristics of Plants

Plants, whether large or small, or growing in deserts or forests, have important characteristics that set them apart from other living things.

- All plants are **multicellular**, or made up of more than one cell.
- Almost all plants can make complex organic molecules from simple chemicals in cell structures called **chloroplasts** (KLOR uh plasts).
- Plants have a substance called **cellulose** (SEL yuh lohs) in their cell walls. Cellulose is made up of chains of glucose units.
- Plants cannot move from place to place.

All plants are multicellular.   Plants are made up of more than one cell. These cells make up different tissues and organs. The most obvious of these are roots, stems, and leaves. Roots anchor a plant in the ground, and they take in water and dissolved substances for the plant to use. Stems hold plants up. Leaves are the primary sites of photosynthesis.

Almost all plants can make their own organic molecules.   Most plant cells have chloroplasts. These structures use light to power the chemical reactions of photosynthesis, the process by which simple inorganic chemicals are used to make complex organic molecules. Any green part of a plant can conduct photosynthesis. Plants that lack chlorophyll take their nutrients from other plants or fungi.

Plants have cell walls.   Each plant cell has a cell wall that surrounds its cell membrane. This wall is made of cellulose.

Plants are not motile.   Plants are unlike animals in that they are not able to move from one place to another. If conditions are not suitable in the place where they are growing, some plants stop growing and lose their leaves. They can begin growing again when conditions improve. Other plants die during such times.

As You Read

In the column labeled *L*, list the four characteristics of plants.

What are a plant's cell walls made of?

Figure 9.1 The places where plants grow, as well as the size to which plants grow, can vary widely.

Needs of Plants

Plants can grow in a variety of different places, including places that most people would find very uncomfortable—Death Valley, for example. In large numbers, plants can form a forest, carpet a lawn, or cover a prairie. To grow, plants need sources of water, minerals, light, and carbon dioxide.

Water The most important need of a plant is a source of water. Almost all plant cells contain water. Plants use water as a raw material for photosynthesis.

Minerals Plants also need a source of certain minerals. For most plants, the soil in which they grow is the source. However, some plants grow on other plants, and their roots never reach the ground. For these plants, fallen leaves and animal wastes provide the needed minerals.

Light Plants also need light for photosynthesis. Different kinds of plants have adaptations that allow them to survive in places with various amounts of light. Some plants grow well in places where the Sun shines directly on them for most of the day. Other plants grow well under the low-light conditions present on the floor of a tropical rain forest.

Carbon Dioxide Carbon dioxide is abundant in the air. Plants use carbon dioxide as a raw material for photosynthesis. They combine carbon dioxide from the air with organic molecules already in their cells to make new organic molecules for growth and energy storage. The process of adding the carbon dioxide is called carbon fixation.

Explain It!

In your Science Notebook, write a paragraph that tells what a plant is.

Include a chart that illustrates how a plant differs from an animal.

Figure 9.2 Mistletoe (left) grows on this tree branch becau_ it. *Welwitschia* (right) produces only two leaves during its e_ which become torn and shredded, collect moisture from th_

Figure It Out

1. To what kingdom do algae belong?
2. What are nonvascular plants?

Figure 9.3 (center) and mosses (right) are The evolution_

After You Read

1. What characteristics do plants have?

2. If you know where a plant grows naturally_ you grow that plant in a pot in your hous_

3. Using your chart, describe how knowin_ help you identify an unknown one. C_ would still like to know in the colum_ information in the column labeled_

Before You Read

In your Science Notebook, make a concept map for the word *plant*. Use the following vocabulary terms: *nonvascular plant*, *vascular plant*, *fern*, *monocot*, and *dicot*.

During Earth's early history, green algae lived in the water. Green algae were once considered plants, but they are now grouped with other similar organisms in Kingdom Protista. There is evidence that these organisms are the ancestors of today's plants.

Bryophytes: Liverworts and Mosses

Bryophytes (BRI uh fites) are considered to be among the least complex living plants. Their life cycles are tied to damp places. Liverworts and mosses are bryophytes.

Liverworts and their relatives are small and grow on the surface of wet soil. Mosses are the small plants that form a green carpet on the floor of many forests. Mosses also grow on trees and rock surfaces.

The cells of both liverworts and mosses take in water by osmosis. Liverworts and mosses lack **vascular tissue**, or the special tissue through which water and other materials move inside a plant. Plants that lack vascular tissue are called **nonvascular plants**. Liverworts and mosses also lack true roots, stems, and leaves.

of modern plants begins with green algae *(left)*. Liverworts
bryophytes, which are nonvascular plants.

Tracheophytes: Vascular Plants

Tracheophytes (TRAY kee uh fites) are true land plants because they have evolved ways to survive independent of wet environments. They are **vascular plants** because they are able to move water from their surroundings through their bodies in vascular tissues. Tracheophytes include club mosses, horsetails, and ferns, as well as gymnosperms and angiosperms.

Ferns are nonflowering vascular plants. Although they produce a variety of leaf shapes, many ferns have leaves that look a great deal like green feathers. Unlike the vast majority of vascular plants, ferns do not make seeds. The reproductive structures that ferns release are dustlike spores.

Gymnosperms (JIHM nuh spurmz) are vascular plants that produce seeds that are not enclosed within a fruit. Pines, firs, redwoods, and sequoias are all gymnosperms that produce seeds in cones. Angiosperms (AN jee uh spurmz) are the most easily recognizable seed-producing plants. Angiosperms are flowering plants. Roses, corn, bamboo, orchids, daisies, and fruit trees are angiosperms that are familiar to most people. In angiosperms, seeds are enclosed in fruits. The fruit protects the seeds as they develop. Because most fruits are edible and the seeds not easily digested, animals often spread seeds far from the parent plants.

Figure 9.4 Ferns are tracheophytes, the next step in the evolution of plants. More than 12,000 species of ferns exist.

As You Read

Use the headings and vocabulary words to add branches to your concept map.

What type of plants are ferns, gymnosperms, and angiosperms?

Although gymnosperms evolved long before flowering plants, today there are many more species of angiosperms than gymnosperms.

Figure 9.5 Angiosperms, the flowering plants, number more than 235,000 species. How do the angiosperms differ from other vascular plants?

Figure 9.6 The first plants to evolve with seeds were the gymnosperms. The sugar pine is a gymnosperm.

CONNECTION: Earth Science

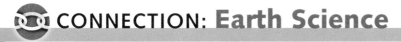

Fossils are the preserved remains or evidence of ancient living things. Fossils of early plants are usually molds. When a plant died, it was buried by sediment that hardened into rock. When the plant itself decayed, the impression it left in the rock was preserved. The rocks in which molds are found can be dated using radioactive elements to determine the age of the mold and the plant that created it.

Monocots and Dicots

Flowering plants are classified into two groups based on the appearance of their seeds. **Monocots** (MAH nuh kahts) have a single seed leaf. If you cut open a corn kernel, which is a seed that may develop into a new plant, you will find one seed leaf. **Dicots** (DI kahts) have two seed leaves. If you cut open a lima bean, you will find a pair of seed leaves—the two halves of the bean.

In addition to corn, other monocots include grasses, bamboo, and orchids. Monocots have parallel veins in their leaves. The flower parts of monocots usually occur in threes or multiples of three.

Dicots have veins in their leaves that look like nets. The flower parts of dicots often appear in multiples of four and five. Roses, maples, and most other broadleaf trees are dicots.

Explore It!

Conduct a plant tour of your home, yard, or neighborhood park, and record your observations, including drawings, in your Science Notebook. Prepare a chart in your Science Notebook with three columns labeled *Ferns*, *Gymnosperms*, and *Angiosperms*. Based on the characteristics of the plants you observe, write the name of each plant in the correct column.

PEOPLE IN SCIENCE: Elizabeth Blackwell 1700–1758

Science does not occur only in the laboratory. Often, people who are not trained in science make important contributions to the body of scientific knowledge.

Herbals are early medical recipe books that describe medical uses for certain plants. Many herbals also include illustrations that help identify the plants described. Many people consider *A Curious Herbal*, done by Elizabeth Blackwell in the first half of the eighteenth century, to be among the most beautiful herbals ever made.

When her husband was imprisoned, Elizabeth, who was a trained artist, decided to produce an herbal to pay off her husband's debts. She moved near a small botanical garden in London and made her drawings from plants that grew there. She sought information about the plants' names and uses from her husband. She made printing plates of her drawings and then hand-colored the plates—all 500 of them. She then sold the plates to bookstores. Elizabeth's herbal is a book that made important contributions to the understanding of plants and their medical uses.

After You Read

1. Use drawings to illustrate how plants have evolved over time. Use arrows to connect your drawings in the correct sequence.

2. Describe the adaptations that allow plants to survive on land.

3. Use your concept map to identify the two main types of plants. In a well-developed paragraph, explain how these two groups are different.

Before You Read

Create working definitions for *photosynthesis* and *cellular respiration*. A working definition is one that develops as you read and think about an idea. Write what you know about these terms before you begin reading. Then add to the definitions as you read and discuss the lesson.

Although plants do not appear to be active, important chemical reactions are occurring in their cells. In fact, many scientists agree that the most important manufacturing process on Earth happens in the cells of plants that are exposed to light. This process is photosynthesis.

Plants as Chemical Factories

Photosynthesis (foh toh SIHN thuh sus) is the main process that producers use to make complex organic molecules out of simple inorganic chemicals. Photosynthesis uses the energy of sunlight, which is absorbed by the green pigment chlorophyll (KLOR uh fihl). The process is often summarized by the equation

$$12CO_2 + 12H_2O \xrightarrow[\text{chlorophyll}]{\text{light energy}} C_{12}H_{22}O_{11} + 12O_2$$

carbon   water   starch   oxygen
dioxide

According to this equation, the raw materials of photosynthesis are carbon dioxide and water. The products are starch and oxygen. The process requires the presence of light energy and chlorophyll, which is contained in the chloroplasts. Photosynthesis takes place in any plant parts that contain chlorophyll.

Chloroplasts are able to trap the energy of light and use it for photosynthesis. Visible light, or white light, is made up of different colors. Each color has a different amount of energy associated with it. Plants use the energy present in red and blue light primarily. The light energy is used to split water molecules, which produces oxygen gas, and to form the new bonds that hold together the atoms in organic molecules. Most plants appear green because they do not absorb this color. Instead, green is reflected off the leaves and other parts of the plant.

Figure 9.7 Plant leaves contain a variety of pigments of different colors. The colors are masked by the green color of chlorophyll, but they show up when cold temperatures and shorter days cause the breakdown of chlorophyll.

1. What are the raw materials of photosynthesis? The products?

2. Hypothesize about what would happen to plants left in the dark.

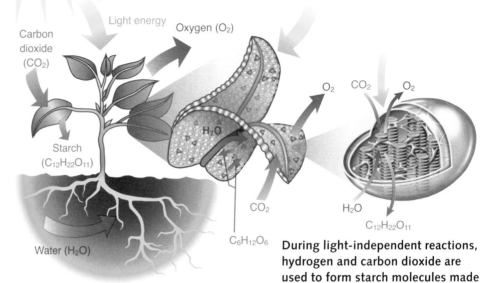

During light-dependent reactions, light energy trapped by the chloroplast decomposes water into hydrogen and oxygen. The oxygen escapes into the air.

Light energy

Oxygen (O_2)

Carbon dioxide (CO_2)

O_2 CO_2 O_2

H_2O

Starch ($C_{12}H_{22}O_{11}$)

CO_2 H_2O

$C_{12}H_{22}O_{11}$

$C_6H_{12}O_6$

Water (H_2O)

During light-independent reactions, hydrogen and carbon dioxide are used to form starch molecules made up of long chains of glucose units.

Figure 9.8 Photosynthesis includes two sets of reactions. One set *(blue)* requires the presence of light; the other *(red)* does not.

As You Read

Review your working definitions for *photosynthesis* and *cellular respiration*.

Where in the plant cell does photosynthesis take place? What process produces the raw materials for cellular respiration?

Cellular Respiration

Like all forms of life, plants need energy to grow and survive. Plants get this energy from photosynthesis and by a process called cellular respiration. **Cellular respiration** (res puh RAY shun) is the process that releases potential energy stored in chemical bonds. Cellular respiration occurs in all living cells. In cellular respiration, the energy stored in organic molecules is released when chemical bonds are broken. The process is often summarized by the equation

$$C_6H_{12}O_6 \;+\; 6O_2 \;\rightarrow\; 6CO_2 \;+\; 6H_2O \;+\; \text{energy}$$

glucose   oxygen   carbon   water
                              dioxide

According to this equation, the raw materials of cellular respiration are glucose and oxygen. The products are carbon dioxide, water, and energy. The equation for cellular respiration is essentially the opposite of the equation for photosynthesis. This means that the products of one process are the raw materials for the other process.

CONNECTION: Math

The equation for cellular respiration is called a balanced chemical equation because the number of atoms of each element is the same on both sides of the equation. Prove that this is true by calculating the total number of atoms of carbon (C), hydrogen (H), and oxygen (O) on each side of the equation.

Plant Hormones

Scientists have discovered that plants make substances called hormones. **Hormones** (HOR mohnz) are special chemicals made in one part of an organism that affect growth and development in other parts of the organism. Very small amounts of hormones are usually able to produce significant effects in cells.

Auxins (AWK sunz) are a group of hormones made by plants. Auxins can cause some plant cells to enlarge and can stop the growth of other plant cells. For example, scientists believe that plants bend toward sunlight because of the action of auxins. Auxins build up on the side of a plant's stem that is away from a source of light. These cells on the dark side of the stem grow longer than the cells on the side of the stem that faces the light source. The longer cells cause the stem to bend toward the light source.

Auxins are also produced at the growing tip of a plant stem. They have the ability to keep potential growth buds on the stem from growing. When the tip of a plant is damaged or removed, the supply of these auxins is stopped. When this happens, the tiny growth buds begin to grow all along the stem, producing more stems and a bushier plant. Gardeners often pinch off or cut off the growing tip for this reason.

Other Chemicals Made by Plants

Plants are also able to make a wide variety of other chemicals. Many of these chemicals are made as defense mechanisms. Tannic acid, for example, is present in the leaves of many plant species. Tannic acid makes the leaves of plants taste bitter. Many insects avoid eating leaves with high levels of tannic acid. Evidence shows that plants can increase the level of tannic acid in their leaves in response to insect attack. For some other organisms, the tannic acid in leaves is a bonus. People drink tea because the tannic acid in tea leaves has a pleasant taste.

The chemical that is the active ingredient in aspirin, acetylsalicylic acid, was originally discovered in the bark of willow trees. Long ago, people noticed that a solution made from willow twigs soaked in water would lower fevers. Scientists fear that other important medicines might be lost if a plant species is destroyed before it can be studied.

Figure 9.9 In an effort to trap more light energy for photosynthesis, many plant stems bend toward the light *(left)*. The bending is caused by auxins. Plants that do not get enough light often grow tall and spindly *(right)*. They look like they are "reaching" for the light they need to survive.

After You Read

1. Describe the processes of photosynthesis and cellular respiration.
2. What functions do plant hormones control?
3. Use your working definitions for *photosynthesis* and *cellular respiration* to describe the relationship between these two processes.

Learning Goals

- Describe three patterns of plant growth.
- Identify ways plants respond to the environment.

New Vocabulary

annual
biennial
perennial
tropism
stimulus
phototropism
thigmotropism

Before You Read

Read the Learning Goals and vocabulary terms for this lesson. In your Science Notebook, draw a T-chart. Label one side *Plant Life Spans* and the other side *Plant Responses.* Place each vocabulary term on the side of the T-chart where you think it belongs.

Just as flowering plants vary in appearance, so too do they vary in life span, or the time between seed germination and death. Some plants live for only a few days and complete their life cycle; others live for hundreds of years. The life spans of plants are determined by both genetic and environmental factors.

Annuals, Biennials, and Perennials

Plants are grouped according to life span into three general categories: annuals, biennials, and perennials.

Annuals (AN yoo ulz) are plants that complete their life cycle—grow from a seed, mature, produce seeds, and die—in one growing season. Many common garden flowers, such as petunias and zinnias, are annuals. Tomatoes and other plants that are killed by frost are also annuals. These plants are grown each year for the fruits or flowers they produce. Their seeds are able to withstand harsh environmental conditions and will sprout when conditions improve.

Biennials (bi EN ee ulz) are plants that complete their life cycle in two years. Many biennials change their appearance dramatically over the two years that make up a normal cycle. Carrots are a common biennial often grown in gardens. During its first growing year, a carrot plant produces green tops and the orange roots that people eat. In its second growing season, a carrot plant produces flowers and seeds. Biennials die after flowering, but the seeds that they produce help their species survive.

Perennials (puh RE nee ulz) are plants that live longer than one or two years. Although some perennials can flower in the first season of growth, many take several years before they are mature enough to produce flowers. Some perennials can live for many years; others are relatively short-lived.

Perennials have various ways of surviving for long periods of time. The stems of peonies die in the winter, but the thick roots store food supplies for next year's growth. Other perennials, such as trees, grow tough, woody stems.

Figure 9.10 Angiosperms can be classified according to their life spans. A petunia *(top)* is an annual. A pansy *(center)* is a biennial. A peony *(bottom)* is a perennial.

Tropisms

Tropisms (TROH pih zumz) are ways in which plants respond to stimuli in their environment. A **stimulus** (STIHM yuh lus, plural: stimuli) is something that produces an action or response. Bending toward light is known as a **phototropism**. As discussed before, this tropism is the result of auxins producing changes in a plant's cells.

Tropisms are said to be positive or negative depending on a plant's response to a particular stimulus. Growing toward light is considered a positive phototropism. Growing away from the pull of gravity is a negative gravitropism. A plant can show both positive and negative tropisms at the same time. When a seed sprouts, its stem grows toward the light and away from the pull of gravity.

Some plants also exhibit **thigmotropism**, or a tropism in response to touch. A positive thigmotropism is most commonly seen in climbing plants, such as peas, grapes, and clematis.

As You Read

Check your predictions about the placement of the vocabulary terms in your T-chart.

What determines whether a plant is an annual, a biennial, or a perennial?

Figure 9.11 The tendrils on a pea plant hold it tight to the object it is climbing on. The tendrils always coil in the same direction.

Figure It Out

1. What type of tropism is this plant exhibiting?

2. How might thigmotropism help a plant survive in the dense growth of a rain forest?

 Extend It!

Working with a partner, design an experiment that would answer one of the following questions:

- Can a plant find its way out of a maze?

- Does a plant know which way is up?

Describe your experiment in your Science Notebook.

After You Read

1. Explain how annuals, biennials, and perennials differ.

2. Describe the stimulus and the plant response in two types of plant tropisms.

3. A homeowner wants to purchase a plant that will grow up a decorative trellis in a flower bed and be attractive for many years. Use a vocabulary word from each side of your T-chart to describe the type of plant that should be purchased.

KEY CONCEPTS

9.1 What Is a Plant?

- Plants are multicellular organisms that capture matter and energy from their surroundings.

- Plants have cell walls composed of cellulose.

- Unlike animals, plants cannot move from place to place.

- Plants can be found in many environments, but all plants need sources of water, minerals, light, and carbon dioxide in order to survive.

9.2 Plant Evolution

- The first plants evolved from green algae that developed adaptations enabling them to survive on land.

- Nonvascular plants lack the specialized tissues that allow water and minerals to move inside the plant. These plants must live in damp places where water is readily available. Mosses and liverworts are examples of nonvascular plants.

- Vascular plants such as ferns, gymnosperms, and angiosperms have specialized vascular tissues that allow water and minerals to move inside the plant. With specialized vascular tissues, these plants can live in different types of environments.

9.3 Plant Chemistry

- Photosynthesis uses energy from the Sun to convert simple inorganic chemicals into complex organic molecules and also releases oxygen gas. Photosynthesis occurs in the cells of the green parts of plants.

- Cellular respiration is the process that releases energy stored in the chemical bonds of organic molecules. This process occurs in the cells of all living organisms.

- Plant hormones such as auxins are chemicals that control a plant's growth and development.

9.4 Patterns of Growth

- A plant can be classified according to its life span as an annual (a plant that completes its life cycle in one growing season), a biennial (a plant that requires two years to complete its life cycle), or a perennial (a plant that lives longer than two years).

- Tropisms are a plant's response to stimuli in the environment. Phototropism is a plant's response to light; thigmotropism is a plant's response to touch.

VOCABULARY REVIEW

Write each term in a complete sentence, or write a paragraph relating several terms.

9.1
multicellular, p. 156
chloroplast, p. 156
cellulose, p. 156

9.2
vascular tissue, p. 158
nonvascular plant, p. 158
vascular plant, p. 159
fern, p. 159
monocot, p. 160
dicot, p. 160

9.3
photosynthesis, p. 161
cellular respiration, p. 162
hormone, p. 163
auxin, p. 163

9.4
annual, p. 164
biennial, p. 164
perennial, p. 164
tropism, p. 165
stimulus, p. 165
phototropism, p. 165
thigmotropism, p. 165

PREPARE FOR CHAPTER TEST

To prepare for the chapter test, create a question from each Learning Goal. Use the information in your Science Notebook to answer each question. Then use these answers to write a well-developed essay about the chapter. Use the Key Concept on the first page of this chapter as your topic sentence.

MASTERING CONCEPTS

True or False
If the statement is true, write "true." If it is false, change the underlined word or words to make the statement true.

1. Plants are multicellular organisms that can <u>capture energy from sunlight</u> but are not motile.

2. A source of <u>minerals</u> is the most important need that plants must meet in order to survive.

3. Plants evolved from <u>bacteria</u> that adapted to life on land.

4. A <u>nonvascular plant</u> can live independent of water because it has special tissues that move water and minerals through the plant.

5. <u>Cellular respiration</u> is the process that green plants use to produce oxygen and organic molecules.

6. The products of photosynthesis are essentially the <u>raw materials</u> for respiration.

7. Plants that complete their life cycle in one growing season are <u>biennials</u>.

8. The response of a plant to stimuli in the environment is called a <u>tropism</u>.

Short Answer
Answer each of the following in a sentence or brief paragraph.

9. Why is vascular tissue important to a plant?

10. What role does chlorophyll play in photosynthesis?

11. Describe two tropisms that affect the stems of a plant.

12. Compare and contrast gymnosperms and angiosperms.

Critical Thinking
Use what you have learned in this chapter to answer each of the following.

13. **Infer** As vascular tissues evolved, plants became much larger. Explain the relationship between these two events.

14. **Compare and Contrast** Compare the processes of photosynthesis and cellular respiration.

15. **Apply Concepts** People often grow plants near windows in their houses. Explain why the plants should be turned every week.

Standardized Test Question
Choose the letter of the response that correctly answers the question.

16. Which of the following are basic needs of plants?

   I. sunlight

   II. water

   III. carbon dioxide

   A. I only

   B. I and II only

   C. II and III only

   D. I, II, and III

Test-Taking Tip

If you don't understand the directions, you can usually ask the teacher to explain them. When the directions are clear enough for you to understand the question completely, you are less likely to pick the wrong answer.

Chapter 10

Nonflowering and Flowering Plants

KEY CONCEPT Plants have evolved with methods of reproduction that enable them to live in a variety of places on Earth.

Tropical rain forests teem with plant life. High above the ground are towering trees and climbing vines. The forest floor is carpeted with lush, green plants of astounding diversity. Yet, a tropical rain forest is only one of the many places on Earth where plants are found.

How are plants able to survive in the various places on Earth? How do they reproduce and ensure the continuation of their species? You will learn the answers to these questions in this chapter.

Think About Ways in Which Plants Reproduce

You may not have been to a rain forest, but you are familiar with many different types of plants. Plants are part of your environment at home and at school. Take a moment to observe the plants in your environment.

- What characteristics can you observe?

- In your Science Notebook, select one plant that you are familiar with and describe its characteristics and how you think it reproduces. Include a diagram to support your description.

NSTA

SciLINKS
THE WORLD'S A CLICK AWAY

www.scilinks.org
Vascular and Nonvascular Plants
Code: WGB10

10.1 Bryophytes: Nonvascular Plants

Learning Goals

- Identify the characteristics of bryophytes.
- Explain how bryophytes reproduce.
- Describe the features of liverworts and mosses.

New Vocabulary

alternation of generations
sexual reproduction
gamete
zygote
sporophyte generation
spore
gametophyte generation
bryophyte
vascular system

Before You Read

Use the headings of this lesson to form questions. Write the questions in your Science Notebook. As you read, write answers to the questions.

From fossil evidence, scientists believe that bryophytes began to live on land about 480 million years ago. The first true land plants, they remain nearly unchanged in appearance to this day. The study of bryophytes, and other early groups of plants, helps in understanding how plants became so successful on land. A basic pattern that occurs in the life cycles of all plants is particularly easy to observe in bryophytes.

The Life Cycle of Plants

All plants have a life cycle, or stages of growth and reproduction, that includes a pattern called alternation of generations. **Alternation of generations** means that plants spend one part of their life cycle as diploid individuals and the other part as haploid individuals. Recall from Chapter 4 that the terms *diploid* and *haploid* refer to the number of sets of chromosomes in cells. Diploid ($2n$) cells have two sets of chromosomes, and haploid (n) cells have one set.

Figure 10.1 summarizes alternation of generations. Cell division that follows meiosis reduces the chromosome number of the cells produced from $2n$ to n. In most organisms, meiosis leads directly to the production of cells used in sexual reproduction. In **sexual reproduction**, two sex cells unite to begin the life of a new individual. In most organisms, the sex cells, or **gametes**, are called eggs and sperm. The female parent produces eggs, and the male parent produces sperm.

An egg and a sperm unite in fertilization. Because each gamete has a haploid number of chromosomes, the union produces an offspring with diploid cells. The fertilized egg is called a **zygote** (ZI goht). Offspring of sexual reproduction are genetically different from their parents.

In the plant life cycle, the $2n$ stage is called the **sporophyte** (SPOR uh fite) **generation** because the plant reproduces with spores. A **spore** is a cell that is able to begin growing into a new plant without joining with another cell. The n stage is called the **gametophyte** (guh MEE tuh fite) **generation** because the plant produces gametes.

Figure It Out

1. Which stage begins with fertilization? With meiosis?

2. What is another name for sex cells?

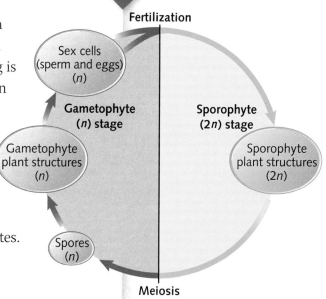

Figure 10.1 Alternation of generations consists of a sporophyte stage and a gametophyte stage.

In plants, spores are produced by cell division that involves meiosis, and thus they are haploid (*n*). The spores develop into haploid gametophytes that produce the haploid gametes.

Characteristics of Bryophytes

As you may recall from Chapter 9, **bryophytes** are nonvascular plants. A nonvascular plant stays small because it lacks a **vascular system**, or internal system of tubes that conducts water and nutrients throughout the plant. Such plants get water by taking it in through the parts that grow aboveground. The water enters and moves through these tiny plants by osmosis. Bryophytes, which include liverworts and mosses, are most frequently found growing in clumps in damp, shady places.

Like the earliest land plants, all bryophytes require water for sexual reproduction. Their sperm must swim to and fertilize their eggs. The gametophytes have a closely packed growth habit, which means that the sperm do not have far to swim. Both the sporophyte and gametophyte generations are relatively easy to identify in most bryophytes.

Liverworts

About 8,000 species of liverworts have been identified. The best-known liverwort species has a gametophyte with a flattened, leaf-shaped body from which umbrella-like stalks rise. Shown in **Figure 10.2**, only the gametophyte of this liverwort is easy to see. Gametes are produced in structures found under the "umbrellas." Male and female gametes are produced on separate umbrella-like stalks.

The sporophyte generation grows from fertilized eggs underneath the "umbrellas" of female stalks. The sporophytes have no chlorophyll and get all their nourishment from the gametophytes they grow on. The sporophytes produce spores that are carried away and grow into new gametophytes.

Mosses

Mosses vary greatly in appearance, but many look like small green brushes. Mosses often form in large colonies that carpet the areas in which they grow. Moss gametophytes look like they have stems and leaves, but these structures are not true stems or leaves. The leaflike structures are mostly only one single cell thick.

A moss sporophyte, which grows from the tip of a gametophyte, is a slender stalk that ends in a spore capsule. Like the sporophytes of liverworts, moss sporophytes lack chlorophyll and take nutrients from the gametophytes they grown on.

As You Read

Using your answers to the questions you wrote, predict whether a tall plant found in the desert could be a bryophyte. Explain your answer.

Figure 10.2 This liverwort species has gametophytes that look like flattened leaves from which tiny umbrella-like stalks rise. Most liverwort species, however, do not take this form.

Life Cycle of a Bryophyte

Once an egg is fertilized, the zygote develops into a sporophyte that is even smaller than the gametophyte on which it grows. The spores are tiny, almost like dust. When released, spores can be carried great distances by currents of water or wind. Unlike the more familiar seed plants, nonvascular plants are dispersed by spores. **Figure 10.3** shows the life cycle of a moss.

Young sporophytes

❶ A zygote begins the sporophyte stage and develops into an embryo that grows into the stalk and capsule.

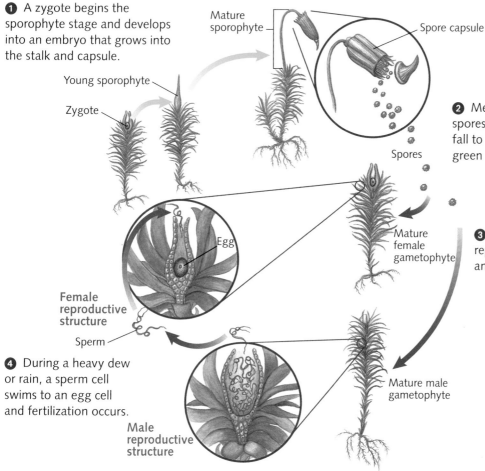

Mature sporophyte

Spore capsule

Young sporophyte

Zygote

Spores

❷ Meiosis occurs in the capsule and spores are produced. When the spores fall to the ground, they grow into the green gametophytes.

Egg

Mature female gametophyte

Female reproductive structure

Sperm

❸ Sex cells are produced in reproductive structures of male and female moss gametophytes.

❹ During a heavy dew or rain, a sperm cell swims to an egg cell and fertilization occurs.

Male reproductive structure

Mature male gametophyte

Figure 10.3 Life Cycle of a Moss. According to the diagram, what structures are produced by the gametophyte generation? By the sporophyte generation?

After You Read

1. What characteristics do bryophytes share?

2. Arrange the following words so that when they are connected with arrows, they correctly illustrate a life cycle that includes alternation of generations: *zygote*, *gametophyte*, *sporophyte*, *spores*, and *gametes*.

3. Design a chart that compares liverworts and mosses. Your chart should have three columns—*Characteristics*, *Liverworts*, and *Mosses*—and three rows: *Vascular Tissue Present*, *Where They Grow*, and *Relative Size of Generations*.

Explore It!

Place about 5 cm of sand in the bottom of one plastic cup and 5 cm of peat moss in the bottom of a second plastic cup. Predict in your Science Notebook which material will absorb the most water. Test your prediction by using a dropper to add water to each cup. Record the amount of water absorbed by each. Was your prediction correct? Based on your observations, do you think it would be better to add peat moss or sand to a flower bed to help keep the soil moist?

Learning Goals

- Identify the characteristics of club mosses, horsetails, and ferns.
- Recognize stages in the reproductive cycles of club mosses, horsetails, and ferns.

New Vocabulary

tracheophyte
rhizome
sporangium

Before You Read

In your Science Notebook, create a chart with three columns. Label the first column *Club Mosses*, the second column *Horsetails*, and the third column *Ferns*. As you read about these three kinds of plants, add information that describes their characteristics.

Hundreds of millions of years ago, when the seas receded and more of Earth's land became exposed, new types of plants evolved. These plants had adaptations that allowed them to survive on dry land. One important adaptation was a system of internal tissues through which water and other nutrients could be carried to all of a plant's parts.

Characteristics of Tracheophytes

Tissues that carry water and nutrients inside an organism's body are called vascular tissues. Most plants that have two types of vascular tissues—xylem and phloem. Plants with xylem and phloem are called vascular plants, or **tracheophytes** (TRAY kee uh fites). Most familiar plants are vascular plants. Because they have vascular tissues, these plants are able to form the organs known as true roots, stems, and leaves.

In mosses and liverworts, the gametophytes, or gamete-producing plants, are much larger than the sporophytes, or spore-producing plants. In tracheophytes, however, the sporophytes are the only generation large enough to be seen without magnification. The gametophytes are either very small or, as in most vascular plants, hidden in specialized organs. Like the bryophytes, the most primitive vascular plants still depend on water for sexual reproduction.

Club mosses, horsetails, and ferns are the most primitive vascular plants. Unlike the more familiar conifers and flowering plants, these plant phyla do not produce seeds. All of these groups of plants make drought-tolerant spores in conelike structures. The spores of these primitive tracheophytes disperse, or spread, new individuals.

Club Mosses

The name "club mosses" can be misleading. These land plants are not related to true mosses. Club mosses are small plants (most are under 30 cm in height) that grow on forest floors and look like small evergreen trees. They thrive in very moist places. **Figure 10.4** shows one species of club moss.

Figure 10.4 This club moss has leafy green stems and yellow conelike structures that produce spores.

When a spore lands in a suitable place, it grows into a gametophyte. This tiny plant can grow below the soil surface or on top of the soil. The gametophytes have structures that produce eggs and sperm on the same plant. A continuous film of water on the plant or on the ground is required for a sperm to reach and fertilize an egg. The fertilized egg develops into a new sporophyte.

Horsetails

Horsetails are relatively small, curious-looking plants that are easily recognized. Some species of horsetails produce frilly branches that resemble the tail of a horse. This bushy appearance gives horsetails their name. Horsetails have silica in their stems. Silica is an abrasive substance. Native Americans and early settlers used bunches of these plants to scrub pots and other utensils, giving horsetails the common name "scouring rushes." Horsetails are common to areas such as marshes, shallow ponds, and stream banks.

Horsetails grow from special underground stems called **rhizomes** (RI zohmz). Conelike structures form at the tips of horsetail stems. Spores are produced in these structures. When the spores are ripe, they are released and carried by air currents to new areas. If a spore lands in a place where conditions are favorable, it begins to grow into a tiny gametophyte plant. The gametophytes make either eggs or sperm. A fertilized egg grows into a new sporophyte individual.

As You Read

With a partner, select one section of this lesson. Take turns reading paragraphs to each other. After each paragraph reading, have the listening partner verbally summarize the paragraph. Then make sure you and your partner add notes to your charts.

How did the structure of horsetails lead to their use in scrubbing pots?

Figure 10.5 Slender stalks of horsetails grow in colonies. Conelike structures at the tips of the silica-containing stems are the sites of spore production.

Figure 10.6 Boston ferns are a popular houseplant.

Ferns

In nature, most ferns grow in shady areas. Because they need water for their sperm to swim to eggs during the gametophyte generation, most ferns grow in damp areas. Most fern sporophytes produce leaves called fronds. Almost everyone is familiar with the featherlike fronds. The leaves of other fern species can be very different in shape. Some look like small buttons, and others look a bit like small swords. In addition to differences in frond shapes, fern species also differ in size. Some ferns are very small (about 1 cm in diameter), while others—tropical tree ferns, for example—can grow many meters tall.

Ferns produce fronds from a rhizome. The rhizome is not a root but rather an underground stem with roots. Fern rhizomes can grow for many years and produce new leaves every growing season. The rhizomes can be cut into many pieces, each of which can grow into separate but identical fern plants. Reproduction that does not involve the union of an egg and a sperm, which is common in all plants, is called **asexual reproduction**.

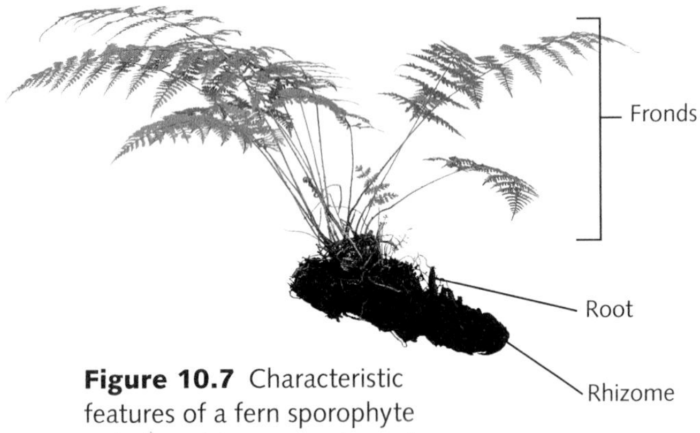

Fern fronds grow in a special way: they unroll. This makes it easy to tell a fern from other kinds of plants. If you look at a young frond as it starts to develop, you can see why it is sometimes called a fiddlehead. As shown in **Figure 10.8**, the unrolling frond looks like the carved top of a violin, fiddle, or cello. As it grows, the frond continues to unroll until it reaches its mature shape and size.

Fern sporophytes produce spores in structures called **sporangia** (spuh RAN jee uh, singular: sporangium), which are usually located on the lower surface of a frond. Some species produce sporangia on separate stalks, while others bear their spores on the top of the fronds. Sporangia are often found in clusters called sori, which are located on the undersides of fronds. In some fern species, the sori form lines. In others, they are arranged around the edge of a frond. Some people even mistake these clusters of sporangia for insects and try to rub them off a frond.

Fronds

Root

Rhizome

Figure 10.7 Characteristic features of a fern sporophyte include a rhizome, roots, and fronds. What is a rhizome?

Figure 10.8 New fronds are called fiddleheads because their shape resembles the carved top of a violin.

☉ CONNECTION: Math

The approximate numbers of living species of bryophytes and tracheophytes are provided here.

Mosses: 20,000; Liverworts: 6,500; Hornworts: 100; Club Mosses: 1,150; Horsetails: 15; Ferns: 11,000 Use this information to prepare a bar graph. Show bryophytes with bars of one color and tracheophytes with bars of another color. Label each bar correctly and include a key. Note: Hornworts are a type of bryophyte.

What conclusion can you reach by looking at your completed graph?

Life Cycle of a Fern

Fern spores form by cell division that begins with meiosis. When the spores are ripe, the sporangia open. The released spores are light and float on the air. If they land in a suitable damp area, they begin to grow. They produce the gametophyte generation of the fern, which is a small, delicate green heart-shaped plant growing on the surface of damp soil. A fern gametophyte is shown in **Figure 10.9**.

The gametophyte produces sperm and eggs in separate structures. The antheridium (an thuh RIH dee um, plural: antheridia) is the male reproductive structure in which sperm are produced. The archegonium (ar kih GOH nee um, plural: archegonia) is the female reproductive structure in which eggs are produced. Sperm need water in order to swim to the archegonia to fertilize the eggs.

Figure 10.9 The fern gametophyte looks like a small green heart growing on the damp soil surface.

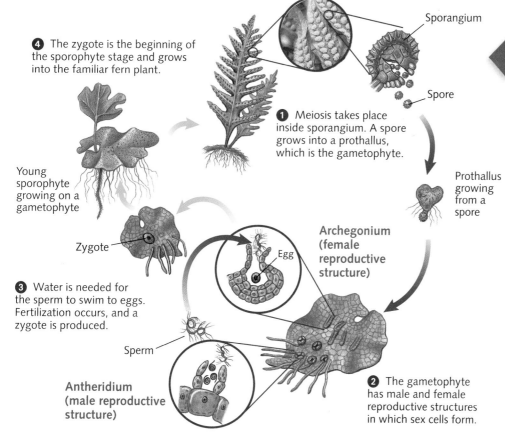

❹ The zygote is the beginning of the sporophyte stage and grows into the familiar fern plant.

Young sporophyte growing on a gametophyte

Zygote

❸ Water is needed for the sperm to swim to eggs. Fertilization occurs, and a zygote is produced.

Sperm

Antheridium (male reproductive structure)

Sporangium

Spore

❶ Meiosis takes place inside sporangium. A spore grows into a prothallus, which is the gametophyte.

Prothallus growing from a spore

Archegonium (female reproductive structure)

Egg

❷ The gametophyte has male and female reproductive structures in which sex cells form.

Figure 10.10 Life Cycle of a Fern. Both stages are photosynthetic and can thus live independently.

After You Read

1. What characteristic do tracheophytes share?

2. Which stage in the life cycle of tracheophytes is most easily recognized?

3. In a well-developed paragraph, explain why club mosses, horsetails, and ferns grow in damp places.

Figure It Out

1. What is the name of the male reproductive structure? The female reproductive structure?

2. What stage begins with the formation of a zygote?

🔄 Extend It!

Ferns were among the plants in Earth's ancient forests. Although they may have died 300 million years ago, these ancient ferns are important sources of energy today. Research how ancient ferns became the coal deposits used for fuel today. Present your research as a sequence chart or a series of drawings to illustrate this process.

Learning Goals

- Identify the characteristics of gymnosperms.
- Describe the four phyla of gymnosperms.

New Vocabulary

gymnosperm
cycad
pollen
ginkgo
conifer
ovule
pollination

10.3 Seed Plants: Gymnosperms

Before You Read

In your Science Notebook, create a concept map based on the word *gymnosperms*. Start by writing anything you already know about gymnosperms. Add information as you read the lesson.

Plants that make seeds are the dominant form of plant life on Earth today. Their ancestors are first seen in the fossil record about 350 million years ago. Seed plants, which have microscopic gametophytes, are not dependent on water for their reproduction. The seeds, not spores, disperse seed plants. Seed plants are generally classified into two major groups: gymnosperms and angiosperms.

Characteristics of Gymnosperms

Gymnosperms (JIHM nuh spurmz) are seed plants that do not produce flowers and produce seeds that are not contained in fruits. Nearly all gymnosperms, such as pine trees and redwood trees, produce seeds in cones. The leaves of most gymnosperms are scalelike or needlelike. Many gymnosperms are called evergreens, because some green leaves are always on their branches. There are four phyla of plants that are classified as gymnosperms: cycads, ginkgoes, gnetophytes, and conifers.

Cycads

Cycads (SI kadz) are gymnosperms that resemble small palm trees. Cycad plants are either male or female. A male cycad plant has cones that produce pollen. **Pollen** grains are the small structures of seed plants in which sperm develop. A female cycad plant has conelike structures that produce eggs.

Figure 10.11 Cycad plants produce either male *(top)* or female *(bottom)* cones.

Figure 10.12 The sago palm *(left)* is a cycad, not a palm. The heavily thorned leaves of *Encephalartos horridus (right)* keep animals from eating this cycad.

Ginkgoes

Ginkgoes (GINK ohz) are gymnosperms with broad, flat leaves that are fan-shaped and usually have two lobes. Like many broadleaf trees, ginkgoes drop their leaves as temperatures fall with the onset of cold weather. The ginkgo has often been called a "living fossil." Fossils that look remarkably like the single existing species of ginkgo have been found in rock layers that date back in time about 270 million years.

Ginkgo trees can live for a long time—a tree that might be almost three thousand years old has been identified. The trees are resistant to many insects, and their very deep roots enable them to withstand strong winds. The leaves turn a beautiful yellow-gold color in the fall before they drop. Like cycads, ginkgo trees are either male or female.

Gnetophytes

Gnetophytes (NEE tuh fites) are found in Asia, North America, Africa, and Central and South America. They include desert shrubs and tropical trees and climbing vines that produce seeds in conelike structures. The shrubs known as Mormon tea (*Ephedra*) in American deserts are gnetophytes.

Conifers

California is home to the world's tallest living thing, largest living thing, and oldest living thing. All three of these record holders are **conifers** (KAH nuh furz), or gymnosperms that produce seeds in cones. The world's tallest living thing is a coast redwood tree called Hyperion. It is more than 115 meters tall. The largest living thing is the General Sherman tree, a giant sequoia. Its volume has been estimated to be about 1,487 m³. The oldest living thing is a bristlecone pine that grows in the White Mountains. According to its rings, it is 4,700 years old!

Figure 10.13 Ginkgoes are often planted on city streets because they can tolerate the smog and air pollution common in many urban areas.

As You Read

Add information to your concept map.

What are the four groups of gymnosperms?

Did You Know?

Like rustlers in the Old West who stole horses, cattle, and sheep, "cycad rustlers" steal cycads. Because cycads are rare and grow slowly, a mature cycad of a desirable species can be worth many thousands of dollars. Today, it is not uncommon for people to put alarms on mature cycad plants in their gardens and greenhouses!

Figure 10.14 Coast redwood trees *(left)* are the tallest conifers. Bristlecone pines *(right)* are conifers with very long life spans.

Figure 10.15 Conifers, the largest group of gymnosperms, form large forests of closely spaced trees, an adaptation for wind pollination.

Extend It!

Scientists are engaged in efforts to save the seeds of many useful and endangered plants in places called seed banks. Conserving seeds in this way will ensure that such plants do not disappear completely from Earth. Use library and Internet resources to find out about seed banks. What organizations manage seed banks? Where are the seed banks located? What kinds of plants are the targets of conservation efforts?

Features of Conifers   There are about 550 species of conifers alive today. The vast majority of conifers are trees, but some are smaller shrubs. All conifers are woody plants. Conifers are the most common trees found in the large forests in colder areas of North America, Europe, and Asia. Conifers include pines, spruces, firs, cedars, yews, redwoods, larches, and junipers.

Most conifers are evergreens, which are plants that have green leaves throughout the year. Some conifers have leaves called needles. Pines, firs, and spruces have clusters of long, thin needles. Other conifers, such as yews, have short, flat leaves. Still others, such as junipers, have small, scalelike leaves. Although most conifers are evergreens, a few species shed their leaves during cold winter weather. In fact, all conifers lose some leaves all the time. A walk in a pine forest will show the soft carpet of brown pine needles that have fallen over the years.

Cones are one of the most obvious features of conifers. Cones are the structures in which conifers produce seeds. In most species, male and female cones form on different branches of the same plant. Male cones produce pollen, which contains the sperm. Female cones produce **ovules** (AHV uhz), which are structures of seed plants in which egg cells are produced. The female cones of most conifers are green when they are young and turn brown and woody when they mature. In junipers, mature female cones have fleshy scales and look like small berries. Some conifers do not produce cones. Instead, their seeds are surrounded by a fleshy protective coating. In yews, the coating is a bright red color that often attracts birds.

Reproduction in Conifers   In order for seed plants to reproduce, pollination must take place. **Pollination** (pah luh NAY shun) is the transfer of pollen grains to the female structures of a plant. Conifers are wind-pollinated. The male cones release large amounts of pollen. This increases the chances that fertilization will occur. The seeds are held in the female cones until they mature. Then, the scales that make up the cones open, and the seeds are dropped. In some species, this process can take place in as little as four months, while in other species, it can take several years.

Usually, conifer seeds that fall from mature female cones are carried away by wind. Seeds are also spread when cones fall to the ground or when birds eat the seeds and deposit undigested ones in their wastes. The cones of some pines will only release their seeds after fire burns the area in which they live. The fire kills adult trees and clears the ground of its thick layer of dropped needles. The seeds sprout because the death of the larger, older trees permits more sunlight to reach the seeds on the ground, which is now clear of debris.

Life Cycle of a Pine

Pines produce male and female cones on the same tree. First, spores form inside the cones through meiosis. The spores develop into male and female gametophyte structures. The male gametophyte is inside a pollen grain, and the female gametophyte is inside an ovule. Wind carries pollen from the male cones to the female cones, where fertilization takes place. Once eggs are fertilized, a cone's seeds develop on the tops of scales.

Figure 10.16 Life Cycle of a Pine.

❶ In cones, cell division that follows meiosis produces spores. In male cones, spores develop into pollen grains that carry the male gametophyte. In female cones, spores develop into ovules that contain the female gametophyte.

Explain It!

In your Science Notebook, explain how seeds are produced and then dispersed in conifers. Include drawings or diagrams to illustrate your explanation.

Young male cone

Young female cone

Meiosis

Meiosis

Scale of male cone

Scale of female cone

Pollination

Ovule

Pollen grain

Fertilization

Figure It Out

1. What process results in the formation of a zygote?

2. What is the female gamete, and where is it produced? What is the male gamete, and where is it produced?

❷ Pines are wind pollinated. Each pollen grain has tiny wings that help wind carry it to a female cone.

Cross section of one ovule

Pine seedling

Sperm in pollen tube

Egg

❸ Inside a female cone, a pollen tube grows from the pollen grain into an ovule. Each sperm passes through a pollen tube and fertilizes an egg. This process may take up to 15 months.

❺ A winged pine seed develops from each ovule. The seed grows into a sporophyte.

Mature female scale with seeds

Seed with embryo

❹ The zygote produced by fertilization grows into an embryo. The embryo is a new, immature sporophyte.

After You Read

1. What three characteristics do most gymnosperms share?

2. Compare and contrast male and female cones.

3. Use the Learning Goals on page 176 to reorganize the information in the concept map you made in your Science Notebook.

Learning Goals

- Describe the characteristics of flowering plants.

- Identify the parts of a flower and their functions.

- Explain how pollination and fertilization are related.

New Vocabulary

angiosperm
sepal
petal
stamen
anther
filament
pistil
stigma
style
ovary

Before You Read

Preview the lesson by looking at the pictures, reading the captions, and reviewing the Learning Goals. In your Science Notebook, write a paragraph predicting what you think the lesson is about, based on your preview.

Flowering plants are probably the most familiar group of plants and are the most diverse group of organisms in the plant kingdom. The first fossil evidence of flowering plants dates to about 140 million years ago. Flowering plants are also called angiosperms.

Characteristics of Angiosperms

Angiosperms (AN jee uh spurmz) are flowering plants that produce seeds that develop inside of fruits. They number more than 250,000 different species and grow almost everywhere. Angiosperms thrive in tropical areas that have few gymnosperms, in deserts that have little water, in temperate areas with seasonal supplies of water, and in cold areas where it becomes warm enough for ice and snow to melt. Some angiosperms even grow in water. Angiosperms provide many of the foods people eat, including fruits, vegetables, and grains.

Structure and Function of Flowers

Flowers are the main reproductive organs of flowering plants. Most flowers have four main parts: petals, sepals, stamens, and a pistil. Each flower part plays an important role in a plant's ability to produce more of its kind. These parts and their functions are shown in **Figure 10.18** on page 181.

Petals and Sepals   In most plants, flowers are produced on the ends of stems. When a flower is developing, it is covered by green leaflike structures called **sepals** (SEE pulz) that surround the bud and protect the delicate tissues inside. When the flower bud is ready to open, the sepals begin to separate at the top and fold downward. **Petals** are the soft parts of the flower that most people think of as the flower itself. A flower's petals are a form of plant "advertisement," attracting animals that the flower needs to complete its life cycle.

Stamens   The **stamens** (STAY munz) are the male reproductive organs of a plant. Each stamen consists of an **anther** (AN thur), which produces pollen grains, and a **filament** (FIL uh mint), which holds the anther above the flower petals. Pollen grains contain the sperm that will fertilize the eggs.

Figure 10.17 Within the tough green sepals, the delicate petals and reproductive parts of this rose flower are protected.

Pistil The **pistil** (PIHS tul), which is made up of the stigma, style, and ovary, is the female reproductive organ of a plant. The **stigma** (STIHG muh) is the sticky tip of a pistil that holds on to any pollen grains that land on it. The **style** connects the stigma to the ovary below. The **ovary** (OH vuh ree) holds the ovules, in which the eggs develop and are fertilized. Once fertilization has occurred, the ovules develop into seeds. The ovary swells and becomes the fruit in many species of flowering plants.

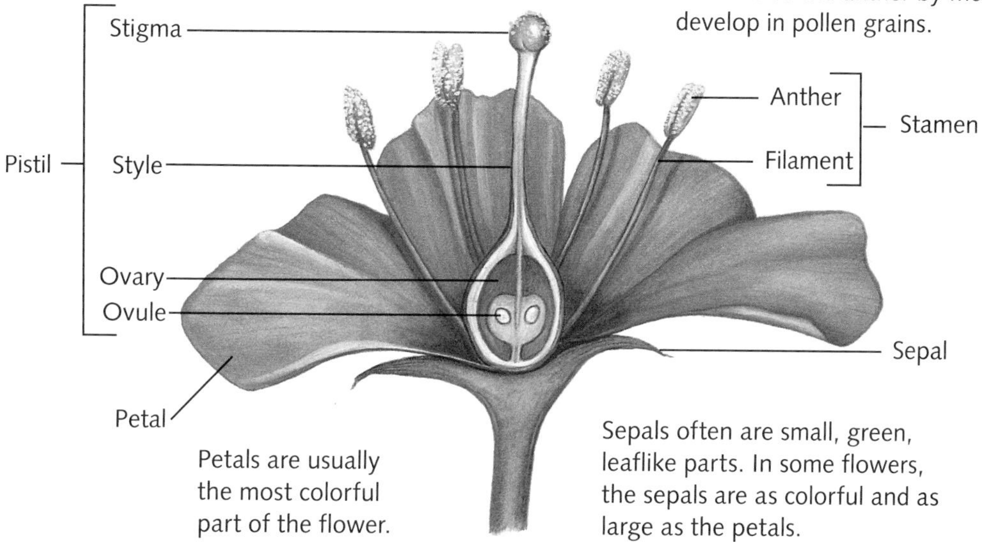

A pistil consists of a sticky stigma where pollen grains land, a long stalklike style, and an ovary. Ovules are the parts of the ovary where eggs are produced.

A stamen consists of an anther and a thin stalk called a filament. Pollen grains form inside the anther by meiosis. Sperm develop in pollen grains.

Stigma
Pistil
Style
Ovary
Ovule
Petal

Anther
Stamen
Filament

Sepal

Petals are usually the most colorful part of the flower.

Sepals often are small, green, leaflike parts. In some flowers, the sepals are as colorful and as large as the petals.

Figure 10.18 Some angiosperms produce flowers with all the major parts shown in this illustration. The stamens and the pistil are reproductive structures. Sepals and petals protect the reproductive structures and attract pollinators.

⊙ CONNECTION: History

Cotton has long been an important and useful plant. Cotton was the main crop in the South before the Civil War. Cotton is picked after the bolls, or fruit of the cotton plant, split open. Just-picked cotton has a number of seeds scattered throughout the long, thin fibers that make up the cotton. Before the invention of the cotton gin by Eli Whitney in 1793, these seeds had to be separated from the cotton fibers by hand. This difficult and often painful task took many hours to complete. Whitney's invention automated the seed-removal process and revolutionized the cotton industry in the United States. Cotton production became more profitable.

Once they are "cleaned" of seeds, the cotton fibers are combed, spun into thread, and woven into cloth. Cotton cloth can be dyed a rainbow of colors and used to make articles of clothing and household goods.

 Explore It!

Use a variety of materials—construction paper, tissue paper, clay, straws, toothpicks, peas, beans, etc.—to build a model of a flower. Your model should have both male and female reproductive structures in addition to petals and sepals. Be sure to include a key that identifies each part of your flower model.

Life Cycle of an Angiosperm

Many flowering plants have both male and female parts in the same flower. However, other plants, such as begonias, have separate male and female flowers on the same plant. Still other plants, such as hollies, produce male and female flowers on separate plants. In order to get colorful holly berries, both male and female holly plants are needed.

With a few exceptions, pollination must take place before seeds can form. Pollen grains can be carried by wind, water, gravity, or animals. Once a pollen grain lands on the female part of the plant, sperm cells and a pollen tube develop. The sperm cells move down the tube, into the ovary, and to an ovule. Fertilization can then take place between a sperm and an egg. The resulting fertilized egg, or zygote, develops into an embryo. **Figure 10.19** illustrates these processes.

As You Read

Read the paragraph you wrote in your Science Notebook about this lesson. Place a check mark next to predictions that are correct. Correct any predictions you now know are wrong.

What is the reproductive structure of angiosperms?

Figure It Out

1. At what point does pollination occur?

2. A plant with flowers is a mature sporophyte. What is the male gametophyte? The female gametophyte?

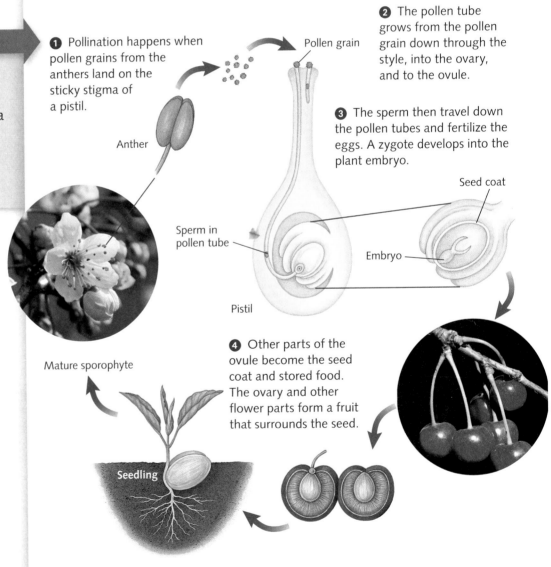

① Pollination happens when pollen grains from the anthers land on the sticky stigma of a pistil.

Anther

Mature sporophyte

② The pollen tube grows from the pollen grain down through the style, into the ovary, and to the ovule.

Pollen grain

③ The sperm then travel down the pollen tubes and fertilize the eggs. A zygote develops into the plant embryo.

Seed coat

Sperm in pollen tube

Embryo

Pistil

④ Other parts of the ovule become the seed coat and stored food. The ovary and other flower parts form a fruit that surrounds the seed.

Seedling

Figure 10.19 Life Cycle of an Angiosperm. This illustration traces the life cycle of a cherry tree.

Agents of Pollination

Wind, water, gravity, and animals are the main agents of pollination. Pollination can occur when wind carries pollen to a stigma. In many plants, however, it is an animal—often an insect—that brings pollen to a stigma. The animal is called a pollinator.

In their search for food, honeybees commonly transfer pollen as they work, as do many other insects. In the United States, bees are the most common pollinators of food plants. Farmers place hives of honeybees in their fields and orchards. When the bees' work is finished, the hives are moved to other fields and orchards.

Some insects are "deceived" into acting as a pollinator by a flower. For example, certain orchid flowers resemble female insects. The orchid gives off a scent that attracts a male insect. The male insect serves as a pollinator when he attempts to mate. Moving in, the male receives a touch of pollen on his head. Later, when another "female" flower attracts the same insect, the flower gets pollinated. Other flowers give off scents that attract pollinators.

Mice, other small rodents, and bats are also agents of pollination. Some birds eat nectar, a sweet liquid made by some flowers, and transfer pollen from anther to stigma. **Figure 10.20** shows some animal pollinators.

CONNECTION: Physics

While many plants depend on other organisms for pollination, some plants depend on physical forces to help them reproduce. Gravity, the force that pulls two objects toward each other, can also pollinate flowers.

When flowers whose filaments hold the anthers above the stigma release pollen grains, the force of gravity pulls the pollen grains down. As the pollen grains fall, they come into contact with the sticky stigma of the flower. With gravity's help, pollen has been delivered to the female reproductive structure, and fertilization can occur. In large fields of corn, pollen is dispersed by wind, but it is also affected by gravity. When released, corn pollen travels relatively short distances as it falls to Earth at the rate of about 30 cm per second.

After You Read

1. What characteristics do angiosperms share?
2. List the parts of a flower. Identify appropriate parts as male or female structures. Describe the function of each part.
3. How is pollination related to fertilization?
4. Rewrite the prediction paragraph in your Science Notebook to include three things you now know about angiosperms.

Figure 10.20 A honeybee looking for nectar *(top)*, a bat approaching a cactus flower *(center)*, and a male bee fooled into thinking the orchid flower is a female bee *(bottom)* are all pollinators.

KEY CONCEPTS

10.1 Bryophytes: Nonvascular Plants

- All plants have life cycles that include a pattern called alternation of generations. Alternation of generations means that a plant spends one part of its life in a stage that produces spores and the other part in a stage that produces sex cells.

- Bryophytes lack vascular tissues for moving water and nutrients throughout the plant. They also have easily distinguished spore-producing and gamete-producing plants.

- Bryophytes depend on water for sexual reproduction and are dispersed by spores.

- The union of a sperm and an egg, or fertilization, produces a zygote, or fertilized egg that develops into a new sporophyte.

- Mosses and liverworts are examples of bryophytes.

10.2 Tracheophytes: Vascular Plants

- Tracheophytes, such as club mosses, horsetails, and ferns, use vascular tissues to move water and nutrients in their bodies.

- Tracheophytes have a tiny gamete-producing stage and a relatively large sporophyte. Primitive tracheophytes are dispersed by spores.

- Horsetails and ferns have underground stems called rhizomes.

- The reproductive structures of ferns include sporangia, antheridia, and archegonia.

10.3 Seed Plants: Gymnosperms

- Seed plants are vascular plants that produce seeds for dispersal. They also have a microscopic gametophyte stage.

- Gymnosperms produce seeds that are not protected by fruit, and they have needlelike or scalelike leaves.

- Cycads, ginkgoes, gnetophytes, and conifers are types of gymnosperms.

- Cones are the reproductive structures of gymnosperms. Male cones produce pollen grains, which are small structures in which sperm develop. Female cones produce eggs.

- Pollination is the process of delivering pollen, which contains the male gamete, to the female parts of the flower so that fertilization can occur.

10.4 Seed Plants: Angiosperms

- Angiosperms are flowering plants that produce seeds protected by fruits.

- Flowers are the reproductive structures of angiosperms.

VOCABULARY REVIEW

Write each term in a complete sentence, or write a paragraph relating several terms.

10.1
alternation of generations, p. 169
sexual reproduction, p. 169
gamete, p. 169
zygote, p. 169
sporophyte generation, p. 169
spore, p. 169
gametophyte generation, p. 169
bryophyte, p. 170
vascular system, p. 170

10.2
tracheophyte, p. 172
rhizome, p. 173
asexual reproduction, p. 174
sporangium, p. 174

10.3
gymnosperm, p. 176
cycad, p. 176
pollen, p. 176
ginkgo, p. 177
conifer, p. 177
ovule, p. 178
pollination, p. 178

10.4
angiosperm, p. 180
sepal, p. 180
petal, p. 180
stamen, p. 180
anther, p. 180
filament, p. 180
pistil, p. 181
stigma, p. 181
style, p. 181
ovary, p. 181

MASTERING CONCEPTS

True or False
If the statement is true, write "true." If it is false, change the underlined word or words to make the statement true.

1. Liverworts and mosses are <u>vascular</u> plants.
2. The male gamete is the <u>sperm</u>, and the female gamete is the <u>egg</u>.
3. Special underground stems from which horsetails grow are called <u>sporangia</u>.
4. The fact that club mosses, horsetails, and ferns live in moist areas aids in <u>pollination</u>.
5. Transfer of pollen grains from the male to the female plant is called <u>fertilization</u>.
6. Cones are the reproductive structures of <u>angiosperms</u>.
7. The seeds of <u>angiosperms</u> are dispersed in fruits.
8. The male part of the flower is the <u>stamen</u>.

Short Answer
Answer each of the following in a sentence or brief paragraph.

9. Describe the life cycle pattern of plants.
10. Why do moss plants remain small?
11. What functions do sepals and petals perform in a flower?
12. What are three ways in which pollination can occur?

Critical Thinking
Use what you have learned in this chapter to answer each of the following.

13. **Infer** A new plant has just been discovered. The plant has small, gray-colored flowers that are hard to see and have no scent. Describe how this new plant might be pollinated.
14. **Compare and Contrast** How are gymnosperms and angiosperms similar? How are they different?
15. **Predict** Pesticides are chemicals designed to kill harmful insects. However, pesticides often kill beneficial insects, as well. What effect would widespread pesticide use have on angiosperms?

Standardized Test Question
Choose the letter of the response that correctly answers the question.

16. Which flower part includes all of the others?
    A. stigma
    B. style
    C. pistil
    D. ovary

> **Test-Taking Tip**
>
> Use scrap paper to write notes. Sometimes making a sketch, such as a diagram or a table, can help you organize your ideas.

PREPARE FOR CHAPTER TEST

To prepare for the chapter test, create a question from each Learning Goal. Use the information in your Science Notebook to answer each question. Then use these answers to write a well-developed essay about the chapter. Use the Key Concept on the first page of this chapter as your topic sentence.

Plant Parts and Functions

KEY CONCEPT The parts of a plant work together to enable the plant to capture energy and matter, grow, and reproduce.

Most plants have roots, stems, and leaves. However, one species of orchids, native to Florida, is a remarkable exception to the majority of plants. An adult ghost orchid does not appear to have either a stem or leaves. Its roots look like a tangled mess of spaghetti. Chlorophyll in its roots captures energy for the plant. When it blooms, its flowers are dramatic and beautiful, appearing like white ghosts floating in the air.

Think About the Structure and Function of Plant Parts

Some plants have parts that are easily recognizable as roots, stems, and leaves.

- Think about the roots, stems, and leaves of some familiar plants. How do you think these plant parts work together to enable the plants to survive? Record your thoughts in your Science Notebook.

- Plant parts are the source of many of the foods you eat. In your Science Notebook, list some of the plant parts you eat. Organize your list by *Roots*, *Stems*, and *Leaves*. When you have completed the chapter, go back to your list and see if you were correct.

NSTA

*SCi**LINKS**.*
THE WORLD'S A CLICK AWAY

www.scilinks.org
Plant Anatomy **Code: WGB11**

11.1 Roots

Learning Goals

- Identify the parts of a plant root.
- List four functions of plant roots.

New Vocabulary

root
taproot
fibrous root
vegetative propagation
root cap
apical meristem
epidermis
root hair
cortex
endodermis
xylem
phloem
hydrotropism

Before You Read

Create a working definition of the term *root*. A working definition is one that develops as you learn more about a topic. Write what you know about this term before you begin reading. Then, revise your definition as you read and discuss the lesson.

All vascular plants have systems. These systems, like all living systems, are made of cells, tissues, and organs that work together to perform specific functions that keep the plants alive. The shoot system of a plant is the aboveground part of the plant. The root system of a plant is generally below the ground. Thus, the roots of most plants are not easily seen. However, the root system is usually about the same size as the shoot system.

Roots are plant organs that have four major functions. Roots anchor and support a plant. Roots absorb water and dissolved minerals from the soil. Roots contain vascular tissues that transport materials to and from the stem. Roots store excess products of photosynthesis in the form of starch.

Types of Roots

Roots come in various shapes and sizes. Roots give a plant the surface area it needs to absorb the water and minerals essential to its survival. The surface area, or total outside layer, of a plant's roots can be as much as 50 times greater than the surface area of its leaves.

Plant roots are adapted to the plant's environment—specifically, to the soil type, temperature, and water. For example, certain desert plants have roots that are more than 20 m long, which enables the plant to obtain water from deep underground. There are two main types of root systems: taproots and fibrous roots.

Taproots Have you ever tried to pull a dandelion out of the ground? If so, you probably found yourself holding a tuft of green leaves and yellow flowers. The top of the dandelion broke off from the root. If you dug up the rest of the dandelion with a shovel and shook off the soil, you would have found the root system. The main root of a dandelion plant looks a bit like a small, pale carrot. It is a taproot.

A **taproot** is a thick, single structure that grows straight into the ground. Smaller, branching roots grow out from its sides. A taproot securely anchors a plant in the soil and serves primarily as a storage organ. It contains starch and sugar made by a plant. Carrots and beets are taproots. Taproots are most commonly found in dicots and gymnosperms. If the top of a dandelion plant breaks off, a new top will grow.

Figure 11.1 The long taproot of a dandelion plant makes it very difficult to remove the dandelion from the soil. How could this plant be removed?

CONNECTION: Geology

Most plants require soil in order to grow. Soil is mostly a mixture of small particles of rock that was broken down through various natural processes. Soil also contains water, air, and the remains of dead plants and animals. As dead organisms decay, various chemicals are released into the soil.

Soils also have naturally occurring mineral nutrients. Different soils have different amounts of rock particles, air, decaying organisms, and mineral nutrients. Nitrogen is one of the most important mineral nutrient plants need to grow. Other nutrients, sometimes called trace elements, are needed by plants in much smaller amounts.

Naturally occurring bacteria are also considered by some scientists to be an important part of soil. Some kinds of bacteria found in soil convert nitrogen in the air into forms of nitrogen that plants can use to grow. Other kinds of bacteria in soil make forms of nitrogen that plants need by breaking down animal wastes and dead organisms.

As You Read

With a partner, discuss your working definition of the term *root.* In your Science Notebook, add facts to the definition and rewrite it if necessary.

Describe the two main types of root systems.

Figure 11.3 One kind of adventitious roots are prop roots. These roots help anchor a plant in the ground and keep a tall and top-heavy plant upright.

Fibrous Roots   If you pulled out some grass plants instead of a dandelion plant, you would notice a different kind of root system. Grass plants have **fibrous roots**. Fibrous roots consist of a great many thin, branching roots that grow from a central point. Fibrous roots look like tufts of stringy hairs. Fibrous roots store less food than taproots do and are most commonly found in monocots. They serve primarily to hold the plant in the ground and provide a large surface area for water and mineral absorption.

Figure 11.2 Lawn grasses have fibrous root systems made up of many small roots that branch out in all directions. Fibrous roots intertwine to make a tough mat of roots.

Adventitious Roots   Roots that grow from the stem are called adventitious roots. These roots have different shapes and functions. Corn plants often produce a type of adventitious roots called prop roots, which help support the plants. Plants that grow in wet places often produce roots that grow upward from the mud and, eventually, above the water. Cypress trees produce these roots, which are called "knees." Knees take in the oxygen that cells in a plant's roots need for respiration. Strong buttress roots can be seen on many tall tropical trees. Buttress roots grow from the base of the tree's trunk. These roots help support the tree as it grows.

Some plants, such as the ghost orchid, produce another kind of adventitious roots called aerial roots. Aerial roots are aboveground roots. These roots hold on to the stems of other plants. Some kinds of ivy, including English ivy, produce aerial roots. The roots support the plant as it grows up a tree or on a wall. Aerial roots do not take nutrients from the plant on which they grow. Instead, they get their nutrition from leaves that fall and decompose near them.

One way to reproduce some plants is to cut a piece of the stem and place it in water or in moist soil. In time, adventitious roots will form from cells in the plant's stem. Coleus and begonias are two kinds of plants that can be reproduced in this way. This is one important method of vegetative propagation. **Vegetative propagation** uses parts of plants to make more plants that are genetically identical to the parent plant. An example of vegetative propagation is shown in **Figure 11.5**. Some roots can also be used to produce a new plant. If such roots are cut apart and then planted, new stems will begin to grow. Dividing a plant's roots is another method of vegetative propagation. Gardeners often use this method to increase the numbers of their plants.

Figure 11.4 English ivy plants attach themselves to vertical surfaces, such as tree trunks, by means of hairy-looking aerial roots.

Uses of Roots

People have used roots in a variety of ways for thousands of years. Some roots are used for food, some for spices, and some for substances such as dyes, medicines, and insecticides. Some plants have roots that absorb toxins left in the soil from pollution along with nutrients from the soil. These plants can be used to "clean" contaminated soil.

Beets, carrots, yams, turnips, and radishes are among the many roots that are eaten. The root of the cassava plant is used to make tapioca, an ingredient in puddings and baby food. The root of the marsh mallow plant was the original source of marshmallow candies. Roots such as licorice, sassafras, and horseradish are used as spices.

ⓘ CONNECTION: **Physics**

Roots grow downward with the force of gravity. However, the water taken in by the roots flows upward, defying gravity. How can that be possible?

Two processes are responsible for the water in roots traveling as far as 100 m to the tops of the tallest trees. The first of these processes is the attraction of water molecules to one another, which is called cohesion. The second process is transpiration, which is the giving off of water from the leaves of a plant.

Because the cohesive force between water molecules is so strong, every molecule that evaporates from a leaf pulls on another molecule of water in the leaf. If each leaf of a tree gives off billions of molecules of water, a tremendous force pulling water up is created. Water molecules also stick to, or adhere, to the sides of the narrow tubes that carry water upward.

Figure 11.5 Roots grew from one of these cut stems of a pyracantha plant. Gardeners often add rooting hormones to the stem to encourage more rapid production of roots. The production of plants from plant parts is called vegetative propagation.

Root Structure and Growth

Plant growth differs from animal growth in an important way. Animals have what is called a closed pattern of growth, while many plants have an open pattern of growth. Animals reach a certain shape and size, and then they stop growing. Many plants continue to grow throughout their lives, producing new cells at their growing tips, which include roots.

Tough cells that make up the root cap cover the end of a root. The tough cells in the root cap are in contact with particles in the soil. The **root cap** protects the more delicate cells that lie just behind the cap. These cells make up the **apical meristem** (AY pih kul • MER uh stem). *Apical* is a word that means "tip" or "end." The cells in a meristem continuously undergo mitosis and cell division. An apical meristem is also located at the tip of each plant stem.

As the cells in the root's apical meristem continue to divide, the newly formed cells enter a zone of elongation. Here the cells become longer, pushing the tip of the root ahead through the soil. The growth in the length of a root is called primary growth.

The **epidermis** (eh puh DUR mus) is the tissue that surrounds the outside of a root. Small **root hairs**, which are extensions of epidermal cells, stick out of the epidermis. Root hairs absorb water, oxygen, and dissolved minerals. They also greatly increase the surface area of the root that contacts the soil. The root **cortex** is found inside the epidermis. The cortex makes up most of the root and is the location where starch made by the plant is stored.

The **endodermis** (en duh DUR mus) is the innermost layer of the cortex. The endodermis surrounds the plant's vascular tissues—xylem and phloem. **Xylem** (ZI lum) is plant tissue that is made up of tube-shaped cells that transport water and dissolved minerals to all parts of a plant. Water and dissolved minerals that enter a root move through the cortex, into xylem cells, and then throughout the plant. **Phloem** (FLOH em) is plant tissue that is similar to xylem in that it is made up of tube-shaped cells. However, phloem transports organic molecules made in a plant's leaves to all parts of the plant. Some of these molecules are stored as starch in the roots.

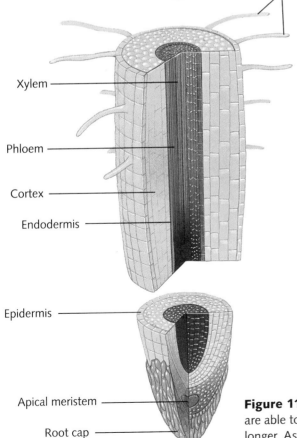

Root hairs

Xylem

Phloem

Cortex

Endodermis

Epidermis

Apical meristem

Root cap

Figure 11.6 Many specialized tissues make up a root. Only cells in the root tip are able to divide. Just behind the area of cell division is an area where cells grow longer. As the number and size of cells increase, the root gets longer and wider.

The arrangement of xylem and phloem differs in the roots of dicots and monocots. Recall that angiosperms, or flowering plants, are classified into two large groups: dicots and monocots. Dicots produce seeds that have two cotyledons, or seed leaves. Monocots produce seeds with a single cotyledon.

The xylem cells in a dicot root are arranged in a star-shaped fashion in the center of the root. Phloem cells are found between the arms of the star. In a monocot root, xylem cells and phloem cells alternate in a ring around the root. **Figure 11.7** shows the arrangement of xylem and phloem tissues in a dicot root and a monocot root.

Roots generally exhibit positive gravitropism and grow downward toward the pull of gravity. Roots have also been said to show positive hydrotropism. **Hydrotropism** is growth toward water. Because it is impossible to separate the growth of a plant's roots as a result of gravity's pull from the growth of a plant's roots in response to water, some scientists believe that hydrotropism may not be an actual tropism.

The growth of plant roots occurs only in damp soil. Roots will not grow in soil that is completely dry. In addition to water, plant roots need oxygen for cellular respiration. Cellular respiration is the process that enables cells to get energy needed for growth.

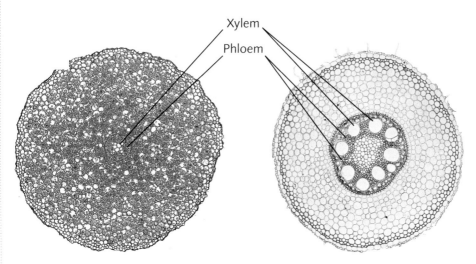

Figure 11.7 You can see the different arrangement of similar structures in these cross sections of a dicot root *(left)* and a monocot root *(right)*. What other types of cells can be seen?

Extend It!

Hydroponics is a method of growing plants without soil. Research the process of hydroponics. Use your research to prepare a list in your Science Notebook of the advantages and disadvantages of growing plants using this process. How does hydroponics affect root growth?

After You Read

1. Review the working definition of *root* in your Science Notebook. Then identify the type of root system that you might eat as part of a meal and the type of root system that you might mow over during the summer.

2. What are the main functions of plant roots?

3. Draw and label the parts of a root. Identify the root you have drawn as a monocot or a dicot. Explain how monocot and dicot roots differ.

Before You Read

Use the headings of this lesson to form questions. Write the questions in your Science Notebook. As you read, write answers to the questions.

Stems are the most visible parts of nearly all plants. **Stems** are part of a plant's shoot system. Most stems grow above the ground. However, the stems of some plants grow underground or along the soil surface. The trunks of trees are likely the largest and most noticeable stems.

Stems have many important functions. Stems support the leaves, cones, fruits, flowers, and even seeds of plants. Stems hold a plant's leaves up toward the sunlight. Stems contain vascular tissues—xylem and phloem—that transport water, dissolved minerals, and organic molecules to and from the roots and leaves. Stems that are green conduct photosynthesis. In some plants, stems store water or products of photosynthesis. This can help the plant survive during times of drought or severe cold.

Types of Stems

Plant stems vary greatly in size and shape. Think about the trunk, branches, and twigs of a tree. All of these structures are stems! As **Figure 11.8** shows, the baobab tree has an enormous trunk and short, stubby branches. The rose plant has thin stems covered with thorns.

There are two classifications of stems: herbaceous stems and woody stems. **Herbaceous** (hur BAY shus) **stems** are green, soft, and flexible stems that usually carry out photosynthesis. Examples of plants with herbaceous stems include basil, petunias, impatiens, carnations, sunflowers, and tomatoes. **Woody stems** are hard, strong, and rigid. Trees, shrubs, and roses have woody stems.

Figure 11.8 Notice the difference between the stems of a baobab tree and those of a rose plant. What are the functions of a stem?

Structure of Stems

A stem is divided into nodes and internodes. A **node** is the point at which leaves are attached to a stem. Some plants produce a single leaf at a node, while other plants produce several leaves at a node. An **internode** is the part of the stem between nodes. A plant that does not receive enough light is easily identified by its longer internodes. The leaves on most plants are arranged in ways that expose the leaves to the greatest amount of light possible, as **Figure 11.9** shows. It is unusual to find a plant whose leaves are arranged in such a way that they shade other leaves under normal growing conditions.

The top angle between a leaf and a stem is called the **axil**. The axil often contains a bud that can grow into a new branch. There is also a bud at the tip of the stem. The stem increases in length when this bud grows. Recall from Chapter 9 that hormones are produced at the tip of a stem. These hormones prevent other buds lower down on the stem from growing. If the bud at the tip of a stem is damaged or removed, the hormones are not produced. The other buds, called axillary buds, begin to grow. The plant produces new stems and grows bushier.

The outside layer of a stem is the epidermis. The epidermis is usually waterproof and may also be covered with a thick, protective layer of dead tissue called bark. In some kinds of trees, the bark is very thick. Just inside the epidermis is a layer of cells that fill in the area surrounding the xylem and phloem. These cells contain chlorophyll. When light strikes this layer, these cells can conduct photosynthesis.

Herbaceous Dicot Stems   The epidermis of a herbaceous dicot stem is covered by a waterproof layer called the cuticle. In some plants, there are pores in the stem that allow the exchange of gases between the stem and the surrounding air. The vascular tissues are arranged in bundles that form a ring near the outside edge of the stem. The cortex is found between the vascular bundles and the epidermis. Tissue known as **pith** fills the rest of the stem within the ring of vascular bundles. An example of a plant with this kind of stem is the pea plant.

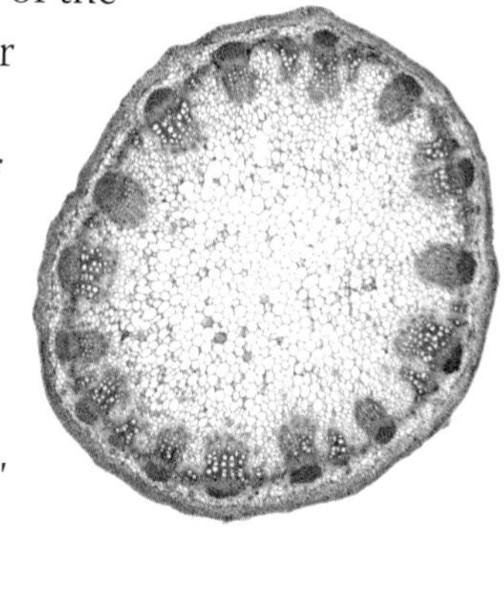

Figure 11.9 Leaves are spaced along a stem in a regular pattern. Note the tiny bud at each spot where a leaf is attached to the stem.

Figure 11.10 A white potato looks like a root, but it is actually a part of an underground stem called a tuber. Tubers store starch. A white potato tuber has buds from which new stems can grow. These buds are commonly called eyes.

Figure 11.11 In the stem of a young herbaceous dicot, separate bundles of xylem and phloem form a ring.

As You Read

Answer the question you wrote in your Science Notebook for each heading. You should be able to answer all three of the questions you wrote. Then, share your answers with a partner and modify them as needed.

Figure It Out

1. What tissue makes up most of a woody dicot stem?

2. What cells function primarily as protection for the stem?

Woody Dicot Stems   Woody dicot stems have two layers of meristematic cells that produce new cells. The first layer is called the **cork cambium**. The cork cambium produces tough cork cells that will become part of the bark of a tree. The cork cells contain a waxy substance that repels water, making the stems waterproof. The cork cells also protect the tree from physical damage.

The second layer is called the **vascular cambium**. This layer produces secondary xylem and phloem cells. A thin layer of secondary phloem cells is found just beneath the bark. Secondary xylem cells fill the inside of a woody stem. The diameter of a tree trunk increases with the growth of layers of secondary xylem cells. Because the amount of growth produced by the vascular cambium differs from season to season, visible rings of tissue can be seen in a woody stem. These rings can be counted to determine the age of a tree. In some tropical areas, trees grow all year long. The wood in the trunks of these trees does not show obvious rings, and so the trees' ages are not easily determined.

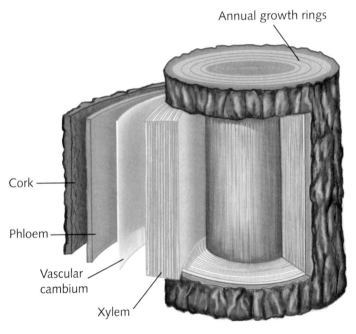

Annual growth rings

Cork

Phloem

Vascular cambium

Xylem

Figure 11.12 The production of secondary vascular tissue increases the size of a woody dicot's stem. Woody stems are composed primarily of dead xylem cells. People put the stems that produce this woody tissue to a variety of uses.

⊙⊙⊙ CONNECTION: Meteorology

The widths of a tree's annual growth rings provide important information about conditions in the tree's environment. In years when moisture is abundant, the rings are wide. In drier years, the rings are narrower. Careful examination of tree rings allows scientists to reconstruct the weather conditions and climate patterns in an area.

Monocot Stems The surface of a monocot stem, such as the stem of a corn plant, is covered by an epidermis. In most monocots, the bundles of xylem and phloem are scattered throughout the stem. However, in some monocots, a ring of scattered bundles of xylem and phloem cells can be observed near the outside edge of the stem. Within a single bundle of vascular tissue, xylem cells can be found facing the center of the stem. Phloem cells can be found facing the outer edge of the stem. There is also an air space in the vascular bundle.

Specialized Stems

Tubers, bulbs, and rhizomes are types of underground stems. A tuber is a fat stem that stores food. The eye of a tuber is a bud that is capable of developing into a new plant. A white potato is an example of a tuber. An onion is an example of a bulb, which is a short, underground stem with thick, food-storing leaves. When planted, a bulb develops into a new plant. Garlic and scallions are other examples of bulbs that are eaten for food. Tulips, daffodils, crocuses, and hyacinths are examples of decorative plants that form bulbs that are not eaten. Rhizomes are stems that grow horizontally beneath or along the ground. As they grow, they produce buds that develop into new plants. Morning glories, strawberries, irises, and ginger plants produce rhizomes. The corkscrew-shaped tendrils on grape vines are another example of a specialized stem. Such stems enable grape vines to cling to a supporting object as they grow.

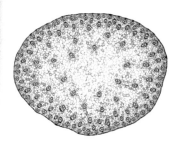

Figure 11.13 The vascular bundles in a monocot stem are scattered throughout the stem.

Figure 11.14 The thick stem of a cactus, such as this saguaro, is an example of a specialized stem. What do you think is the function of such a stem?

 Explore It!

Design an experiment that would demonstrate that some specialized stems store organic nutrients produced by the plant through photosynthesis. In your Science Notebook, list the materials that you will need and the procedure you plan to follow. With approval from your teacher, conduct the experiment. Record your observations in your Science Notebook.

After You Read

1. A plant has a thick stem that is gray in color and very rough. Use answers to your heading questions to identify the function of this plant's stem.

2. What structures are found in all stems?

3. Design a concept map that illustrates the different types of stems.

Learning Goals

- List the functions of leaves.
- Describe the structure of leaves.
- Relate the structure of leaves to their environment.

New Vocabulary

leaf
blade
petiole
cuticle
mesophyll
stoma
guard cell
transpiration

Before You Read

Look at the four main headings within this lesson. Divide a page in your Science Notebook into four sections. Write one heading in each section. Then, write a sentence summarizing what you know about each heading.

You have probably seen solar panels on the roofs of buildings. These large, flat surfaces collect the energy of sunlight. A plant's leaves do the same thing. In fact, leaves can be thought of as the world's oldest solar-energy collectors.

Leaves are part of a plant's shoot system. **Leaves** are plant organs whose main functions include capturing the energy of sunlight, making organic molecules through the process of photosynthesis, and exchanging gases with the environment. Most leaves are flat and have a relatively large surface area that receives sunlight. However, leaves come in a great variety of sizes and shapes.

Leaf Variation

The flat part of a leaf is called the **blade**. Leaves are classified as either simple or compound. A simple leaf is made up of a single blade. Oak trees and apple trees have simple leaves. A compound leaf has a blade that has split into two or more smaller sections. These small sections are called leaflets. Palms, roses, and clovers have compound leaves. If the leaflets of a compound leaf are attached to each other at a single point, the leaf is palmately compound. This name comes from the fact that the leaf form resembles the way the fingers of a hand are attached to the palm. If the leaflets are attached along an extended stalk, the leaf is pinnately compound. Pinnately compound leaves look something like feathers.

The stalk that joins a leaf blade to a stem is called a **petiole** (PET ee ohl). Most leaves are attached to stems by petioles. A petiole is usually small. Some plants, however, have large petioles. The large petioles of celery and rhubarb are often eaten. Not all leaves are joined to stems by petioles. Some leaves, such as grass blades, are joined directly to the stem. The petiole contains vascular tissues—xylem and phloem—that extend from the stem into the leaf and form veins.

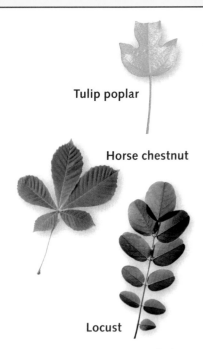

Tulip poplar

Horse chestnut

Locust

Figure 11.15 The leaf of the tulip poplar *(top)* is a simple leaf. The leaves of a horse chestnut *(center)* and a locust *(bottom)* are compound leaves, in which the blade is divided into leaflets. The horse chestnut leaf is palmately compound. The locust leaf is pinnately compound.

Figure 11.16 The maple *(left)* has leaves attached to its stem by a petiole. The St. John's wort *(right)* is a plant that has leaves attached directly to its stem.

The leaves of a dicot, such as a maple tree, have a branching network of veins. The leaves of a monocot, such as an iris, have veins that run parallel to one another. The veins connect with xylem and phloem in the stem. Water enters the leaves through the xylem, and organic molecules are moved out of the leaves through the phloem.

Leaf Structure

The internal structure of a plant leaf can more easily be seen if the cut surface of the leaf is examined under a microscope. The outermost layer of a leaf is the epidermis. A leaf has both an upper and a lower epidermis. The epidermis acts like a clear window to let sunlight pass into the leaf. The upper epidermis is covered with a waxy, waterproof coating called the **cuticle** (KYEW tih kul). The cuticle, which can make real plants look as if they are made of shiny plastic, helps prevent the leaf from losing excess water. The epidermis of some plants has glands and hairs, and sometimes even soft, fuzzy surfaces. **Figure 11.18** shows the structures of a leaf.

Just beneath the epidermis are two layers of cells known collectively as **mesophyll** (MEZ uh fihl). Mesophyll is the photosynthetic tissue of a leaf. Palisade mesophyll lies very close to the upper epidermis, where it receives maximum exposure to sunlight. The cells in this layer are column-shaped and are packed very closely together. They contain many chloroplasts—the structures in plant cells that contain chlorophyll.

Beneath the palisade mesophyll is the spongy mesophyll. The cells in this layer are loosely packed and irregularly shaped. These cells have air spaces that allow carbon dioxide, oxygen, and water vapor to flow freely around the cells. Carbon dioxide enters a leaf through tiny pores in the epidermis and diffuses into the mesophyll cells. Water vapor and oxygen pass out of the mesophyll cells and the leaf through these openings.

Dicot leaf

Monocot leaf

Figure 11.17 The veins in a dicot leaf have a branching pattern. A monocot leaf has parallel veins.

As You Read

Add new information to the summaries you wrote in your Science Notebook for *Leaf Variation* and *Leaf Structure*. Be sure to include the vocabulary terms used in each section.

What is the waxy, waterproof coating of a leaf called?

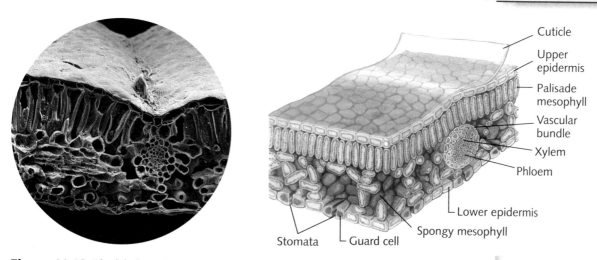

Cuticle
Upper epidermis
Palisade mesophyll
Vascular bundle
Xylem
Phloem
Lower epidermis
Spongy mesophyll
Guard cell
Stomata

Figure 11.18 The labels in the diagram *(right)* identify the parts of a leaf seen in cross section. Match these labels to the leaf parts shown in the micrograph of a cross section of a leaf *(left)*. In which layers does photosynthesis take place? How are the cells in these layers adapted to their function?

1. How do guard cells regulate the size of the opening of a stoma?

2. Why is gas exchange important to a plant?

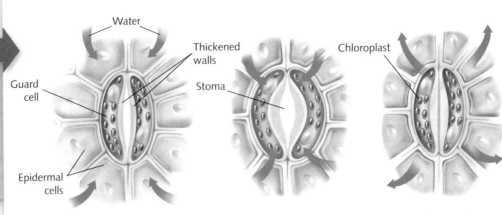

Water

Thickened walls

Chloroplast

Guard cell

Stoma

Epidermal cells

❶ The guard cells have flexible cell walls.

❷ When water enters the guard cells, the pressure causes them to bow out, opening the stoma.

❸ As water leaves the guard cells, the pressure is released and the cells come together, closing the stoma.

Figure 11.19 One guard cell is located on each side of a stoma. Water movement into and out of the guard cells regulates the size of the opening of each stoma.

Explore It!

In your Science Notebook, predict what will happen if the leaves of a plant are covered with a plastic bag. Help your class design an experiment to test your prediction. As your teacher performs the experiment, record the hypothesis, materials, and procedure in your Science Notebook. After two days, carefully observe the inside of the bag. Record and explain your observations.

The tiny pores, or openings, in the lower epidermis of most leaves are called **stomata** (STOH muh tuh, singular: stoma). Stomata control gas exchange between the leaf and the outside environment. The number of stomata in leaves varies greatly. Some plants may have as many as several million stomata on a single leaf. Plants such as water lilies that float on the surface of water have leaves with stomata in the upper epidermis. Other plants have leaves with stomata in both the upper and lower epidermis. Two **guard cells** lie on either side of a stoma. Guard cells are epidermal cells that contain chloroplasts. Guard cells are able to change shape as they absorb and release water. As a result, the guard cells open and close the stoma. **Figure 11.19** illustrates this process. The opening and closing of the guard cells regulates the amount of carbon dioxide, water vapor, and oxygen that enters and leaves the leaf.

The Movement of Nutrients in Plants

The veins in a leaf contain xylem and phloem cells. These cells join with the xylem and phloem in the stems and roots of a plant. Water and mineral nutrients enter a root and move into the xylem. The water and minerals move up the xylem in the plant's stem and enter the leaves. In the leaves, some of the water is used in the process of photosynthesis, which occurs primarily in the palisade mesophyll. Recall that during photosynthesis, water molecules are split in the presence of sunlight and chlorophyll, and oxygen gas is given off.

Most of the water passes out of leaves as water vapor through stomata in a process called **transpiration** (trans puh RAY shun). Because water molecules attract other water molecules, water that leaves the stomata by transpiration actually "pulls" water up from the roots. Organic molecules made in the leaves move around the plant in the phloem. Water drawn into the phloem by osmosis helps push nutrients through the phloem.

Specialized Leaves

The shapes and sizes of leaves vary depending on the plant's environment. Many species of cactus, for example, lack typical leaves. Their leaves are the spines that protect these plants from being eaten. Other plants that live in dry environments often have thick, fleshy leaves that are covered with a thick cuticle, which helps prevent water loss.

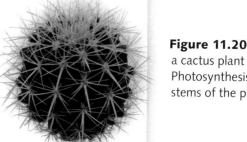

Figure 11.20 The leaves of a cactus plant are its spines. Photosynthesis occurs in the stems of the plant.

Plants that live in areas that receive a great deal of rain often have long, pointy leaves. The points are called drip tips, and they help the leaves shed water. Some leaves are vividly colored and surround the small flowers on a plant. These leaves are called bracts. The flowers of dogwood trees and poinsettia plants have showy bracts that help the plants attract insects that pollinate the small flowers.

Tendrils that help some plants hold on to supports as they climb are modified leaves. These tendrils look like small, delicate whips that are able to coil tightly around the structures on which they climb. Tendrils can be seen on garden pea plants.

Perhaps the most incredible leaves are those found on plants that trap insects. The leaves on a Venus's-flytrap snap shut when an insect touches hairs in the trap. Digestive juices then help break down the insect. Pitcher plants have modified leaves that look like pitchers and act as traps for catching insects. Any insect that gets caught inside these plants is decomposed by the plants' digestive juices.

Figure 11.21 Drip tips help the leaves of plants in a rain forest shed water *(top)*. The colorful leaves on this poinsettia plant look like flower petals *(bottom)*. The actual flowers are the small structures in the center.

Figure 11.22 The Venus's-flytrap *(left)* snaps shut to trap an insect. Pitcher plants *(right)* make a sweet-tasting liquid in glands on the rims of their traps. This liquid attracts insects.

After You Read

1. Describe three functions of leaves.

2. What structures in a leaf make up its transport system?

3. Select one specialized leaf described in your Science Notebook, and explain how its structure is related to its environment.

4. Use the information recorded in your Science Notebook about movement of water and organic molecules in plants to write a summary paragraph.

Learning Goals

- Describe the structure of a seed.
- Identify ways that seeds can be dispersed.

New Vocabulary

seed
seed coat
embryo
cotyledon
plumule
epicotyl
hypocotyl
radicle
germination
dormancy

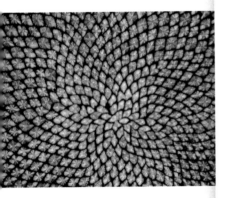

Before You Read

In your Science Notebook, create a T-chart. Label one column *Seed Structure* and the other column *Seed Dispersal*. As your first entry, record what you already know about each topic. Then, in the appropriate column, record important information discussed in this lesson. Try to use the vocabulary terms in your entries.

Seeds can be thought of as a plant's insurance that there will be another generation of its kind. A **seed** is a structure from which a new plant grows. Seeds look quite different from one another. Seeds can be as fine as dust or as large as coconuts. Orchid plants produce seeds that are so small they are hardly recognizable as seeds. Because these seeds are so fine, they cannot contain a great deal of food for a developing plant. In fact, orchid seeds depend on certain species of fungi to provide the food they need to grow. Orchids produce millions and millions of seeds, which helps increase their chances of survival.

Although seeds vary widely in appearance, they all have the same basic structure. A seed consists of a seed coat, an immature plant, and stored food. A seed develops from a fertilized ovule and contains a new plant and all that is necessary for its development.

Structure of a Seed

The **seed coat** is a tough protective coat that envelops the entire seed. Seed coats vary in thickness from the thin, brown, papery covering of a peanut to the thick covering of a coconut. A seed coat protects the delicate tissues of the young plant in the seed. It keeps these tissues from losing water and protects them from certain kinds of physical damage. The seed coat generally opens only after the seed has fallen in a moist place and has begun to take in water. The "skins" of peanuts, corn kernels, and lima beans are seed coats.

Enclosed within the seed coat is the embryo. The **embryo** (EM bree oh) is the immature plant that begins to grow when environmental conditions are just right. The embryo develops from a fertilized egg. The embryo eventually gives rise to the leaves, stem, and roots of a new plant.

Figure 11.23 The seeds of a sunflower *(top)* show an intricate spiral arrangement with symmetry. Sunflower seeds are eaten by humans, birds, and other animals. A useful oil can be extracted from sunflower seeds. Vanilla *(bottom)* comes from the seed pods of the vanilla orchid. Each seed pod contains millions of tiny seeds that are used to flavor many kinds of food.

Figure 11.24 Peanuts are one of the few kinds of seeds that develop underground. Peanuts are sometimes called "ground nuts." After the peanut plant flowers and is fertilized, the stem bends toward the ground and pushes its tip below the soil surface.

In addition to the seed coat and the embryo, a seed contains stored nutrients. In many seeds, the stored nutrients make up a large part of the seed. The seedling plant will use the stored nutrients until it is able to manufacture its own organic molecules through photosynthesis. In some seeds, the stored food surrounds the embryo. In other seeds, the stored food is contained in one or more **cotyledons** (kah tuh LEE duns), or seed leaves. **Figure 11.25** shows examples of seeds with different numbers of cotyledons.

Dicot Seeds   The seeds of dicots have two cotyledons. If the seed coat is removed with care, the two cotyledons can be gently separated at the seam where they are joined together. Inside are two tiny leaves called the **plumule** (PLOOM yool). These tiny leaves are usually folded, and they lie on one of the cotyledons. Both cotyledons are attached to the embryo just below the plumule. A short stem is located just above the point at which the cotyledons are attached. This stem is called the **epicotyl** (EH pih kah tul). The epicotyl will become the stem of the mature plant. Another stem, called the **hypocotyl** (HI puh kah tul), is located just below the point of attachment. At the base of the hypocotyl is a region called the **radicle** (RA dih kul). The primary root of the plant begins to grow from the radicle.

🔍 **Explore It!**

Use a hand lens to observe bean seeds that have been soaked in water. Carefully remove the seed coat, and then separate the two halves of each seed. Use the hand lens to identify and observe the parts of the seed. In your Science Notebook, draw and label the parts of the seed you have observed.

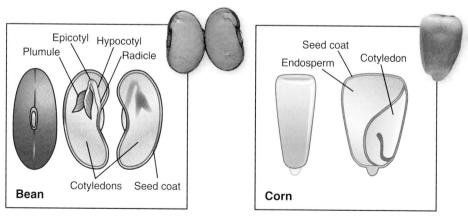

Figure 11.25 A bean is a dicot; corn is a monocot. In bean seeds *(left)*, all of the nutrients are stored in the cotyledons. In seeds such as corn kernels *(right)*, most of the nutrients are stored in a tissue called endosperm.

Figure It Out

1. How are these seeds alike? How are they different?

2. What part of the embryo becomes the plant stem? What part becomes the primary root?

As You Read

Review the information you have recorded in the column labeled *Seed Structure*. Define and describe a seed using the following terms: *seed, seed coat, embryo,* and *cotyledon.*

Monocot Seeds The seed of a monocot has one cotyledon. Corn is an example of a monocot. In corn seeds, the embryo does not depend upon the single cotyledon for food. Instead, food for the embryo is stored as endosperm, and the delicate new leaves are protected in a cylinder of tissue. Endosperm is formed during fertilization and is rich in important nutrients. The cotyledon helps transfer nutrients to the embryo.

Gymnosperm Seeds Gymnosperms also have cotyledons in their seeds. Pines produce seeds with eight cotyledons. When a pine seed begins to grow, the cotyledons that resemble smaller versions of pine needles are in a ring around the baby plant's stem, as **Figure 11.26** shows.

Figure 11.26 Pine seeds have eight tiny cotyledons. The pine cotyledons look like the small leaves or needles on adult trees.

Seed Dispersal

Seeds that fall close to the parent plant often have to compete for available nutrients and other resources, such as light and water. Plants have evolved many ways to ensure that seeds are transported from where they are formed. The transportation of seeds is called seed dispersal. Seeds are dispersed by water, wind, humans, and other animals.

Coconuts produce some of the largest seeds in the plant kingdom. These seeds are encased in tough husks made of strong fibers with air spaces in between them. Thus coconut seeds can float in water and are dispersed from one place to another on ocean currents. Other plants that live in or near water produce seeds that can be dispersed by moving water and waves.

Dandelion seeds have small fluffy threads attached to them. The threads help the wind carry the seeds aloft. Other seeds, such as those produced by maple trees, have two winglike structures. These structures act much like propellers to move the seeds from tree to ground.

Animals and humans play a part in seed dispersal. Animals carry fruits to other locations and leave behind the seeds. They also excrete indigestible seeds in their body wastes. Humans and animals can pick up and carry seeds that have sharp barbs on their seed coats. The barbs stick in an animal's fur or on a human's clothing, and the seeds get carried to another place. When the seeds eventually fall off, they can grow where they land. The invention of self-sticking fabric tape is said to have its origin in observations of the tiny hooks that attach tick seeds to the fur of animals.

Extend It!

Your teacher will divide the class into groups of three to four students and assign each group a method of seed dispersal. Work with your group to design a seed that can be dispersed in that way. Use the materials provided by your teacher to construct your group seed. Materials might include cotton balls, tissue paper, hook-and-loop fasteners, foam packing material, clay, pipe cleaners, and glue. Once you have constructed your seed, devise an experiment to determine how successful your design is.

Seed Germination

The process by which the embryo in a seed begins to develop into a new plant is called **germination** (jur muh NAY shun). Germination begins when growing conditions are favorable. Water, oxygen, and favorable temperatures are common requirements of germination. Absorbed water causes the cotyledons to swell, cracking open the seed coat and activating the embryo. **Figure 11.27** illustrates seed germination.

Seeds often undergo a period of inactivity called **dormancy** (DOR mun see) before they begin to grow. During dormancy, the embryo can survive long periods of bitter cold, extreme heat, or drought. The embryo will not begin to grow until dormancy is broken. The embryos in many tree seeds, for example, must go through a period of cold temperatures or through several winters before they will begin to grow. Some seeds need darkness to break dormancy, while others need exposure to light. The seeds of some plants have begun to grow after periods of dormancy lasting as long as 1,000 years! Having some knowledge of the conditions that seeds need to break dormancy is essential to growing plants from seeds.

Explain It!

The seeds of some plants, such as prairie wildflowers and conifers, will not germinate until dormancy is broken by exposure to fire. Research plant species such as the bishop pine or lupines that depend on exposure to fire to germinate. In your Science Notebook, explain why this method of ending dormancy is an important adaptation for survival.

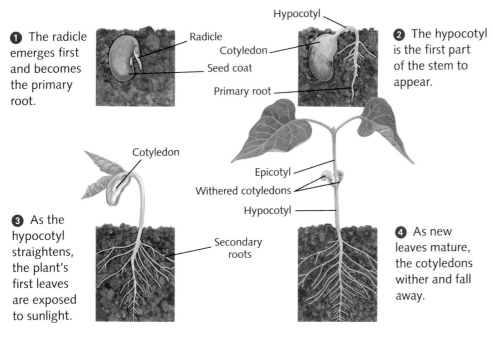

1 The radicle emerges first and becomes the primary root.

Hypocotyl
Radicle
Cotyledon
Seed coat
Primary root

2 The hypocotyl is the first part of the stem to appear.

Cotyledon

Epicotyl
Withered cotyledons
Hypocotyl

3 As the hypocotyl straightens, the plant's first leaves are exposed to sunlight.

Secondary roots

4 As new leaves mature, the cotyledons wither and fall away.

Figure 11.27 The events of seed germination vary in different kinds of plants. This illustration shows how a bean seed germinates to begin the growth of a new plant.

CONNECTION: Math

Many seeds absorb water before germinating. Determine the amount of water absorbed by ten bean seeds prior to germination using a graduated cylinder and the technique of water displacement. In your Science Notebook, describe the technique of water displacement, and record all of your measurements.

After You Read

1. Use the information in your T-chart to draw and label the parts of a seed.

2. How might each of the following seeds be dispersed to a new location: A seed inside a sweet berry? A lightweight seed that spins? A seed that floats?

3. In a well-developed paragraph, define and describe the process of germination. Accompany your paragraph with a correctly labeled drawing.

Summary

KEY CONCEPTS

11.1 Roots

- Roots anchor and support a plant, absorb water and minerals from the soil, and contain vascular tissues that transport materials to and from the stem.
- Two major types of root systems are taproots and fibrous roots.
- The epidermis surrounds the vascular tissues—xylem and phloem—that transport water, minerals, and organic molecules throughout the plant.

11.2 Stems

- Stems provide support for a plant, hold leaves up toward the sunlight, and transport material between a plant's leaves and roots.
- Plants can have woody stems that are hard and rigid or herbaceous stems that are more flexible and carry out photosynthesis.
- The arrangement of vascular tissues differs in monocot and dicot stems.
- Some stems are specialized to carry out other functions.

11.3 Leaves

- Leaves are designed to capture the Sun's energy and carry out photosynthesis.
- Leaves are plant organs whose structure is closely related to their function and well adapted to their environment.
- The flat part of a leaf is called the blade. Leaves are classified as either simple or compound depending on the number of blades. A simple leaf is made up of a single blade. A compound leaf consists of several leaflets.

11.4 Seeds

- Seeds contain a protective coating, stored food, and an immature plant that grows and develops when conditions are favorable.
- Seeds can be dispersed to new environments by water, wind, humans, and animals.
- Germination is the process by which the embryo in a seed begins to develop into a new plant. Water, oxygen, and favorable temperatures are common requirements of germination. Germination begins when dormancy ends and growing conditions are favorable.

VOCABULARY REVIEW

Write each term in a complete sentence, or write a paragraph relating several terms.

11.1
root, p. 187
taproot, p. 187
fibrous root, p. 188
vegetative propagation, p. 189
root cap, p. 190
apical meristem, p. 190
epidermis, p. 190
root hair, p. 190
cortex, p. 190
endodermis, p. 190
xylem, p. 190
phloem, p. 190
hydrotropism, p. 191

11.2
stem, p. 192
herbaceous stem, p. 192
woody stem, p. 192
node, p. 193
internode, p. 193
axil, p. 193
pith, p. 193
cork cambium, p. 194
vascular cambium, p. 194

11.3
leaf, p. 196
blade, p. 196
petiole, p. 196
cuticle, p. 197
mesophyll, p. 197
stoma, p. 198
guard cell, p. 198
transpiration, p. 198

11.4
seed, p. 200
seed coat, p. 200
embryo, p. 200
cotyledon, p. 201
plumule, p. 201
epicotyl, p. 201
hypocotyl, p. 201
radicle, p. 201
germination, p. 203
dormancy, p. 203

True or False

If the statement is true, write "true." If it is false, change the underlined word or words to make the statement true.

1. The <u>epidermis</u> is the protective tissue found in the leaves, stems, and roots of plants.

2. <u>Phloem</u> transports water throughout a plant.

3. <u>Stems</u> support a plant and transport water and nutrients between plant parts.

4. <u>Leaves</u> are responsible for anchoring a plant and absorbing water and minerals from the soil.

5. Leaf <u>stomata</u> control the exchange of gases with the environment.

6. Seeds can be <u>germinated</u> by animals, humans, wind, and water.

7. The arrangement of some plant tissues of monocots and dicots is <u>different</u>.

8. Water vapor is given off by plants in a process called <u>photosynthesis</u>.

Short Answer

Answer each of the following in a sentence or brief paragraph.

9. Name and describe the functions of the two types of vascular tissues found in plants.

10. How do the functions of taproots and fibrous roots differ?

11. A plant's stem is soft, flexible, and green. What are the probable functions of this stem?

12. What functions are the leaves and stem of a cactus specialized to perform?

Critical Thinking

Use what you have learned in this chapter to answer each of the following.

13. **Apply Concepts** During late winter, holes are drilled into some sugar maple trees in order to collect syrup. What type of tissue is being "tapped"?

14. **Analyze** Why are the roots of a plant usually the first part of the plant to appear from a germinating seed?

15. **Infer** Why are seeds a good food source for animals?

Standardized Test Question

Choose the letter of the response that correctly answers the question.

16. Where does most of the photosynthesis occur in a typical plant?

    A. xylem

    B. roots

    C. leaves

    D. stems

Test-Taking Tip

Remember that qualifying words such as *most* can mean that all the choices could be considered true. Ask yourself which choice best answers the question.

PREPARE FOR CHAPTER TEST

To prepare for the chapter test, create a question from each Learning Goal. Use the information in your Science Notebook to answer each question. Then use these answers to write a well-developed essay about the chapter. Use the Key Concept on the first page of this chapter as your topic sentence.

Not-So-Delicious Apples

THE FRUIT KNOWN as the Red Delicious apple first appeared in an Iowa field in the 1880s. This round, juicy apple was called the Hawkeye. People loved it and later renamed it the Delicious apple. Although it was very popular, breeders thought they could improve sales of the apple by improving it. The improvements changed the apple's original shape, color, and taste almost completely.

Over the years, breeders crossed Delicious apples with apples that had a less rounded shape and a redder color. They renamed it the "Red" Delicious. Careful breeding made it crispy and firmed its skin so it would not bruise easily. Breeders also changed the apple to make it last longer on the store shelf.

Unfortunately, breeding can go too far. Breeding for one desirable trait can accidentally breed out other desirable traits. For example, apples with genes for a deep red color may also have genes that make them less sweet or less juicy. Breeders may end up with beautiful, red apples that people would rather look at than eat! This is what happened to the Red Delicious. At some point, it became a beautiful, crispy, firm, long-lasting apple that just didn't taste as good as the original.

Once a new variety has been bred, how do growers produce only that variety of apples? The answer lies in a process called grafting. Most apples you buy in stores came from trees propagated by grafting rather than trees grown from seeds.

In grafting, growers cut pieces from the tips of branches of trees that produce a particular variety of fruit. These branch tips will become grafts. Growers then attach the grafts to the ends of the branches of a rootstock chosen for its hardiness.

Research and Report

There are hundreds of varieties of apples. The Red Delicious is just one of them. With a partner, research one of the apple varieties. Find its origin, its characteristics, how it is used, and how it was developed. Include a drawing of the apple, as well as a map that shows where farmers grow it today.

The rootstock's branches and their new branch tips grow together. Year after year, each tree produces the same variety of fruit as the tree from which the grafts came.

Occasionally, however, slight variations may occur as a result of gene mutations. Branches in which these mutations occur are called sports. A sport might form apples with a desirable trait, such as a redder color, more sweetness, or a slightly different shape. Growers can graft branch tips from a sport onto a rootstock to produce apples with the desirable new trait. And, this is how an 1880s Iowa farmer's discovery began its journey to becoming the Red Delicious apple.

⚬ CAREER CONNECTION ORGANIC FARMER

MANY PEOPLE WANT healthier, safer foods. They want fruits and vegetables free of chemicals. They don't want to eat plants or animals that have been genetically altered. They also want to make sure their food is produced in a way that protects the environment. Organic farmers provide the foods that such people want to buy.

Like other farmers, organic farmers must produce good crops. However, they make sure their crops get nutrients and are protected from diseases and pests in different ways. They do not use chemical fertilizers or chemical poisons to help keep foods safer and reduce pollution.

To grow crops naturally, organic farmers must know how their crops interact with the soil and with other living organisms. Instead of using pesticides, they might attract beneficial insects that eat crop pests. To enrich the soil, they might rotate crops. Organic farmers also use natural fertilizers, such as compost.

Organic farmers love living on the land, and they enjoy providing fresh, healthy food to their own families and to other people.

Arctic Bank

ON A SMALL, windswept island at the edge of the Arctic Ocean, scientists are busily burying something of great importance. They are storing this treasure in a vault deep in a mountainside. The vault has thick, concrete walls. Huge steel doors stand at the entrance. Outside, a sturdy fence surrounds the vault, and polar bears roam the grounds. What is being hidden in the vault? Seeds. Millions of seeds.

Why bury millions of seeds in an arctic vault? The reason is that burying the seeds could prevent the extinction of many plant species. Scientists know that regional wars have already wiped out some unique types of plants. They worry that even greater damage could be done by a global disaster, such as a large-scale war, a natural catastrophe, or a disease that attacks a type of food crop. The result could be the disappearance of many important plants—and the seeds that could grow into new ones.

The seed bank being constructed on Norway's Svalbard Islands is the scientific community's response to this potential disaster. Many of the world's nations will cooperate to stock the seed bank. The bank will serve as insurance against losing crop species and as a means of preserving Earth's plant diversity. It will contain seeds of every food crop grown today—even unique plants that grow in limited areas. So if a type of plant disappears from the world outside the vault, it won't be lost forever.

The Svalbard Islands site is an excellent choice of location. There, the vault is very safe. The seeds are well preserved by constant cold temperatures, always below freezing. If the electricity were to fail, permafrost around the vault would keep the seeds cold.

Although there are other seed banks on Earth, the one on the remote island off Norway will be the largest. When all two million seeds are safely in place, this vault will protect a treasure more valuable than gold—the continuation of plant species.

The Fastest Vine in the South

YOU CAN RIP IT OUT of the ground. You can pour pesticides on it. You can even set fire to it. Yet you still might not kill it. That's the problem with kudzu, a fast-growing vine that is creeping across America.

Kudzu is an invasive species—one that doesn't grow here naturally. Because it's native to Japan, it has no natural enemies in North America. It just grows wild—and fast. This vine can grow up to 30 cm per day!

Wherever kudzu takes root, it smothers everything around it. It grows over other plants, killing them. It twists itself around trees, often uprooting them.

Kudzu was first brought to the United States in 1876. People in the South were soon planting the fast-growing vines to provide decoration and shade on porches. Today, it has spread as far north as Massachusetts and as far west as Texas and Iowa.

To begin to eliminate kudzu, a combination of mowing, pesticides, and burning is needed. Scientists are also considering the use of a beetle native to Asia that attacks kudzu roots. But they will not import the beetle until they know that it will not attack other native species. They are cautious about solving one problem by creating another!

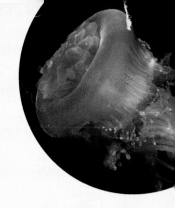

Unit 5

Animals

Sponges, Cnidarians, Worms, and Mollusks

KEY CONCEPT Invertebrate animals can be classified based on their physical characteristics.

You put on your diving gear and jump into the cool, clear ocean water. As you go deeper, you notice an amazing variety of fishes and other animals that you do not recognize.

Sponges, corals, and jellyfish are all types of organisms you might find during your underwater exploration. These unusual animals have special characteristics that allow them to live in Earth's great ocean depths.

Think About Identifying Animals

Animals inhabit Earth's land, air, and water. Each animal has a specific role to play in its environment and special characteristics that help it survive.

- Spend some time observing and thinking about the animals you see on a regular basis. Then, look through books and magazines for pictures of animals that may be less familiar because they live in places not easily observed.

- In your Science Notebook, make a T-chart. Label one column *Land* and the other column *Water*. List at least ten animals that live in each type of environment. At the bottom of each column, identify some characteristics that those animals must have in order to live successfully in that environment.

www.scilinks.org
Animals **Code:** WGB12

Learning Goals

- Identify the common characteristics of animals.
- Describe the difference between vertebrates and invertebrates.
- Compare the three types of animal symmetry.

New Vocabulary

sexual reproduction
asexual reproduction
vertebrate
invertebrate
radial symmetry
bilateral symmetry
asymmetrical

Before You Read

In your Science Notebook, write your own definition of the word *animal*. Make a drawing of an animal, and list three physical characteristics that make it an animal.

Think of some of the living things you see on your way to school each day, such as people, birds, dogs, cats, and insects. What do all of these organisms have in common? You would be correct if you said that they are all different types of animals. The animal kingdom includes an amazing variety of living things, from tiny insects to giant jellyfish to giraffes to people like you.

Characteristics of Animals

All animals have characteristics that separate them from bacteria, fungi, protists, and plants. One of these characteristics is nutrition. Unlike plants, animals depend on other livings things for food. Some animals eat plants, some eat other animals, and some eat both plants and animals.

All animals are multicellular, which means they are made up of many cells. Your own body is made up of between 10 trillion and 100 trillion cells. An animal cell has a nucleus and other organelles, is surrounded by a cell membrane, and lacks a cell wall.

Most animals are motile, which means they are able to move from place to place. Moving around enables animals to escape from enemies and find important things, such as food, water, shelter, and mates.

Animals usually reproduce by sexual reproduction. **Sexual reproduction** occurs when a sperm cell and an egg cell join to form a new organism. Some animals, however, can reproduce asexually. **Asexual reproduction** occurs when an individual produces a new individual identical to itself.

Figure It Out

1. What physical characteristics do all of the animals shown have in common?

2. Identify ways in which these animals are different from one another.

Figure 12.1 From *(left to right)*, the monarch butterfly, sea slug, rhinoceros, and dolphin are all members of the animal kingdom.

Groups of Animals

The animal kingdom can be divided into two large groups: vertebrates and invertebrates. A **vertebrate** (VUR tuh brayt) is an animal that has a backbone. Humans, snakes, horses, tigers, and dogs are all examples of vertebrates. An **invertebrate** (ihn VUR tuh brayt) is an animal that does not have a backbone. Grasshoppers, earthworms, octopuses, and jellyfish are all invertebrates. Invertebrates make up about 95 percent of all animal species.

Animals can also be divided into groups based on their body plan, the way the features of the body are arranged. **Figure 12.2** shows the three basic body plans of animals. Animals with **radial** (RAY dee ul) **symmetry** have bodies that are arranged in a circle around a central point, the way spokes are arranged around the hub of a bicycle wheel. These animals have a top and a bottom, but no front, back, or head. If you were to draw an imaginary line across the top of a jellyfish, you would see that both halves look the same. You could draw the line in any direction and still see two halves that look alike.

Animals with **bilateral** (bi LA tuh rul) **symmetry** have bodies with two similar halves. They can be divided into right and left halves by drawing an imaginary line down the length of the body. Animals with bilateral symmetry have a front side, back side, head end, and tail end. If you draw an imaginary line down the middle of a lizard's body, you will see the same features on both sides of the line.

The simplest animals have no symmetry and are called **asymmetrical**. The prefix *a-* means "without." Sponges are asymmetrical.

CONNECTION: Geography

One of the characteristics common to most animals is their ability to move freely from one place to another. People of all cultures use animals such as horses, camels, elephants, llamas, and donkeys to transport goods. Each culture must use animals that are well adapted to the local climate.

After You Read

1. List three common characteristics of all animals.

2. How is a vertebrate different from an invertebrate? How does this difference explain the fact that the bodies of invertebrates are usually smaller than those of vertebrates?

3. Review the animal drawing in your Science Notebook. Identify the type of symmetry the animal has, and explain how this type of symmetry will help it survive in its environment.

As You Read

Look at the animal you drew in your Science Notebook.

Is your animal a vertebrate or an invertebrate? Explain your answer.

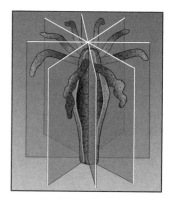

Radial symmetry

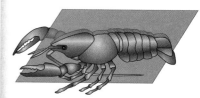

Bilateral symmetry

Asymmetry

Figure 12.2 The body plan of an animal helps it survive in its environment. What type of body plan do humans have?

12.2 Sponges

Before You Read

In your Science Notebook, create a concept map with the word *Sponge* in the center. Draw a circle outside the smaller circle. In the outside circle, write or draw any information that you predict describes the characteristics of a sponge.

Examine a kitchen sponge closely and you will see that it is filled with pores, or little holes. These pores are what give sponges their phylum name Porifera (puh RIH fuh ruh), which means "pore bearer." Your artificial kitchen sponge was most likely made by humans, but natural sponges are animals that live in water.

Characteristics of Sponges

Sponges are the simplest group of invertebrates. They are asymmetrical and are found in a variety of colors, shapes, and sizes. Most species of sponges live in the ocean, but a few species live in freshwater. As adults, sponges are not able to move. Because they show little or no movement, sponges were once classified as plants. Since sponges cannot conduct photosynthesis like plants do, they are now classified as animals.

A sponge has a hollow body shaped like a sac. The inside of the sponge's body contains two layers of cells. Moving water carries algae, tiny animals, and oxygen through the pores into the sponge. The sponge's cells remove the food and oxygen from the water in a process called **filter feeding**. The cells also release waste products into the water. The water and waste products leave the sponge through a larger opening at the top of its body.

Figure 12.3 The red beard sponge *(left)*, blue vase sponge *(center)*, and giant barrel sponge *(right)* vary greatly in color, size, and shape.

Structure of a Sponge

A sponge has a skeleton that supports its cells. Many sponges have skeletons made of small, sharp structures called **spicules** (SPIH kyewlz). Spicules are made of either a glasslike or a chalky substance. In other sponges, the skeletons are made of a softer, rubberlike substance. These sponges are the ones used as bath sponges. In still other sponges, the skeletons are made of both spicules and the rubberlike substance. A sponge's spiny skeleton helps protect it from predators. Some sponges release poisonous chemicals as a defense mechanism against predators that try to eat them.

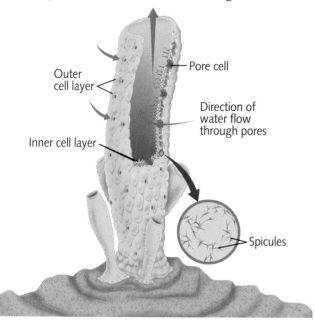

Outer cell layer

Pore cell

Direction of water flow through pores

Inner cell layer

Spicules

Figure 12.5 A sponge has no tissues, organs, or organ systems. The body of a sponge consists of two layers of cells surrounding a central cavity. The red arrows indicate the flow of water into and out of the sponge.

Figure 12.4 The skeleton of the glass sponge is made of thousands of glasslike spicules. What is one function of the spicules?

Reproduction in a Sponge

Sponges reproduce both asexually and sexually. Sponges reproduce asexually by **budding**. In budding, a small bulge grows on the side of a sponge. When this bulge falls off the parent, it grows into a new sponge. The ability of an organism to regrow body parts is called **regeneration** (rih je nuh RAY shun). Regeneration is common among simple animals.

Most sponges produce both sperm cells and egg cells, but at different times of the year. During sexual reproduction, sperm are released by the sponge and carried through the water to the eggs of another sponge. The fertilized egg develops into a swimming larva. The larva attaches itself to a rock or the ocean floor and develops into an adult sponge.

After You Read

1. Review the characteristics of sponges you included in the concept map in your Science Notebook. Why are sponges classified as animals?

2. How does a sponge take food and oxygen into its body?

3. How does a sponge reproduce?

Before You Read

Look at the photographs and diagrams in this lesson. In your Science Notebook, write two questions that this lesson might answer. Your questions should relate to something about cnidarians that you would like to know.

Have you ever had this experience? You and a friend are walking along the beach. You see a jellyfish and start to poke it with your foot. Your friend yells out, "Be careful! Don't touch that jellyfish because it might sting you!" A jellyfish is a type of animal called a cnidarian (ni DARE ee un). Like sponges, cnidarians are invertebrate animals that live in water.

Characteristics of Cnidarians

The phylum name Cnidaria comes from the Greek word for "nettle." A nettle is a plant with stinging cells along its surface. Cnidarians are animals that have stinging cells along their bodies. Being stung by a jellyfish or other cnidarian can be a very painful experience.

Cnidarians are more complex than sponges. A cnidarian's body has a hollow central cavity with an opening called the mouth. Most cnidarians have armlike structures called tentacles that surround the mouth. The **tentacles** (TEN tih kulz) contain stinging cells that help cnidarians capture food. The stinging cells paralyze or kill prey, and the tentacles then help get the prey into the mouth to be digested in the body cavity.

Cnidarian Body Plans

Cnidarians have two basic body plans: the polyp and the medusa. The **polyp** (PAH lup) is shaped like a vase and usually does not move about. A sea anemone is a type of polyp. The **medusa** (mih DEW suh, plural: medusae) is shaped like a bell and is free to swim around. A jellyfish is a type of medusa. Both polyps and medusae have radial symmetry.

Figure 12.6 A sea anemone *(left)*, jellyfish *(center)*, and the hydras *(right)* are cnidarians.

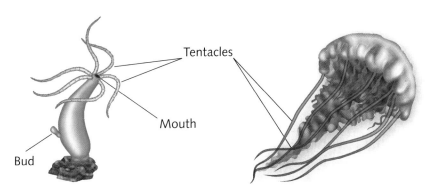

Tentacles

Mouth

Bud

Figure 12.7 Cnidarians have two basic body plans: the vase-shaped polyp *(left)* and the bell-shaped medusa *(right)*. The medusa's mouth is centered under the "bell."

Figure It Out

1. How does a cnidarian obtain its food?

2. Compare the location of the mouth in the polyp and in the medusa.

Reproduction in a Cnidarian

Cnidarians reproduce both asexually and sexually. Like sponges, cnidarians reproduce asexually by forming buds that fall off the parent and grow into new polyps. **Figure 12.7** shows a bud on the polyp body plan. Cnidarians can also reproduce sexually by producing eggs and sperm. Sperm are released into the water and fertilize eggs. Many cnidarians have a life cycle that includes both a polyp stage and a medusa stage.

Types of Cnidarians

There are three main classes of cnidarians: hydras, jellyfish, and sea anemones and corals. Hydras live in freshwater and spend their entire lives as polyps. Jellyfish spend most of their time as medusae.

Sea anemones and corals are polyps that live attached to rocks on the ocean floor. They are brightly colored and look like flowers. Sea anemones are larger than corals and do not have hard outer skeletons. Corals live in large groups called colonies and have hard outer skeletons. When the coral polyps die, their skeletons remain behind. Over time, the skeletons build up and form large, rocklike structures called reefs. Coral reefs provide homes, protection, and food for many sea animals.

Extend It!

In a small group, research the Great Barrier Reef in Australia and other coral reefs throughout the world. In your Science Notebook, write a letter to your local newspaper explaining why people should work to protect the world's coral reefs.

Figure 12.8 The Great Barrier Reef, off the coast of Australia, is the world's largest coral reef. It is 2,000 kilometers long and can be seen from space.

After You Read

1. What are the characteristics of cnidarians?

2. How are the two body forms of cnidarians the same? How are they different?

3. How do cnidarians reproduce?

4. Read the questions and answers you wrote in your Science Notebook. What changes would you make now that you have finished this lesson?

Learning Goals

- Describe how worms are different from cnidarians.
- Identify the three main phyla of worms.
- Compare the characteristics of flatworms, roundworms, and segmented worms.

New Vocabulary

parasite
anus

12.4 Worms

Before You Read

In your Science Notebook, create a chart with three columns. Label the first column *Flatworms*, the second column *Roundworms*, and the third column *Segmented Worms.* As you learn about these three kinds of worms, add to your chart information that describes their characteristics.

A flash of lightning wakes you up as a rainstorm moves into the area. The next morning, you notice that the streets and sidewalks are covered with earthworms! The rain has flooded the earthworms' underground homes, forcing them to escape by crawling to the surface.

Earthworms are just one kind of worm in the animal kingdom. There are three main phyla of worms based on shape: flatworms, roundworms, and segmented worms. Worms are soft-bodied invertebrates.

Worms differ from cnidarians in several ways. They have bilateral symmetry with a definite head and tail. Sense organs and nerves are found in the head. They have long, narrow bodies with no arms or legs. They can live in water and on land.

Flatworms

Worms with a flattened shape are called flatworms. They are grouped in the phylum Platyhelminthes (Pla tih hel MIHN theez). The word *Platyhelminthes* comes from two Greek words: *platy,* meaning "flat," and *helminth,* meaning "worm." Flatworms are the simplest of all worms. There are three main classes of flatworms. The planarian is a free-living, freshwater flatworm. Most planarians are only a few centimeters long and are harmless to humans.

Two classes of flatworms are parasites. A **parasite** is an organism that lives inside or on another organism. Parasites take food from this organism. The two classes of parasitic flatworms are the flukes and tapeworms. Flukes live in the body tissues of animals and cause sickness. Tapeworms live in the intestines of humans and other vertebrates. The tapeworms absorb the digested food of the animals in which they live.

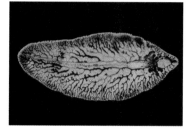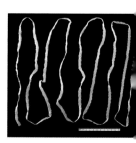

Figure 12.9 The planarian *(left),* the fluke *(center),* and the tapeworm *(right)* are all types of flatworms.

Figure 12.10 The trichina worm *(left)*, the *Ascaris lumbricoides (center)*, and the hookworm *(right)* are all types of parasitic roundworms.

Roundworms

If you have a cat or dog, you may at one time have had to get medicine from a veterinarian to protect your pet from heartworms. Heartworms are a type of parasitic roundworm. Roundworms make up the largest phylum of worms. Many different types of roundworms are parasites. Trichina, *Ascaris*, and hookworm are all types of parasitic roundworms that can infect humans. There are about 50 species of roundworms that are human parasites.

Roundworms belong to the phylum Nematoda (ne muh TOH duh). The word *nematode* means "threadlike." Roundworms have long, thin bodies that come to a point at both ends. They live on land or in water. Most free-living roundworms are tiny. There can be as many as one to ten million roundworms living in one cubic meter of soil or water! Some roundworms are helpful to farmers because they eat insects that feed on plants. Roundworms also act as decomposers by breaking down the bodies of dead plants and animals, which helps build rich soil.

Roundworms have complete, yet simple, digestive systems with two openings—a mouth and an anus. Food enters the body through the mouth and wastes leave the body through the **anus** (AY nus). This makes the digestive system of the roundworm more complex than that of the sponge or cnidarian.

Figure 12.11 This dog heart is infected with heartworms.

Figure It Out

1. Why is *nematode* a good name for this roundworm?

2. How might the heartworms keep the heart from working properly?

As You Read

Use the information you have learned to describe flatworms and roundworms.

What is a parasite?

CONNECTION: Health

Parasitic roundworms can cause serious disease in humans and other animals. *Trichinella* is a parasitic roundworm that lives in the muscle tissue of pigs and other mammals. People should never eat pork or game animals such as deer and wild hogs unless the meat is well cooked.

Explain It!

In your Science Notebook, write a public service announcement for a local television or radio station explaining the importance of preventing heartworm disease in dogs and cats.

Segmented Worms

A segmented worm has a body divided into many segments, or sections. The phylum Annelida (an NEL ud uh) contains the segmented worms, the most complex of the worms. The word *annelid* means "little rings" and describes the bodies of these worms. Earthworms and leeches belong to this phylum. Segmented worms can live in soil, freshwater, or the ocean.

Earthworms   The earthworm is the most familiar of the segmented worms. The body of an earthworm is divided into at least 100 segments. The earthworm has a slimy outer layer of mucus that helps it glide through the soil. It also has tiny, stiff hairs called bristles on its skin that help it move along the ground.

As earthworms burrow through the soil, they eat dead plants and animal bodies. They digest some of this food, but the rest is passed out through the anus into the soil. This produces fertilizer for the soil. Earthworm tunnels loosen the soil and allow air to move into it. This air helps plants grow.

Leeches   Leeches are another type of segmented worm. Some leeches are free-living and eat small invertebrates. Other leeches are parasites that attach to the skin of a vertebrate and feed on its blood.

Figure 12.12 Earthworms *(top)* are the most common annelid worms. Hundreds of earthworms can live under the surface of a square meter of soil. Most leeches *(bottom)* live in freshwater.

CONNECTION: Chemistry

Thousands of years ago, parasitic leeches were used by the ancient Egyptians for a medical procedure called bloodletting. It was believed that sick people contained "bad blood" and that leeches could remove it. Over time, doctors realized that illness was caused by many other factors and that leeches might be weakening and hurting patients. But, once again, leeches are popular with doctors.

Leech saliva contains a chemical called hirudin that keeps blood from clotting. Surgeons who reattach body parts such as fingers and toes find that hirudin helps blood flow through the body parts, which aids the healing process after surgery. Leech saliva also contains a chemical that acts like a natural anesthetic to lessen pain. Leeches can also be used to drain blood from a wound.

After You Read

1. What are the three main phyla of worms?

2. How are flatworms, roundworms, and segmented worms different from one another?

3. Look at the chart in your Science Notebook. Add to the chart to describe the characteristics of segmented worms.

12.5 Mollusks

Before You Read

In your Science Notebook, divide a page into three columns. Label the first column *Snail*, the second column *Clam*, and the third column *Octopus*. Draw a picture of each animal in the correct column. Predict some characteristics of each, and write them under the appropriate drawing.

Learning Goals

- Identify the common characteristics of mollusks.

- Describe how mollusks are divided into classes.

- Compare the characteristics of gastropods, bivalves, and cephalopods.

New Vocabulary

radula

If you have watched a snail move across the ground, collected shells on the beach, or eaten oysters, clams, or octopus, you are familiar with mollusks. The word *mollusk* comes from the Latin word meaning "soft." Mollusks are animals in the phylum Mollusca (mah LUS kuh).

Characteristics of Mollusks

Mollusks are soft-bodied invertebrates that usually have hard inner or outer shells. Mollusks live on land, in freshwater, and in the ocean. Like worms, mollusks have bilateral symmetry.

Mollusks are divided into three main classes based on whether or not they have a shell, the kind of shell they have, and the kind of foot they have. The foot of a mollusk is the part of its body used for movement. The three main classes of mollusks are the snails and slugs, the two-shelled mollusks, and the tentacled mollusks.

Snails and Slugs

Gastropods (GAS troh pahdz), which include snails and slugs, are mollusks with a single shell or no shell at all. The word *gastropod* means "stomach foot." Most gastropods move by using a foot found on the same side of their body as their stomach.

A snail is a gastropod with a single shell. A snail's head contains two pairs of tentacles. The longer pair of tentacles has eyes at each tip. The shorter pair helps the snail detect smells. A snail's head also contains a mouth and a jaw.

Most snails eat plants. A snail has a structure in its mouth called a radula. The **radula** (RA juh luh) is like a file that helps a snail scrape off small pieces from plants that it can easily swallow. As a snail moves, it leaves a trail of mucus behind. This mucus helps the snail's foot glide over different types of surfaces.

Slugs are gastropods that do not have shells. They protect themselves from predators by hiding under rocks and logs during daylight hours. Some slugs have chemicals in their bodies that are poisonous to predators.

Figure 12.13 A snail *(top)* and a sea slug *(bottom)* are mollusks called gastropods.

Explore It!

Working with a partner, use a hand lens to examine the shells of an oyster, a clam, a scallop, and a mussel. In your Science Notebook, describe how they are alike and how they are different. Add diagrams of the shells to support your descriptions.

As You Read

Review the characteristics of each mollusk you drew. Make corrections or additions to your list as you learn more about these animals.

What are the three main classes of mollusks?

Figure 12.14 The clam *(left)*, the scallop *(center)*, and mussels *(right)* are all types of bivalves. The shells of bivalves are held together by hinges and strong muscles.

Two-Shelled Mollusks

Mollusks that have two shells held together by strong muscles are called bivalves. The word *bivalve* means "two shells." Oysters, clams, scallops, and mussels are bivalves. All bivalves live in the ocean or in freshwater. Unlike a gastropod, a bivalve does not have a radula. Bivalves get their food by filter feeding, like sponges. As water moves into the shell of a bivalve, food particles stick to a mucus layer on its body.

Some bivalves stay in one place throughout their lives, and others move freely. Oysters and mussels attach themselves to rocks under the water. Clams and scallops move freely through the water. They clap their shells together, forcing water out of the shell. The movement of the water pushes them forward. Clams and scallops can also use their feet to bury themselves in sand or mud.

Many people enjoy bivalves as food. Bivalves can be cooked or eaten raw. Some bivalves, such as oysters, are valued for their ability to produce pearls. When a grain of sand gets into an oyster shell, it acts as an irritant. In response, the oyster secretes a substance to coat the sand particle so it will not rub against the oyster's body. Over time, the sand becomes completely covered with many layers of secretions, and a pearl is formed. Pearls are used to make jewelry.

Figure 12.15 A pearl was produced by this oyster. Sometimes, an oyster will contain more than one pearl.

CONNECTION: Physics

Newton's third law of motion states that for every action, there is an equal and opposite reaction. The movement of a clam or a scallop demonstrates Newton's third law of motion. As bivalve shells snap shut, water is forced out in one direction, moving the bivalve in the other direction. Octopuses and squids move in a similar way, using jets of water.

Tentacled Mollusks

The most complex of the mollusks are the tentacled mollusks, also called the cephalopods (SEHF uh loh pahdz). The word *cephalopod* means "head-footed." Cephalopods have large, well-developed heads. They also have tentacles that they use to capture food and to move themselves. The octopus, the squid, and the chambered nautilus are different types of cephalopods. A cephalopod may have an outer shell, an inner shell, or no shell at all.

Octopuses and Squids   The octopus has no shell. It has eight arms lined with suction cups to help it capture food. The octopus lives mostly on the ocean floor, usually crawling in search of food. The squid has an inner shell that runs along its back and keeps its body stiff. A squid has eight arms of equal size, plus two longer arms called tentacles. Squids are fast swimmers and usually stay in open water.

Both the octopus and the squid can move quickly to hunt or to avoid predators. They can squirt jets of water out of their bodies to move them through the water. This type of movement is called jet propulsion. Octopuses and squids also produce a dark-colored ink that they can release into the water. This ink helps them hide so that they can escape from predators.

Chambered Nautilus   The chambered nautilus has a coiled outer shell with many chambers, or rooms. By taking in or giving off gas from these chambers, the nautilus can move to different depths in the ocean. As a nautilus grows, it gains more living space by building new chambers connected to the old ones. It moves to hunt for food, but not as quickly as the octopus or the squid.

Figure 12.17 The chambered nautilus is a type of cephalopod that has an outer shell. How is it similar to the octopus and the squid?

Figure 12.16 The octopus *(top)* and squid *(bottom)* are examples of cephalopods.

After You Read

1. Review the characteristics of the mollusks you drew in your Science Notebook. In a well-developed paragraph, describe the three characteristics that are common to all mollusks.
2. How are mollusks divided into classes?
3. How are gastropods, bivalves, and cephalopods different from one another?

KEY CONCEPTS

12.1 Introduction to Animals

- Vertebrates are animals with backbones. Invertebrates are animals without backbones.
- The bodies of animals can show radial symmetry, bilateral symmetry, or asymmetry.

12.2 Sponges

- Sponges belong to the phylum Porifera. Their bodies are covered with pores.
- Sponges remove food and oxygen from water as it enters their pores.
- Sponges can reproduce asexually by budding. Sponges can reproduce sexually by the joining of a sperm cell and an egg cell.

12.3 Cnidarians

- Cnidarians have a hollow body cavity with one opening, called the mouth.
- Cnidarians have stinging cells on their tentacles. They use their tentacles to catch prey.
- Cnidarians have two body plans: the polyp and the medusa.

12.4 Worms

- Flatworms are members of the phylum Platyhelminthes. They have flat bodies and live in freshwater.
- Roundworms are members of the phylum Nematoda. Some nematodes are free-living. Some are parasites that live off other organisms.
- Segmented worms are members of the phylum Annelida. Earthworms and leeches are annelids.

12.5 Mollusks

- Mollusks are members of the phylum Mollusca. They have soft bodies, and many are protected by inner or outer shells.
- Most mollusks have a thick muscular foot or tentacles.

VOCABULARY REVIEW

Write each term in a complete sentence, or write a paragraph relating several terms.

12.1
sexual reproduction, p. 210
asexual reproduction, p. 210
vertebrate, p. 211
invertebrate, p. 211
radial symmetry, p. 211
bilateral symmetry, p. 211
asymmetrical, p. 211

12.2
filter feeding, p. 212
spicule, p. 213
budding, p. 213
regeneration, p. 213

12.3
tentacle, p. 214
polyp, p. 214
medusa, p. 214

12.4
parasite, p. 216
anus, p. 217

12.5
radula, p. 219

PREPARE FOR CHAPTER TEST

To prepare for the chapter test, create a question from each Learning Goal. Use the information in your Science Notebook to answer each question. Then use these answers to write a well-developed essay about the chapter. Use the Key Concept on the first page of this chapter as your topic sentence.

True or False
If the statement is true, write "true." If it is false, change the underlined word or words.

1. <u>Asexual reproduction</u> occurs when a sperm cell and an egg cell join to form a new individual.

2. Animals without backbones are called <u>vertebrates</u>.

3. Sponges remove food and oxygen from water by a process called <u>filter feeding</u>.

4. Cnidarians use <u>tentacles</u> to capture food.

5. An <u>earthworm</u> is a parasitic segmented worm that attaches to the skin of a vertebrate and feeds on its blood.

6. A snail is a type of <u>cephalopod</u>.

Short Answer
Answer each of the following in a sentence or brief paragraph.

7. What are the common characteristics of all animals?

8. Describe how sponges reproduce asexually and sexually.

9. Compare the two body plans of cnidarians.

10. How are parasites harmful to humans and other animals?

11. What methods do cephalopods use to move from one place to another?

Critical Thinking
Use what you have learned in this chapter to answer each of the following.

12. **Compare and Contrast** Identify ways in which sponges are similar to and different from cnidarians.

13. **Apply Concepts** Sponge farmers cut parent sponges into smaller pieces and throw the pieces back into the water. Why do they do this?

14. **Relate** A chemical is used to kill unwanted insects in a flower garden. This chemical also kills the earthworms. How might this affect the plants growing in the garden?

Standardized Test Question
Choose the letter of the response that correctly answers the question.

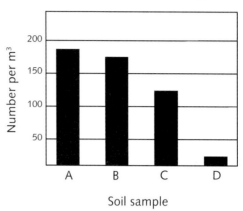

Earthworms and Soil Type

15. The graph shows the number of earthworms per cubic meter in different types of soil. What can you infer about soil sample D?

A. It is the perfect temperature for earthworms.

B. It contains plenty of food for earthworms.

C. It is unhealthy for earthworms.

D. It provides earthworms with protection from predators.

Test-Taking Tip

Resist the urge to rush, and don't worry if others finish before you. Use all the time you have. If you are able, clear your mind by closing your eyes and counting to five or taking another type of short break. Extra points are not awarded for being the first person to finish.

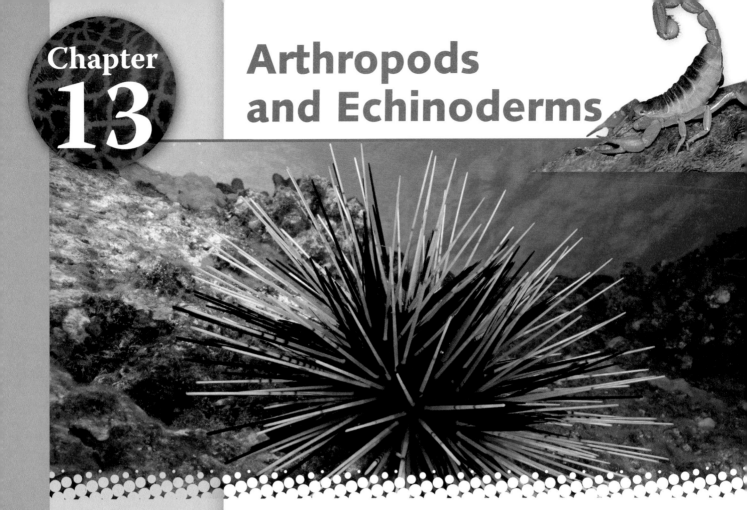

KEY CONCEPT Arthropods and echinoderms can be classified based on their physical characteristics.

Are you afraid of scorpions? Many people are. But did you know that some people keep scorpions as pets? Scorpions are members of the phylum of invertebrates called the arthropods.

Another phylum of invertebrates is the echinoderms. Echinoderms include the beautiful but poisonous Hawaiian sea urchin shown in the photo, as well as sand dollars, sea cucumbers, and sea stars. In this chapter, you will learn about these land and sea invertebrates.

Think About Making Observations

Each time you make an observation about your surroundings, you may think of questions that you would like to have answered. As you find the answers to these questions, you will learn more about the things around you.

- Insects make up the largest class of arthropods. In your Science Notebook, make a list of five insects you have observed around your home or yard and a separate list of five insects you have observed around your school.

- Look at your two lists. What insects did you add to both lists? What insects did you add to only one list? What do your observations suggest?

NSTA

SCI LINKS
THE WORLD'S A CLICK AWAY

www.scilinks.org
Invertebrates **Code: WGB13**

13.1 Arthropods

Before You Read

In your Science Notebook, draw a Venn diagram. Label the circle on the left *Spider*. Label the circle on the right *Crab*. In the left circle, list three physical characteristics of a spider. In the right circle, list three characteristics of a crab. In the overlapping area, list three physical characteristics that a spider and a crab have in common.

Learning Goals

- Name three characteristics of all arthropods.
- Describe the characteristics of arachnids and crustaceans.
- Compare the characteristics of centipedes and millipedes.

New Vocabulary

appendage
exoskeleton
molting
antenna

Recall Vocabulary

regeneration (p. 213)
parasite (p. 216)

If you have ever been bitten by a mosquito or frightened by a spider, you have been bothered by an arthropod. The arthropods (AR thruh pahdz) are the largest phylum of animals on Earth. So far, more than one million species of arthropods have been identified. Insects, spiders, ticks, and crabs are all examples of arthropods.

Arthropods have been on Earth for a long time. Some arthropod fossils are more than 500 million years old. Arthropods called trilobites (TRI luh bites) inhabited Earth's oceans for millions of years, where they lived successfully and in great numbers. Today, none of these early arthropods exist. They are extinct.

Figure 13.1 Trilobites were some of Earth's earliest arthropods.

Characteristics of Arthropods

All arthropods share some of the same structural characteristics. They all have jointed appendages. An **appendage** (uh PEN dihj) is a structure such as a leg, an arm, or an antenna that grows out of an animal's body. The phylum name Arthropoda comes from the Greek words meaning "jointed legs." An appendage with joints allows an animal to be more flexible as it moves or grabs prey. Think about how the joints in your fingers allow you to do many things.

Like segmented worms, arthropods also have segmented bodies. Because of this, some scientists believe that segmented worms and arthropods have a common ancestor.

All arthropods have a hard outer covering called an **exoskeleton**. The exoskeleton acts like a suit of armor to protect the arthropod. The exoskeleton does not grow as the animal grows. It must be shed and replaced with a new, larger exoskeleton. This process is called **molting**.

Figure It Out

1. How is an exoskeleton useful to an arthropod?

2. Infer why an arthropod would be in danger while it is molting.

Figure 13.2 In order to get bigger, arthropods must molt, or shed their exoskeletons.

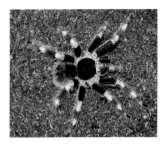

Figure 13.3 The Brazilian white knee tarantula *(left)*, deathstalker scorpion *(center)*, and American dog tick *(right)* are all examples of arachnids.

Arachnids

Spiders, scorpions, ticks, and mites belong to the class Arachnida (uh RAK ni duh). All arachnids have eight legs and two body regions: the head-chest region and the abdomen. The head-chest region contains the arachnid's sense organs, heart, and appendages. The abdomen contains the arachnid's reproductive, respiratory, and digestive organs.

Spiders

Does seeing a spider make you fearful or curious? Of the more than 35,000 species of spiders, only about a dozen are dangerous to humans. In North America, you need to look out for only the two species shown in **Figure 13.4**—the female black widow and the brown recluse.

Spiders usually feed on insects. Some large spiders can feed on small vertebrates such as mice, birds, lizards, and frogs. Spiders catch prey in different ways. Many make webs of a thin, strong, flexible substance called silk. Silk is released from glands in the spider's abdomen. Some spiders jump out and catch their prey by surprise. All spiders kill their prey by using fangs to inject poison into the prey's body.

Figure 13.4 The female black widow spider *(top)* is black with a red, hourglass-shaped mark on the lower side of her abdomen. The brown recluse spider *(bottom)* is brown with a violin-shaped mark on its head.

Legs A spider's four pairs of walking legs are located on the head-chest region.

Eggs Some spiders lay their eggs and never see their young. Others carry an egg sac around with them until the eggs hatch.

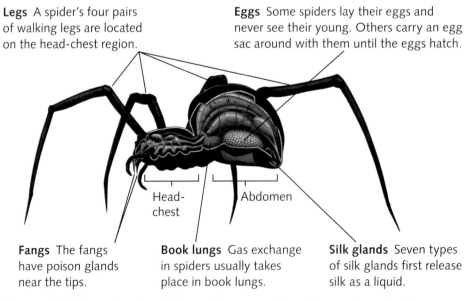

Head-chest

Abdomen

Fangs The fangs have poison glands near the tips.

Book lungs Gas exchange in spiders usually takes place in book lungs.

Silk glands Seven types of silk glands first release silk as a liquid.

Figure 13.5 The external and internal structures of a spider are shown in this diagram. How many pairs of legs does a spider have?

Spiders use special organs called book lungs to get their oxygen. They reproduce sexually. The female's eggs are fertilized by the male's sperm. Female spiders wrap their eggs in a silken sac or cocoon, where the eggs remain until they hatch. Internal and external structures of a spider are shown in **Figure 13.5**.

Scorpions, Ticks, and Mites

Scorpions live mainly in hot climates such as deserts, tropical rain forests, or grassy prairies. Scorpions are active at night. During the day, they hide under rocks or logs to stay cool. If you go camping, be careful to check your shoes in the morning. Sometimes scorpions hide inside them to stay warm.

A scorpion has a stinger at the end of its abdomen. It uses the stinger to inject venom into its prey. The sting of a scorpion is painful to humans, but it does not usually cause death.

A tick has a different body shape than that of other arachnids. The head and the abdomen of a tick are joined together. Ticks are parasites that live on animals and plants. Some ticks that attach to humans can cause disease. Lyme disease and Rocky Mountain spotted fever are two human diseases caused by ticks.

If chiggers have ever made you itch, you have experienced a mite. Like ticks, chiggers and other mites are parasites. Ear mites cause the ears of dogs and cats to itch. Dust mites feed on dead skin cells found in house dust and on bedding.

Figure 13.6 Scorpions are the only arthropods that give birth to live young. The young scorpions stay on the mother's back until they molt for the first time. How is this helpful to the young?

 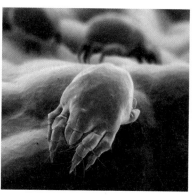

Figure 13.7 The deer tick *(left)* transmits Lyme disease to humans. The dust mite *(right)* feeds on dead skin cells.

⚡ Extend It!

Many people suffer from dust allergies caused by dust mites. Research the topic of allergies to dust mites. Find out what people can do in their homes to help reduce the number of dust mites. Record your findings in your Science Notebook.

🔵 CONNECTION: Health

People who have a dust allergy may be allergic to the exoskeletons and droppings of dust mites. Dust mites are found in large quantities in pillows, mattresses, carpeting, and upholstered furniture.

Figure 13.8 The krill *(left)*, lobster *(center)*, and pill bug *(right)* are all examples of crustaceans.

As You Read

Look at your Venn diagram. Make any additions or corrections to the list of characteristics shared by a spider and a crab and characteristics that make them different.

To what class do crabs belong?

Crustaceans

Have you ever eaten shrimp, crabs, lobsters, or crayfish? All of these animals belong to the class of arthropods called Crustacea (krus TAY shee uh). Most crustaceans live in water. Some, such as pill bugs, live in moist soil. Crustaceans are an important food source for many living things. Whales eat tiny crustaceans called krill.

Crustaceans have five or more pairs of appendages. Each body segment of a crustacean has appendages. Lobsters, crabs, and crayfish have claws as their first pair of appendages. These claws help crustaceans catch prey and protect themselves. Crustaceans also have two pairs of antennae. **Antennae** (singular: antenna) are sense organs used for taste, touch, smell, and balance.

Crustaceans obtain their food in many ways. Some eat dead plants and animals. Others eat prey they have captured with their claws. Some crustaceans eat small, plantlike organisms.

Most crustaceans take in oxygen through gills. Gills are organs used to absorb oxygen from water. Most crustacean species exhibit male and female sexes. They reproduce when the female's eggs are fertilized by the male's sperm.

Did You Know?

The stone crab shown in **Figure 13.9** lives in the warm water off the coast of Florida. If one of its claws is removed, the stone crab will regenerate, or regrow, the claw. Because stone crab claws are sold as food, people who fish for them remove one claw and return the animals to the water. The same stone crab can be caught again and have another claw removed in about one year. This helps prevent their extinction from overfishing.

Figure 13.9 Lobsters, crabs, and crayfish have claws. The stone crab *(left)* can regrow its claws if they are removed. While most crustaceans live in the ocean, the crayfish *(right)* lives in freshwater.

Centipedes and Millipedes

Centipedes and millipedes have long, wormlike bodies with many body segments. They also have many legs. Centipedes have one pair of legs on each body segment. Millipedes have two pairs of legs on each body segment.

Although the word *centipede* means "100 legs" in Latin, most centipedes have between 30 and 50 legs. Centipedes have flat bodies. They are carnivores, or animals that feed on the flesh of other animals. Centipedes actively hunt for their food and kill it using poisonous claws attached to the first segment of their bodies. They move quickly and inject their prey with deadly venom. Centipedes eat insects, snails, slugs, and worms.

Although the word *millipede* means "1,000 legs" in Latin, most millipedes have far fewer legs. A millipede has more body segments than a centipede does. Unlike centipedes, millipedes are not carnivores. They eat dead or decaying plant material. Millipedes are slow-moving animals that do not have claws.

If a millipede is attacked, it will roll up into a ball. The animal has stink glands that can give off a poisonous and bad-smelling substance. This helps drive away predators.

Figure 13.10 The giant desert centipede *(top)* is the largest centipede in the United States. It is usually about 18 to 20 centimeters in length and can live for up to five years. A millipede *(bottom)* has two pairs of legs per segment. A millipede's legs move in a wavelike motion as it walks.

⬤⬤⬤ CONNECTION: Art

In your Science Notebook, draw diagrams of what you think a centipede and a millipede look like. Make your diagrams resemble the diagram of a spider in Figure 13.5. Label each diagram. Using a marker, indicate characteristics that distinguish one animal from the other.

After You Read

1. What are the three common characteristics of all arthropods?

2. What are two ways in which a spider catches its prey?

3. How are antennae important to an arthropod?

4. According to your Venn diagram, in what ways are a spider and a crab alike? In what ways are they different?

Did You Know?

Several millipede species live with ants. The millipede lives in an ant nest and helps clean mold and dead plant material from the nest. When the ants move to a new location, the millipede goes with them.

Learning Goals

- List two reasons why insects are so successful.
- Describe the body structure of insects.
- Distinguish between complete and incomplete metamorphosis.
- Describe how insects are helpful and harmful.

New Vocabulary

thorax
metamorphosis
larva
pupa
nymph
camouflage

Before You Read

In your Science Notebook, create a concept map for *Insects*. Preview the lesson, and predict some characteristics of insects to include in your map.

You may have noticed that the largest group of arthropods was not discussed in the previous lesson. This group is the insects. Flies, ants, grasshoppers, and butterflies are all insects. There are more species of insects—more than one million—than of all other animals put together.

Figure 13.11 The grasshopper *(left)*, ant *(center)*, and butterfly *(right)* belong to the largest group of animals on Earth—the insects.

The Success of Insects

Scientists believe that insects appeared on Earth about 400 million years ago. They developed into an amazing variety of shapes, colors, and sizes. It is estimated that there are 200 million insects for every person on Earth! Insects are found in all of Earth's environments.

Specialized body parts are one reason that insects became such a successful group of animals. Different species of insects, for example, have different kinds of mouth parts. This allows insects to eat many types of food. Many insects are also able to fly. Thus, they can more easily hunt for food, escape from predators, and move to new places to live.

Figure 13.12 A light body allows a water strider *(left)* to walk on the surface of water. The long tube at the end of a butterfly's mouth *(center)* allows it to drink nectar from flowers. The wings on a ladybug *(right)* allow it to move easily from one place to another.

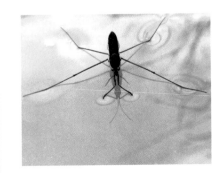

Insect Structure

Insects have three main body sections: a head, a thorax, and an abdomen. The **thorax** is the middle, or chest, section of the insect's body. The head contains the eyes and antennae. Insects have two kinds of eyes: simple and complex. Simple eyes contain only one lens and are used to detect light. Compound eyes contain many lenses. They help the insect detect movement.

The insect's wings and legs are attached to the thorax. Insects have six legs. In jumping insects, such as grasshoppers, one pair of legs is larger than the others. Most insects have one or two pairs of wings.

The abdomen of an insect contains many of its organs. The insect has a system of tubes that carry oxygen through the exoskeleton and into the body. These tubes also remove carbon dioxide from the body.

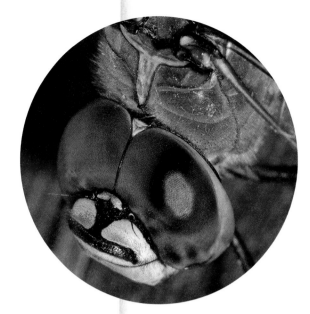

Figure 13.13 The compound eyes of this dragonfly help it detect movement. This allows the dragonfly to catch its prey and avoid predators.

Legs Insects have six legs. By looking at an insect's legs, you can sometimes tell how it moves about and what it eats.

Eyes Grasshoppers have two compound eyes and three simple eyes.

Antennae Insects have one pair of antennae, which is used to sense vibrations and food in the environment.

Abdomen   Thorax   Head

Wings Most grasshoppers have two pairs of wings. Both pairs contribute to flight. As a grasshopper flies, the wings push down to give the insect "lift" and move it forward.

Tubes These tubes carry oxygen throughout the insect's body.

Nervous System Grasshoppers, like other insects, have a complex nervous system that includes a brain and nerve cells.

Figure 13.14 The external and internal structures of a grasshopper are shown in this diagram. How does a grasshopper's number of legs compare with that of a spider?

 Explore It!

Use a hand lens to examine the insects given to you by your teacher. How would you describe the exoskeletons? How many legs, wings, body sections, and antennae does each insect have? Record your observations in your Science Notebook.

Growth and Development of Insects

All insects reproduce sexually. In most species, the male's sperm fertilize the female's eggs inside her body. As a young insect grows, it undergoes a series of changes called **metamorphosis** (me tuh MOR fuh sihs). There are two types of metamorphosis: complete and incomplete.

Most insects, including flies, beetles, bees, moths, and butterflies, go through complete metamorphosis. There are four stages in complete metamorphosis. The first stage is an egg. The egg hatches into the second stage, a larva (LAR vuh, plural: larvae). The **larva** is the wormlike stage of an insect. A larva spends most of its time eating.

Next, the larva enters the third stage, called the pupa stage. During the **pupa** stage, the larval tissues are broken down and replaced by adult tissues. The pupa stage of a moth is protected by a structure called a cocoon. The last stage is the adult insect that comes out of the cocoon.

❶ Insects begin life as a fertilized egg. The egg hatches into a larva.

❷ Larvae eat huge amounts of food to supply the energy needed for tremendous growth.

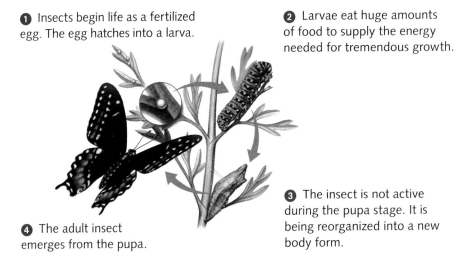

❸ The insect is not active during the pupa stage. It is being reorganized into a new body form.

❹ The adult insect emerges from the pupa.

Figure 13.15 During complete metamorphosis, insects change from an egg to an adult.

Incomplete metamorphosis has three stages. The first stage is an egg. The egg hatches into the second stage, a young insect called a **nymph**. A nymph looks much like the adult insect. The nymph grows into the last stage, the adult. A grasshopper goes through incomplete metamorphosis.

Explain It!

Imagine that you are a butterfly. In your Science Notebook, describe your experience as you develop from an egg into an adult. Describe each stage of your development. Include diagrams if you wish.

Figure It Out

1. Why does a larva eat so much?

2. Compare the activity of an insect in the larva stage with the activity of the insect in its pupa stage.

As You Read

Use what you have learned in this lesson to make corrections or additions to your concept map.

What are the two types of insect metamorphosis?

Figure 13.16 During complete metamorphosis, a Monarch butterfly larva (left) becomes a pupa (center). A grasshopper nymph (right) grows from an egg during incomplete metamorphosis.

Insect Behavior

Insects have a wide range of behaviors. Some behaviors help insects survive. Most insects live by themselves. They compete with each other for food, water, and mates.

Some insects, such as ants and bees, live and work together in colonies. Each member of an insect colony has a specific job. In an ant colony, the female worker ants build and defend the nest and gather food. They also care for the young ants. The queen ants lay the eggs, and the male ants fertilize the eggs.

Insects have many defenses against predators. Bees and wasps defend themselves with painful stings. Stinkbugs taste or smell bad to predators. Some insects have body colors or shapes that help them blend in with their surroundings. This adaptation is called **camouflage** (KA muh flahj).

CONNECTION: Chemistry

One way insects can communicate with each other is by releasing special chemicals called pheromones. An insect that finds food may drag its abdomen along the ground and release a pheromone as it returns to the colony. Other insects can follow this scent back to the food source.

Figure 13.17 Worker ants *(left)* protect the larvae in the colony. A bee *(center)* uses its stinger for protection. A praying mantis *(right)* uses camouflage to blend in with its surroundings.

PEOPLE IN SCIENCE Karl von Frisch 1886–1982

Karl von Frisch was an Austrian scientist who studied insect behavior. He won the Nobel Prize in Medicine and Physiology in 1973 for his studies of bees. In his early work, von Frisch showed that honeybees can see colors. He also found that bees can distinguish among dozens of similar smells.

Von Frisch also found that honeybees use dances to give other bees information about the distance and direction to a food source. Bees use a round dance to show other bees that a food source is a short distance (up to about 50 meters) away from the hive. Bees use a zigzag or waggle dance to show other bees that food is farther away. Bees will also angle their bodies during this dance to show other bees the direction in which they should travel to find the food.

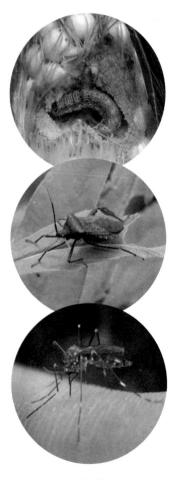

Figure 13.19 The corn earworm *(top)* and the squash bug *(center)* are insects that damage crops. The mosquito *(bottom)* is a parasite that can transmit diseases to humans.

Insects and Humans

What is your reaction when you see an insect? Many people do not like insects. They believe it is better to kill them than to leave them alone. Although some insects are harmful to people, many are helpful.

Bees produce honey and beeswax. They also help plants reproduce by carrying pollen from one plant to another. Honeybees pollinate many commercial crop plants. Without pollination, plants such as apples, peaches, cherries, pumpkins, squash, and tomatoes would not produce the fruits we eat.

Silk comes from the cocoons of silkworms. Silk thread is the strongest natural fiber. It is also elastic, which allows fabric made from it to spring back into shape when stretched.

Insects also eat other insects that can destroy plants. Ladybugs feed on aphids that can destroy many crops, for example. Some insects feed on dead plant and animal material. The dung beetle breaks down cow droppings into substances that make the soil good for planting.

Substances from some insects have been used in medicines for arthritis. Blowfly larvae, for example, are used to treat infections. These larvae feed on dead tissue and secrete a substance that helps the healthy tissue heal.

Some insects are harmful to humans. Species such as grasshoppers and beetles can do major damage to crops. Mosquitoes, flies, and fleas can carry small organisms that cause disease. A mosquito is a parasite that can transmit malaria and West Nile virus to humans.

Figure 13.18 Examples of useful insects include the silkworm *(left)*, dung beetle *(center)*, and blowfly larvae *(right)*.

After You Read

1. What are two reasons why insects are so successful?

2. How are complete metamorphosis and incomplete metamorphosis alike? How are they different?

3. What are two ways that insects protect themselves from predators?

4. How are insects helpful to humans? How are they harmful?

5. Use your concept map to describe three characteristics that all insects have in common.

13.3 Echinoderms

Learning Goals

- Identify the common characteristics of echinoderms.
- Describe the body structure of a sea star.
- Compare the characteristics of sea stars, sea urchins, sea cucumbers, and sand dollars.

New Vocabulary

endoskeleton
water-vascular system
tube feet

Recall Vocabulary

radial symmetry (p. 211)

Before You Read

Use the lesson title and headings to create a lesson outline. Label the title with the Roman numeral *I*, and label the headings with the letters *A*, *B*, and *C*. Under each heading, record important information from the lesson.

If you have walked on the beach, it is likely that you have seen a sea star. A sea star belongs to a phylum of animals called the echinoderms (ih KI nuh durmz). Sea stars, sea urchins, sea cucumbers, and sand dollars are echinoderms. Echinoderms live in Earth's oceans. Examples of echinoderms are shown in **Figure 13.20**.

Figure 13.20 A sea star *(left)*, a sea urchin *(center)*, and a feather star *(right)* are all members of the phylum Echinodermata.

Characteristics of Echinoderms

Like humans, echinoderms have an **endoskeleton**, or internal skeleton. The endoskeleton is made of calcium-containing plates from which the spines protrude. The plates and spines are covered with a thin layer of skin for protection. The word *echinoderm* comes from the Greek words meaning "spiny skin."

The adult bodies of echinoderms have radial symmetry. They often consist of five or more arms coming out of a central body. The larval stages of echinoderms have bilateral symmetry. All echinoderms have a **water-vascular system**. This is a system of water-filled tubes that runs through the echinoderm's body. This system functions in obtaining food and oxygen, removing wastes, and helping the echinoderm move.

Thousands of tube feet are connected to the water-vascular system. **Tube feet** act like suction cups. They help the echinoderm move and hold on to captured food.

Figure 13.21 The tube feet of an echinoderm help it gather food and move.

As You Read

Share your outline with a partner, and make any necessary additions or corrections.

How does a sea star obtain and eat its food?

Figure It Out

1. What is the function of the sea star's tube feet?

2. Compare the physical characteristics of a sea star with those of a spider.

🔍 Explore It!

Design and conduct an experiment to show what types of surfaces are best to use with suction cups. In your Science Notebook, describe how an echinoderm's tube feet are similar to suction cups.

Figure 13.23 This sea star is regenerating the rest of its body from one arm.

Sea Stars

Sea stars are commonly called starfish. Sea stars are not fish, but they do have a star shape. A sea star has five or more arms extending from a central body. The undersides of the arms are covered with tube feet.

Sea stars are hunters that eat mollusks such as clams, oysters, and mussels. A sea star uses its arms and tube feet to catch prey. The sea star grabs an oyster with its arms. It holds on to the closed shell with its tube feet and pulls. When the shell opens, the sea star pushes its stomach through its mouth and into the opening in the shell. Digestive chemicals break down the oyster's body, and the sea star has a tasty meal!

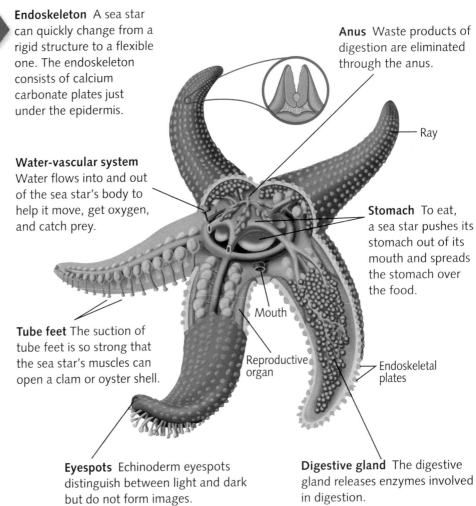

Endoskeleton A sea star can quickly change from a rigid structure to a flexible one. The endoskeleton consists of calcium carbonate plates just under the epidermis.

Anus Waste products of digestion are eliminated through the anus.

Ray

Water-vascular system Water flows into and out of the sea star's body to help it move, get oxygen, and catch prey.

Stomach To eat, a sea star pushes its stomach out of its mouth and spreads the stomach over the food.

Tube feet The suction of tube feet is so strong that the sea star's muscles can open a clam or oyster shell.

Mouth

Reproductive organ

Endoskeletal plates

Eyespots Echinoderm eyespots distinguish between light and dark but do not form images.

Digestive gland The digestive gland releases enzymes involved in digestion.

Figure 13.22 This diagram shows the external and internal structures of a sea star.

People who fish for mollusks do not like the sea stars that eat them. Years ago, sea stars found in mollusk beds were chopped into many pieces and thrown back into the ocean. People did not realize that, like other invertebrates, sea stars can regenerate body parts, as **Figure 13.23** shows. Chopping them up only resulted in more sea stars and fewer mollusks!

Other Echinoderms

Brittle stars look like sea stars, but their arms are long and whiplike. If you tried to pick up a brittle star, its arms would break off. This is a useful adaptation for getting away from predators. Brittle stars eat dead or decaying plant and animal bodies. Brittle stars move by waving their arms against the ocean bottom.

Sea cucumbers have long, flexible bodies. They look very different from sea stars and brittle stars. Sea cucumbers have rows of tube feet on their underside. This allows them to crawl along the ocean bottom. They are filter feeders that use tentacles to sweep food into the mouth. Feather stars are also filter feeders. These echinoderms have short bodies and many arms that look like feathers.

Long, sharp spines cover and protect the bodies of sea urchins. Some sea urchins have poison in their spines, which can deliver painful stings. Sea urchins have five teeth that scrape off and chew algae for food. Sea urchins use their spines and tube feet to move from one place to another.

Sand dollars have no arms. They look like large coins. Their bodies are flat and covered with very short spines. The spines help them burrow into the sand. Sand dollars usually eat tiny particles of food that float in the water.

Figure 13.24 A brittle star *(top)* looks like a sea star with longer arms. A sea cucumber *(bottom)* uses its tentacles to gather food.

Figure 13.25 A sea urchin *(left)* is protected by long spines. Sand dollars *(right)* use their short spines to burrow into the sand.

After You Read

1. What are three common characteristics of echinoderms?

2. Describe the function of an echinoderm's water-vascular system.

3. Explain how sea stars and brittle stars are alike. How are they different?

4. Review the information in your outline. In a well-developed paragraph, describe the adaptations echinoderms have that allow them to catch prey and escape from predators.

Did You Know?

The sea cucumber has an unusual defense system. If attacked by a predator, the sea cucumber will eject its internal organs. The organs are sticky, and they surround the predator while the sea cucumber escapes. The missing organs are quickly regenerated.

Summary

KEY CONCEPTS

13.1 Arthropods

- All arthropods have jointed appendages, segmented bodies, and a hard exoskeleton.

- Arachnids, including spiders, scorpions, ticks, and mites, have two body regions and eight legs.

- Shrimp, crabs, lobsters, and crayfish are crustaceans. Crustaceans have five or more pairs of appendages and two pairs of antennae. Many crustaceans have claws.

- Centipedes and millipedes have long, wormlike bodies with many segments and many legs. Centipedes are active hunters. Millipedes eat dead or decaying plant material.

13.2 Insects

- Insects have three main body sections: a head, a thorax, and an abdomen. They have six legs.

- As insects grow, they undergo a series of changes called metamorphosis. Complete metamorphosis has four stages: egg, larva, pupa, and adult. Incomplete metamorphosis has three stages: egg, nymph, and adult.

- Some insects live by themselves. Other insects, such as ants and bees, live in colonies.

- Insects can be helpful by making products humans use and by eating other insects that destroy crops. Insects can be harmful by eating crops and by causing disease.

13.3 Echinoderms

- Echinoderms have spiny skin and an endoskeleton.

- Echinoderms use a water-vascular system and tube feet to catch prey and to move.

- Both the sea star and the brittle star have five or more arms extending from a central body. Sea stars eat clams, oysters, and mussels.

- Sea cucumbers have long, flexible bodies. Feather stars have many featherlike arms. Sea urchins are covered with spines. Sand dollars have no arms and look like large coins.

VOCABULARY REVIEW

Write each term in a complete sentence, or write a paragraph relating several terms.

13.1
appendage, p. 225
exoskeleton, p. 225
molting, p. 225
antenna, p. 228

13.2
thorax, p. 231
metamorphosis, p. 232
larva, p. 232
pupa, p. 232
nymph, p. 232
camouflage, p. 233

13.3
endoskeleton, p. 235
water-vascular system, p. 235
tube feet, p. 235

PREPARE FOR CHAPTER TEST

To prepare for the chapter test, create a question from each Learning Goal. Use the information in your Science Notebook to answer each question. Then use these answers to write a well-developed essay about the chapter. Use the Key Concept on the first page of this chapter as your topic sentence.

True or False

If the statement is true, write "true." If it is false, change the underlined word or words.

1. An <u>exoskeleton</u> is a structure such as a leg or an arm that grows out of an animal's body.

2. <u>Antennae</u> are sense organs used for taste, touch, smell, and balance.

3. The <u>abdomen</u> is the middle, or chest, section of the insect's body.

4. The second stage of complete metamorphosis is called a <u>pupa</u>.

5. A <u>water-vascular system</u> is a system of water-filled tubes running through an echinoderm's body.

6. Spines cover and protect the bodies of <u>sea stars</u>.

Short Answer

Answer each of the following in a sentence or brief paragraph.

7. What are three common characteristics of all arthropods?

8. Describe the process of molting.

9. Explain why insects are so successful.

10. Compare the processes of complete metamorphosis and incomplete metamorphosis.

11. How does a sea star eat a clam?

Critical Thinking

Use what you have learned in this chapter to answer each of the following.

12. **Compare and Contrast** What are three ways in which arthropods and echinoderms differ from each other?

13. **Apply Concepts** Stone crab hunters remove one claw from a crab and return the crab to the water to grow a new claw. Explain why the crab hunters leave the crabs with one good claw.

14. **Classify** Your friend thinks that a spider is an insect. How would you explain why spiders are not classified as insects?

Standardized Test Question

Choose the letter of the response that correctly answers the question.

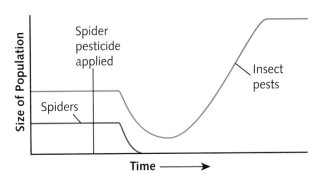

Effects of a Pesticide on Spiders and Insects

15. A pesticide is a chemical that is used to kill certain arthropods. The graph shows how a spider pesticide affects a spider population and an insect population. Which is most likely the reason for the change in the insect population after a spider pesticide is used?

   A. The insects eat the pesticide as food.

   B. The insects eat the spiderwebs as food.

   C. The pesticide causes the insects to reproduce more quickly.

   D. The spiders die from the pesticide and do not eat the insects.

Test-Taking Tip

Avoid changing your answer unless you have read the question incorrectly. Usually, your first choice is the correct choice.

Fishes, Amphibians, and Reptiles

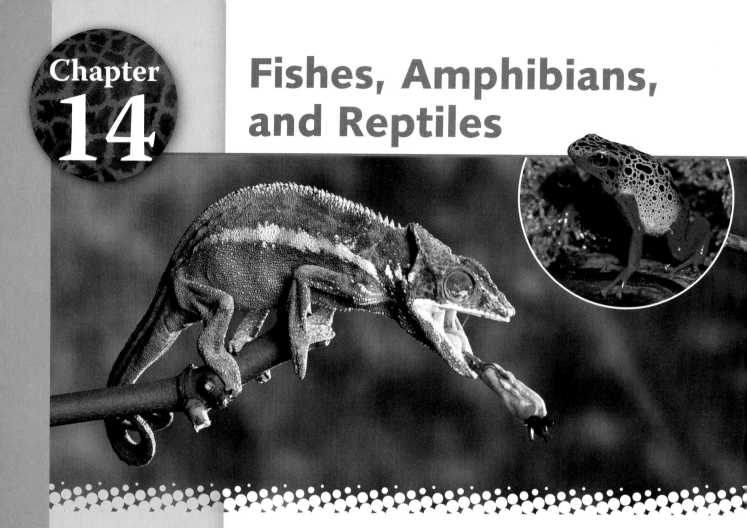

KEY CONCEPT Fishes, amphibians, and reptiles can be classified based on their physical characteristics.

The chameleon lizard sits quietly on a branch, hungry for its next meal. With lightning speed, the chameleon's tongue shoots out of its mouth. The sticky tongue captures a fly and pulls it into the chameleon's mouth.

A chameleon is a type of animal called a reptile. This chapter is about reptiles and two other kinds of animals, the fishes and the amphibians. The blue poison arrow frog is an amphibian. It has a deadly chemical in its skin that people use to make poison arrows. This chemical produces almost instant death in some animals.

Think About Classifying Animals

When things are classified, they are put into groups based on their characteristics. Scientists classify animals into groups.

- Look at the headings of the lessons in this chapter. Think about what you already know about fishes, amphibians, and reptiles.

- In your Science Notebook, describe the characteristics you are familiar with for each of these groups of animals. List some animals that you know belong in each group.

NSTA

SCi LINKS
THE WORLD'S A CLICK AWAY

www.scilinks.org
Vertebrates Code: WGB14

14.1 What Is a Vertebrate?

Learning Goals

- Describe the common characteristics of chordates.
- Describe the common characteristics of vertebrates.
- Distinguish between ectothermic and endothermic animals.
- Discuss the origin of the vertebrates.

New Vocabulary

vertebra
cartilage
notochord
ectotherm
endotherm

Recall Vocabulary

vertebrate (p. 211)
endoskeleton (p. 235)
exoskeleton (p. 225)

Before You Read

Read the lesson title, headings, and Learning Goals. Look at the photos and diagrams. Predict what you think you will learn in this lesson. Write two or three sentences that describe your predictions in your Science Notebook.

Try this: Place your hand on your back at waist level and press gently. Move your hand up and down about five centimeters from that point. You should feel bony protrusions, or "bumps." The bumps are on the bones that make up your spinal column. The bones that go up and down your back are called **vertebrae** (VUR tuh bray, singular: vertebra). These bones are joined together with **cartilage** (KAR tuh lihj), a tissue that is softer than bone. Cartilage is very flexible and strong. This strong column of bone and cartilage is your backbone.

Humans aren't the only animals with backbones. Fishes, frogs, snakes, and birds also have backbones. An animal with a backbone is called a vertebrate. Vertebrates belong to the phylum Chordata (kor DAH tuh). Members of the phylum Chordata are called chordates.

In addition to the vertebrates, two other groups of animals are classified as chordates. Examples of these animals, the tunicates and lancelets, are shown in **Figure 14.1**. Tunicates are also called sea squirts. They live in the ocean attached to objects and are filter feeders. Lancelets, which are also filter feeders, look like small, thin fish that lack fins. They bury themselves in the sand with their heads sticking out to catch prey.

Characteristics of Chordates

All chordates—tunicates, lancelets, and vertebrates—have certain characteristics in common. They all have a hollow nerve cord. The nerve cord is a tube of nerves located near the animal's back. The nerve cord is protected by a notochord. The **notochord** (NOH tuh kord) is a flexible, rodlike structure made up of large, fluid-filled cells and stiff, fibrous tissues that supports the animal's back. In vertebrates, the notochord is replaced by the backbone.

At some point in their lives, all chordates have several pairs of pharyngeal (fuh RIN jee ul) pouches just behind the mouth. In many vertebrates, including humans, the pharyngeal pouches become parts of the ears, the jaws, and the throat. The pharyngeal pouches of fish develop into sets of gills. Fish use their gills to take in oxygen and give off carbon dioxide.

Figure 14.1 Tunicates *(top)* and lancelets *(bottom)* are two types of chordates that are not vertebrates.

1. What are the functions of an endoskeleton?

2. Compare these two endoskeletons. How are they alike, and how are they different?

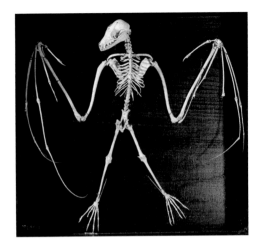

Figure 14.2 The endoskeletons of a frog *(left)* and a bat *(right)* support their bodies.

Characteristics of Vertebrates

The backbone of a vertebrate runs down the middle of its back. The vertebrate's backbone is part of its endoskeleton. An endoskeleton is an internal skeleton that supports and protects the body. It also gives the muscles a place to attach. The endoskeleton of vertebrates includes the skull and the ribs. The skull protects the brain, while the ribs protect the heart, lungs, and other internal organs.

As you may recall from Chapter 13, arthropods have an exoskeleton. As an arthropod grows, it must leave its old exoskeleton and produce a new, larger one. By contrast, the endoskeleton of a vertebrate grows as the body of the vertebrate grows. Animals with endoskeletons can grow larger than animals with exoskeletons or no skeletons.

Some vertebrates, such as fishes, amphibians, and reptiles, have body temperatures that change with the environment. This kind of animal is called an **ectotherm** (EK tuh thurm). Look at the lizard in **Figure 14.3**. It will sit in the sunlight on a cool morning to warm itself. When it gets too hot, the lizard will move to a cooler spot in the shade.

Unless you have a fever, your body temperature stays the same most of the time. Birds and mammals are called endotherms. An **endotherm** (EN duh thurm) is an animal that maintains a constant body temperature.

As You Read

Look at the sentences you wrote in your Science Notebook. Add information to expand or correct your predictions.

What is an endotherm?
What is an ectotherm?

Figure 14.3 This lizard is an ectotherm. Its body temperature will change with the temperature of the environment. The lizard will use the morning sunlight to heat up its body. It will move to the shade if it gets too hot. What is an animal that has a constant body temperature called?

🔴 CONNECTION: Health

A device called a thermostat helps control the temperature of your house. A part of your brain called the hypothalamus acts like a thermostat for your body. It keeps your body temperature the same whether the air temperature is hot or cold. The hypothalamus is found in the center of the brain and is about the size of an almond. This region of the brain is part of the endocrine system of the human body.

Origin of Vertebrates

Scientists use fossils to determine how long animals have been on Earth. Fossil evidence suggests that vertebrates have been on Earth for more than 500 million years. The first vertebrates developed from small, fishlike chordates that lived in water. Many different kinds of fishes developed from these early vertebrates.

Over time, some new fish species developed and had new characteristics that allowed them to move from water to land. One of these new characteristics was strong fins that could be used to crawl on land. They also had lungs to bring oxygen into their bodies. These animals were the first amphibians. **Figure 14.4** shows that reptiles developed from ancient amphibians. Mammals and birds developed from ancient reptiles.

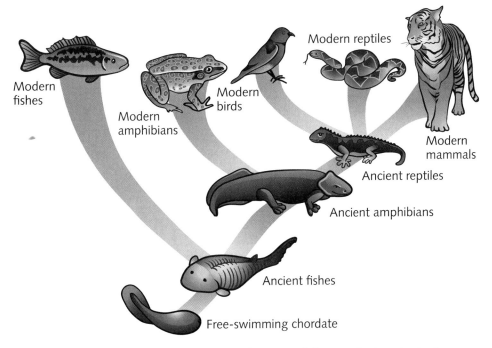

Figure 14.4 This diagram shows the development of the vertebrates. Ancient fish species that lived more than 500 million years ago could be the ancestors of all vertebrates on Earth today.

Explore It!

Use beads and string to make a model of a vertebrate backbone. Tie a large knot at one end of a piece of string. Slide the beads onto this string. Tie another large knot at the other end of the string. Gently bend the string of beads at different places and record your observations in your Science Notebook.

Extend It!

Tunicates and lancelets are members of the phylum Chordata. Use the library or the Internet to find out about these animals. Record your findings in your Science Notebook. Include labeled diagrams.

After You Read

1. Describe three common characteristics of all chordates.

2. What is a vertebrate?

3. How is an ectotherm different from an endotherm?

4. Look at the sentences you have written in your Science Notebook. Use the information to summarize this lesson.

Learning Goals

- Describe the main characteristics of fishes.
- Distinguish among the three groups of fishes.
- Identify the functions of the parts of bony fishes.

New Vocabulary

scavenger
swim bladder

Figure 14.5 The anemone fish, or clownfish, hides in the tentacles of a sea anemone. The sea anemone's tentacles are poisonous to other fish but not to the anemone fish.

Extend It!

Lampreys were accidentally introduced into the Great Lakes during the early 1800s. Use the library or the Internet to research the effects of the lampreys. Include a description of the efforts being made to control the lamprey population.

Before You Read

Write three facts that you already know about fish in your Science Notebook. Leave six lines of space below each fact. As you read the lesson, add at least two more facts to each fact you already know.

Have you ever been inside a pet store? Pet stores usually have aquariums, or large water-filled glass containers that hold many species of fish. Fishes come in many colors, shapes, and sizes. There are more than 25,000 species of fishes. Fish are an important source of food for many people.

Characteristics of Fishes

Most fishes share four common characteristics. In addition to being ectotherms, fishes have scales, fins, and gills. The scales on a fish protect it from predators. The fins help the fish move through the water. Gills allow the fish to bring oxygen into its body. The gills also remove carbon dioxide from a fish's body.

Fishes can be divided into three main groups. These groups are the jawless fishes, the cartilaginous fishes, and the bony fishes.

Jawless Fishes

The group known as jawless fishes include the lampreys and the hagfishes. Both have soft bodies covered with a slimy skin instead of scales. Lampreys and hagfishes look very different from most fishes. Most lampreys are parasites. A lamprey has sharp teeth that hook onto the body of a fish so that the lamprey can suck out the fish's blood and other body fluids.

A hagfish is a scavenger. A **scavenger** (SKA vun jur) is an animal that feeds on dead or dying animals. The hagfish has tentacles around its mouth. It is almost blind and uses its tentacles to find food.

Figure 14.6 The lamprey *(left)* is a parasite that feeds on other fishes. The hagfish *(right)* uses its tentacles to find food.

CONNECTION: Earth Science

Because they are made of cartilage, shark skeletons rarely turn into fossils. The cartilage decomposes quickly, and the tissue is not replaced by minerals. Shark teeth are hard and turn into fossils easily. Fossil teeth are often the only evidence of ancient sharks. Some ancient sharks grew quite large—more than 20 meters long—and their fossil teeth measure 10 centimeters or more.

Cartilaginous Fishes

Move the outer part of your ears or nose gently with your fingers. The ears and nose hold their shapes, but they are flexible. This is because they contain cartilage. Cartilaginous (kart uhl AJ uh nuhs) fishes have skeletons made of cartilage instead of bone. This group of fishes includes sharks, rays, and skates.

Although many people think of sharks as fast-swimming, scary predators, this is not the case. Most sharks are actually shy animals that prefer to be left alone. Fewer than ten species of sharks are considered dangerous to humans. More people die each year from bee stings than from shark attacks.

Most sharks are carnivores that spend much of their time hunting prey. They have sharp teeth and powerful jaws. Most sharks have five to 15 rows of teeth in each jaw. As the teeth in the front rows break off, new teeth from the back replace them. Sharks replace their teeth all their lives. A shark may go through 30,000 teeth in its lifetime!

Rays and skates have flat, wide bodies with long, thin tails. Rays and skates travel over the ocean floor in search of food. Most are harmless to humans, but some have poisonous spines on their tails to scare away predators. Some species can stun small fish with an electric shock. Stepping on one of these rays or skates could be a very unpleasant experience!

Figure 14.7 The skeleton of the great white shark *(top)* is made of cartilage. The shark jaw *(bottom)* contains many rows of sharp teeth. What is cartilage?

Did You Know?

Sharks do not blink their eyes. They have upper and lower eyelids that do not move or close over their eyes. When biting prey, some sharks protect their eyes with a thin third eyelid.

Figure 14.8 The blue-spotted stingray *(left)* and the skate *(right)* are cartilaginous fishes that search for food on the ocean bottom.

Bony Fishes

If you have ever eaten a whole trout or catfish, you know why they are called bony fishes. Bony fishes have skeletons made of bone instead of cartilage. About 95 percent of all fish species are bony fishes.

There are three main groups of bony fishes. The first group is the lobe-finned fishes. People thought that lobe-finned fishes had been extinct for millions of years. Then, in 1938, a lobe-finned fish called a coelacanth (SEEL uh kanth) was found in South Africa. Coelacanths have fins that look like paddles.

The second group of bony fishes is the lungfishes. Lungfishes have both lungs and gills. This characteristic allows them to live in shallow waters that dry up in the summer. They cover themselves with mucus to keep moist until water returns.

Figure 14.9 Two types of bony fishes are the lobe-finned fishes and the lungfishes. The coelacanth *(left)* is a lobe-finned fish. The African lungfish *(right)* is a lungfish.

The third group of bony fishes is the ray-finned fishes. Ray-finned fishes have fins that are supported by long bones called rays. These fishes are the ones you most commonly see and eat. Catfish, tuna, and trout are ray-finned fishes. So are perch, salmon, and bass. The sea horse is also an example of a ray-finned fish. The sea horse is unusual because it swims upright, moves slowly, has a long snout, and broods its young in the stomach pouches of the males.

Figure 14.10 The flying fish *(left)*, the largemouth bass *(center)*, and sea horses *(right)* are all examples of ray-finned fishes.

External Structure A bony fish has a head, body, and tail. The eyes, nostrils, gills, and mouth are found on the head of the fish. A bony fish has different types of fins. These fins help the fish move through the water to hunt for food. They also help the fish avoid predators such as larger fish or people.

The body of a bony fish is covered with scales. The scales are covered with slimy mucus. This mucus keeps the scales moist to help the fish swim better. It also helps the fish slip away from predators.

Internal Structure As **Figure 14.11** shows, a bony fish has an endoskeleton that protects its internal organs. Muscles attach to the fins of the fish to help it move through the water. The gills of the fish help it take oxygen from the water. They also remove carbon dioxide from the fish's body. A bony fish also has a **swim bladder**. The swim bladder acts and looks like a small balloon.

Figure it Out

1. What are some of a bony fish's internal organs?

2. Hypothesize about what could be related to the structure and arrangement of a bony fish's fins.

Rainbow trout

Swim Bladder A swim bladder is an air-filled sac. The fish can add air to the swim bladder to rise. It can release air from the swim bladder to sink.

Scales Scales are covered with slippery mucus. This allows the fish to move through the water with less friction.

Kidney

Urinary bladder

Reproductive organ

Stomach

Intestine

Liver

Heart

Gills Gills are thin, blood-vessel-rich tissues where gases are exchanged.

Fins Fins help the fish move through the water. Different kinds of fishes have different kinds of fins. Fishes that live near coral reefs have small fins. This helps them move through small spaces. A tuna has large fins. This helps it move quickly in open water.

Figure 14.11 The external and internal structures of a bony fish are shown here.

After You Read

1. Explain the three common characteristics of all fishes.

2. How is a cartilaginous fish different from a bony fish?

3. What is a swim bladder?

4. Referring to your Science Notebook, name three facts about fishes.

Learning Goals

- Identify the common characteristics of amphibians.
- List the major groups of amphibians.
- Describe metamorphosis in frogs.
- Explain how frogs survive in water and on land.

New Vocabulary

hibernation
estivation
nictitating membrane
tympanum

Before You Read

In your Science Notebook, write each vocabulary term in this lesson. Leave some space below each term. Using your own words, write a definition for each term as you read about it.

Have you ever caught a frog? If so, you might have found it to be moist and slippery. Frogs have cool, slimy skin. Frogs, toads, and salamanders belong to the class Amphibia (am FIH bee uh). *Amphibia* means "double life." Most amphibians live part of their lives in water and part on land.

There are three main groups of amphibians. They are the frogs and toads, the salamanders and newts, and the caecilians (see SIL yuhns).

Characteristics of Amphibians

All amphibians are ectothermic vertebrates. Many have smooth, moist skin. Their eggs do not have hard outer shells like birds' eggs do. Thus, the eggs must be laid in water or in other moist areas. Amphibians use their lungs, gills, and skin to breathe. Their webbed feet help them move through the water.

Like insects, young amphibians change into adults through the process of metamorphosis. **Figure 14.13** shows the process by which a frog egg becomes an adult frog. In frog metamorphosis, fertilized eggs hatch into tadpoles. Tadpoles look like small fish. They have gills and fins. As the tadpoles grow into adult frogs, they lose their tails. The adult frogs use their lungs and skin to breathe.

Figure 14.12 The leopard frog *(top)*, the golden toad *(center)*, and the spotted salamander *(bottom)* are all types of amphibians.

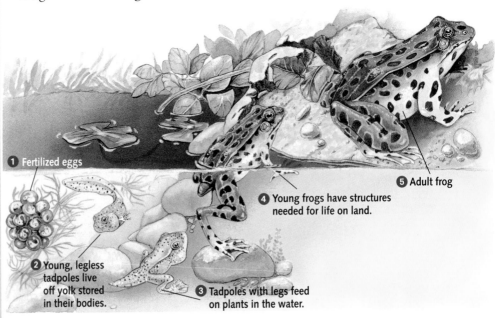

❶ Fertilized eggs

❷ Young, legless tadpoles live off yolk stored in their bodies.

❸ Tadpoles with legs feed on plants in the water.

❹ Young frogs have structures needed for life on land.

❺ Adult frog

Figure 14.13 A frog begins life in water. As it changes to an adult, it moves to land.

Figure 14.14 The caecilian *(left)*, the barred tiger salamander *(center)*, and the fire-bellied newt *(right)* are amphibians that eat insects, worms, and insect larvae.

Caecilians

Caecilians are amphibians that do not have legs. They look very much like earthworms. Most caecilians are blind, but some have tiny eyes. Some caecilians have bony scales. Most species burrow into the soil to live and find their food. A few species live in freshwater. Since they cannot see, caecilians use tentacles near their noses to find food. Caecilians live in the tropical areas of South America, Africa, and Asia.

Salamanders and Newts

Salamanders and newts are amphibians with legs and tails. Like frogs, they have smooth, moist skin. Found worldwide, most salsmanders and newt spend most of their lives on land. A few species live all their lives in freshwater. Many are poisonous. Most salamanders of North America live under rocks and logs in the woods. Like the caecilians, they eat small invertebrates.

Explain It!

In your Science Notebook, write a letter to the governor of your state. Explain why it is important to protect amphibian habitats in your area. Describe a plan to keep the amphibians from becoming extinct. Form small groups with your classmates, and discuss individual letters. Then, using the most important ideas from each group's letters, create one clearly written letter to send to the governor.

CONNECTION: Environmental Science

How long does it take for a group of animals to become extinct? You might think that the answer is a long time, but this is not always true. Today, many amphibians are in danger of becoming extinct. Some scientists believe that almost one-third of Earth's amphibian species are in danger.

Many amphibian habitats are being destroyed. The destruction of rain forests leaves many amphibians without a place to live. Amphibians are very sensitive to chemical pollution. They have very thin skin, and their eggs do not have hard shells. An amphibian can absorb poison just as easily as it absorbs oxygen. This can kill the amphibian or cause its offspring to be deformed. The frog in the photo has three back legs. This deformity is likely the result of poison absorbed by its parents.

Amphibians are important because they eat many harmful insects. Scientists around the world are now working on a plan to save Earth's amphibians. Protecting amphibian habitats and reducing pollution levels will help save many species of amphibians.

Deformed frog

Figure 14.15 The red-eyed tree frog *(top)* and the American toad *(bottom)* belong to the largest group of amphibians.

⊕ CONNECTION: Health

You may have been told that you get warts by touching a toad. A wart is a hard, rough growth on the surface of the skin. Touching the skin of a toad does not cause warts. A wart is caused by a virus.

Frogs and Toads

Frogs and toads are similar in many ways, but they also have some differences. Frogs spend most of their lives in or near water. Toads spend more time on land. A toad might be found in a backyard or a garden. Frogs have smooth skin, and toads have bumpy skin. Frog eggs are usually found in a ball. Toad eggs are often found in a long chain.

Frogs and toads are ectotherms. Their body temperature changes with the environment. During the winter, they bury themselves in mud or leaves until the temperature gets warmer. This period of winter inactivity is called **hibernation** (hi buhr NAY shun). Other animals, such as bees and snakes, also hibernate in the winter. In the summer months, frogs and toads bury themselves to escape the heat. This period of inactivity during intense heat is called **estivation** (es tuh VAY shun).

External Structure of a Frog Frogs are the most common amphibians. They live near most lakes and ponds. The outside of a frog is adapted to life both in water and on land. A frog's eye has a third eyelid called a **nictitating** (NIK tuh tayt ing) **membrane**. This membrane is clear so the frog can see through it. A nictitating membrane protects each of a frog's eyes and keeps them moist.

Frogs do not have ears. Instead, they have a round structure called a **tympanum** (TIHM puh nuhm) located just behind each eye. These structures are a frog's eardrums. The tympanum allows the frog to hear well in water and on land. Male frogs use sound to attract females. Females call to the males to let them know if they want to mate.

Frogs have two sets of limbs. The hind, or back, legs are strong and muscular. They help the frog jump around on land. Webbed feet help the frog swim. The front legs of a frog are small. They help the frog land after it has jumped.

As You Read

Make corrections or additions to the vocabulary definitions you have written.

How are hibernation and estivation alike? How are they different?

Figure 14.16 The tympanum is the round structure located behind each of a frog's eyes *(left)*. A leopard frog *(right)* uses its strong, muscular legs to jump through the air.

As **Figure 14.17** shows, a frog has eyes on the top of its head. This helps the frog see above the water when its body is below the surface. In this way, a frog can remain almost completely hidden while it hunts for and then captures insects for a tasty meal! The tongue of a frog is long and sticky. It is attached to the front of the frog's mouth.

Internal Structure of a Frog Frogs have many internal organs you may recognize, including a heart, a liver, and an intestine. Adult frogs use lungs to breathe air. Frogs also exchange gases through their skin. This skin must remain moist. If their skin dries out, most amphibians will die.

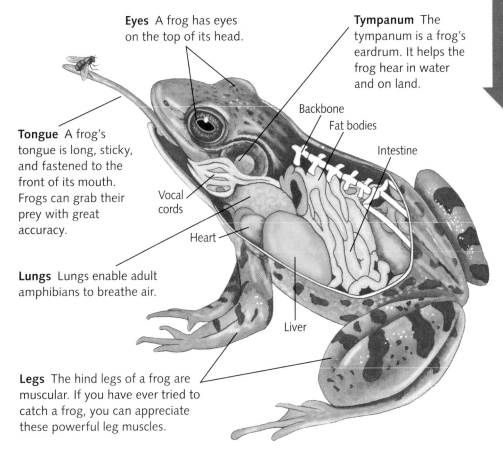

Eyes A frog has eyes on the top of its head.

Tympanum The tympanum is a frog's eardrum. It helps the frog hear in water and on land.

Backbone

Fat bodies

Intestine

Tongue A frog's tongue is long, sticky, and fastened to the front of its mouth. Frogs can grab their prey with great accuracy.

Vocal cords

Heart

Lungs Lungs enable adult amphibians to breathe air.

Liver

Legs The hind legs of a frog are muscular. If you have ever tried to catch a frog, you can appreciate these powerful leg muscles.

Figure 14.17 The external and internal structures of a frog are shown here.

Figure 14.18 The golden dart frog is the world's most poisonous amphibian.

After You Read

1. What are three common characteristics of amphibians?
2. Describe what happens during the metamorphosis of a frog.
3. How can frogs survive in water and on land?
4. Using the definitions in your Science Notebook, write a well-developed paragraph that describes what frogs do to survive hot summer and cold winter temperatures.

Learning Goals

- Identify the common characteristics of reptiles.
- List examples from each major group of reptiles.
- Describe how the amniotic egg allows reptiles to live away from water.

New Vocabulary

amniotic egg

Figure 14.19 The thick skin of the marine iguana allows it to successfully live on land.

Figure 14.20 Reptiles are able to survive exclusively on land because of their amniotic eggs. What is the function of the amnion?

Before You Read

Reword the headings of this lesson so that they form questions. Write the questions in your Science Notebook. As you read, write answers to these questions.

Characteristics of Reptiles

Reptiles are adapted to life on land. They have thick, scaly skin that keeps them from drying out. Since reptiles cannot exchange gases through their skin, they have well-developed lungs. Like the amphibians, reptiles are ectothermic vertebrates. The three main groups of the class Reptilia are the turtles and tortoises, the alligators and crocodiles, and the snakes and lizards.

The Amniotic Egg   The eggs of fishes and amphibians do not have hard shells. These eggs dry out easily, so they must be laid in water. A reptile egg, however, has a hard outer shell. Reptile eggs can be laid on dry land. Most reptiles lay their eggs in protected places beneath sand, soil, gravel, or bark. A reptile egg is called an amniotic (am nee AH tihk) egg. An **amniotic egg** surrounds the developing embryo with food and a tough shell and will not dry out on land. A reptile hatches by breaking its shell with a thorny tooth on its snout. This egg tooth drops off shortly after hatching.

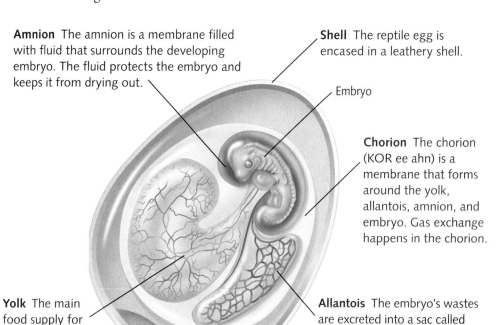

Amnion The amnion is a membrane filled with fluid that surrounds the developing embryo. The fluid protects the embryo and keeps it from drying out.

Shell The reptile egg is encased in a leathery shell.

Embryo

Chorion The chorion (KOR ee ahn) is a membrane that forms around the yolk, allantois, amnion, and embryo. Gas exchange happens in the chorion.

Yolk The main food supply for the embryo is the yolk. It is attached to the embryo.

Allantois The embryo's wastes are excreted into a sac called the allantois (uh LAN tuh wus). When a reptile hatches, it leaves the allantois behind.

Turtles and Tortoises

There are about 250 species of turtles and tortoises. Turtles live mainly in or near the water. They are found in both freshwater and seawater. Tortoises live mainly on land.

Turtles have flat shells, while tortoises have round, domed shells. The shells of the turtle and the tortoise are connected to their bodies. These reptiles cannot leave their shells. Turtle and tortoise shells are very strong. The shell of a box turtle can hold up to 200 times its own weight! Shells provide protection for turtles and tortoises. Most turtles are able to pull their limbs, tails, and heads into their shells for protection against predators.

Figure 14.21 The painted turtle *(left)* and the red-footed tortoise *(right)* are examples of another group of reptiles.

Figure It Out

1. What is the function of a turtle's or a tortoise's shell?

2. Describe how the shells of the painted turtle and the red-footed tortoise are different from each other.

Turtles do not have teeth. They have beaks that are hard like bird beaks. Turtles eat plants and animals. The alligator snapping turtle shown in **Figure 14.22** has a small, wormlike structure on the bottom of its mouth. The turtle sits still with its mouth open on the bottom of a river, a lake, or a pond. When a fish swims up to eat the "worm," the turtle snaps its jaws shut and eats the fish.

The loggerhead turtle might be called the best navigator in the animal kingdom. After it hatches, the young turtle moves to the ocean. As it grows, it travels through thousands of kilometers of ocean water. The turtles will mature and mate. When a female is ready to lay her eggs, she will return to the same beach where she was born. How does she find her way? Scientists think that the loggerhead turtle uses ocean waves and Earth's magnetic field to move in the right direction.

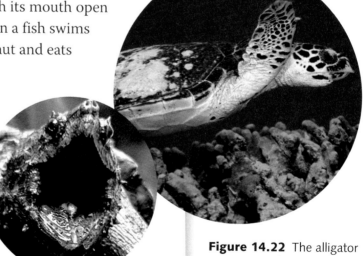

Figure 14.22 The alligator snapping turtle *(left)* lures fish with a wormlike structure in its mouth. The female loggerhead turtle *(right)* will return to her birthplace to lay her eggs.

Figure 14.23 Alligators and crocodiles look very similar. The alligator *(left)* shows only its upper teeth when its mouth is closed. The crocodile *(right)* shows both its upper and lower teeth when its mouth is closed.

Alligators and Crocodiles

Alligators and crocodiles are the largest living reptiles. They can move quietly through the water to search for food, often with only their eyes visible on the surface. They can spend their days lying in the sunlight along the bank of a river and then floating motionless in the water, looking much like a log.

Do you know how to tell the difference between an alligator and a crocodile? As **Figure 14.23** shows, the alligator has a wide head with a rounded snout. The crocodile has a narrow head with a triangle-shaped snout. When its mouth is shut, only the alligator's upper teeth are visible. You can see both the upper and lower teeth of the crocodile.

Both alligators and crocodiles live in tropical areas. They spend most of their time in the water. Alligators live primarily in North America and Asia. Crocodiles live mostly in North America, Africa, Asia, and Australia.

Unlike most reptiles, alligators and crocodiles care for their young. A female crocodile will stay and guard her nest until her babies hatch from their eggs. The hatching babies call to their mother, and she carries them to the water. The babies then start feeding on crabs, shrimps, and insects.

Figure 14.24 This baby crocodile will be carried by its mother to the water so it can feed.

🔗 CONNECTION: **Chemistry**

For several crocodile species, the temperature inside the nest determines the sex of the young. If the temperature of the nest is 30°C, females will hatch from the eggs. If the temperature is above 34°C, males will hatch from the eggs. If the temperature is between these values, the nest will contain both males and females.

Figure 14.25 The gecko *(left)* is a lizard. The copperhead *(center)* is a snake that injects its prey with venom. The ball python *(right)* is a snake that squeezes its prey to death. To what category of eaters do all snakes and most lizards belong?

Lizards and Snakes

Lizards and snakes are the most common reptiles. They are closely related and share some characteristics. They live mostly in warm areas. They have dry skin covered with scales. As lizards and snakes grow, they shed their old skins. These skins are replaced with new ones.

All snakes and most lizards are carnivores. Large snakes can eat very large prey. The major difference between lizards and snakes is that snakes do not have legs, whereas most lizards have four legs. A snake uses its tongue to find prey and to get information about the world around it. The tip of its tongue senses chemicals. This helps the snake catch prey and avoid predators.

Poisonous snakes have hollow fangs that inject venom into their prey. Some snakes wrap around their prey and squeeze it to death. However, most snakes are neither poisonous nor constrictors. Instead, they get their food by grabbing it with their mouths and swallowing it whole. Their prey includes rodents, amphibians, fish, insects, eggs, and other reptiles. Snakes are afraid of people. People are usually bitten by snakes only when they frighten the reptiles.

Snakes sometimes scare their predators by spitting venom at them. The spitting cobra, for example, can spit its venom over a distance of more than two meters!

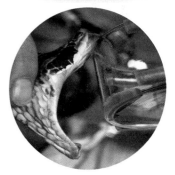

Figure 14.26 The rattlesnake *(top)* uses its tongue to sense chemicals. Venom of poisonous snakes is collected *(bottom)* for medicinal purposes.

CONNECTION: Medicine

Snake venom can be used to make an antivenin that helps people who are bitten by snakes. Snakes are "milked" to collect the venom from their fangs. Scientists are also researching ways in which snake venom can be used to cure diseases.

After You Read

1. What are three common characteristics of reptiles?
2. Describe the differences between an alligator and a crocodile.
3. How do snakes kill their prey?
4. Use the information in your Science Notebook to describe how the amniotic egg allows a reptile to live away from water.

KEY CONCEPTS

14.1 What Is a Vertebrate?

- A chordate is an animal with a nerve cord.
- A vertebrate is a chordate with a backbone.
- A vertebrate has an endoskeleton that protects the brain, heart, lungs, and other internal organs.
- An ectotherm is an animal whose body temperature changes with the temperature of the environment. An endotherm has a body temperature that stays the same.

14.2 Fishes

- Fishes are ectotherms. They have scales, fins, and gills.
- The three main groups of fishes are the jawless fishes, the cartilaginous fishes, and the bony fishes.
- The largest group of fishes is the bony fishes. Bony fishes have scales and swim bladders. Swim bladders help bony fish rise and sink in the water.

14.3 Amphibians

- Most amphibians live part of their lives in water and part on land.
- Frogs and toads begin life in the water as tadpoles. They become land-dwelling animals as they undergo metamorphosis.
- The three main groups of amphibians are the caecilians, the salamanders and newts, and the frogs and toads.

14.4 Reptiles

- Reptiles have dry, scaly skin and an amniotic egg. These adaptations allow them to live only on the land.
- The three major groups of reptiles are the turtles and tortoises, the alligators and crocodiles, and the lizards and snakes.

VOCABULARY REVIEW

Write each term in a complete sentence, or write a paragraph relating several terms.

14.1
vertebra, p. 241
cartilage, p. 241
notochord, p. 241
ectotherm, p. 242
endotherm, p. 242

14.2
scavenger, p. 244
swim bladder, p. 247

14.3
hibernation, p. 250
estivation, p. 250
nictitating membrane, p. 250
tympanum, p. 250

14.4
amniotic egg, p. 252

PREPARE FOR CHAPTER TEST

To prepare for the chapter test, create a question from each Learning Goal. Use the information in your Science Notebook to answer each question. Then use these answers to write a well-developed essay about the chapter. Use the Key Concept on the first page of this chapter as your topic sentence.

True or False

If the statement is true, write "true." If it is false, change the underlined word or words.

1. <u>Vertebrae</u> are bones that go up and down your back.

2. An animal that has a backbone is called a(n) <u>invertebrate</u>.

3. Most fish use <u>lungs</u> to bring oxygen into their bodies.

4. The skeleton of a shark is made of <u>cartilage</u>.

5. Animals that are not active in the winter go through <u>estivation</u>.

6. The head of a(n) <u>crocodile</u> is wide with a rounded snout.

Short Answer

Answer each of the following in a sentence or brief paragraph.

7. What are three common characteristics of all chordates?

8. How is an ectotherm different from an endotherm?

9. Describe what happens to a frog as it goes through metamorphosis.

10. Describe the features of fish that help them survive in nature.

11. How do the shells of the turtle and the tortoise differ from one another?

Critical Thinking

Use what you have learned in this chapter to answer each of the following.

12. **Compare and Contrast** How do caecilians differ from earthworms?

13. **Apply Concepts** The poisonous coral snake has yellow, red, and black bands around its body. The harmless king snake has bands of red, black, and yellow. How is its coloring a useful trait to the king snake?

14. **Make Predictions** What would happen to a frog if it could not move from the land to the water? Explain your answer.

Standardized Test Question

Choose the letter of the response that correctly answers the question.

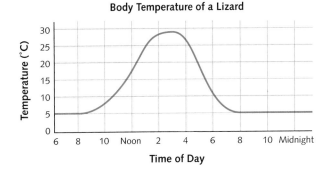

Body Temperature of a Lizard

15. The graph shows the body temperature of a desert lizard from 6 A.M. to midnight. Where do you think the lizard was from 8 A.M. to noon?

   A. sitting on a rock in the sunlight

   B. sitting in the shade

   C. burying itself underground

   D. none of the above

Test-Taking Tip

If "none of the above" is one of the choices in a multiple-choice question, be sure that none of the choices is true.

Chapter 15

Birds and Mammals

KEY CONCEPT Birds and mammals can be classified based on their physical characteristics.

The bald eagle is a majestic animal. It glides gracefully through the air. Its blackish-brown body and white head, neck, and tail distinguish it from all other animals. So recognizable is this animal that it is part of the emblem of the United States.

A bald eagle is a type of animal called a bird. This chapter is about birds and another class of animals, mammals. The giraffe is a familiar mammal, easily identified by its long neck, which allows it to eat leaves from tall trees. The lion, another mammal, is the only natural enemy of the adult giraffe.

Think About Making Inferences

When you make an inference, you make a logical conclusion based on something you observe.

- Look at the photographs of birds given to you by your teacher. Concentrate on the feet and legs of each of the birds.

- In your Science Notebook, make inferences about the type of environment each bird may live in based on the structure of its feet and legs.

NSTA

SCiLINKS
THE WORLD'S A CLICK AWAY

www.scilinks.org
Birds **Code: WGB15A**
Mammals **Code: WGB15B**

15.1 Birds

Learning Goals

- Describe the common characteristics of birds.
- Distinguish among the different types of birds.
- Discuss the origin of birds.

New Vocabulary

feather
contour feather
down feather
preening
sternum

Recall Vocabulary

molting (p. 225)

Before You Read

In your Science Notebook, draw a three-column chart to organize facts about birds. Label the first column *Characteristics*. Label the second column *Origins*. Label the third column *Types*. Preview the Learning Goals, headings, subheadings, and visuals. In each column, write one or two facts that you already know about each topic.

Like fishes, amphibians, and reptiles, birds are vertebrates. Birds have an important characteristic that sets them apart from these animals.

Characteristics of Birds

Birds are endotherms, or animals that have a constant body temperature. Birds can be active no matter what the temperature is in their environment. Thus, they are found in almost all places on Earth. Like reptiles, birds lay eggs that have hardened shells. Birds share other common characteristics.

Bird Feathers   All birds have feathers. A **feather** is made of dead cells that contain the same material found in human fingernails. Feathers come in many colors, shapes, and sizes. Feathers help keep a bird warm. They also help the bird fly, and they cover and protect the bird's skin.

Birds have two main kinds of feathers: contour feathers and down feathers. A **contour feather** is hard and stiff. It gives the bird its shape and protects the bird. A **down feather** is small and fluffy. It is close to the bird's skin and keeps the bird warm. People use down feathers in coats, jackets, blankets, and sleeping bags. Over time, feathers on a bird become worn and damaged. Damaged feathers are replaced through a process called molting. Birds lose only a few of their feathers at a time.

Birds must take care of their feathers. In a process called **preening**, birds apply an oil to their feathers. This oil comes from a gland in the bird's tail. The oil keeps the feathers from drying out and breaking. The oil also makes the feathers waterproof.

Figure 15.1 A bird preens itself *(left)* to protect its feathers. A contour feather *(center)* is strong and stiff. It protects the bird's skin. A down feather *(right)* is small and fluffy. It keeps the bird warm.

Figure 15.2 The large bone on this bird's skeleton is called the sternum. Muscles attached to the sternum allow a bird to fly.

Explore It!

Use a hand lens to look at a variety of contour and down feathers. In your Science Notebook, draw the different types of feathers. Make a list of the characteristics that the feathers share. Also describe how they are different. How does the structure of these feathers help them perform their functions?

Bird Wings All birds have front limbs that are wings. Most birds are able to fly. Birds that can fly have powerful muscles attached to their breastbones. The breastbone is called the **sternum** (STUR num). Look at **Figure 15.2**. The sternum on this bird skeleton is the large bone above the bird's legs. The muscles attached to the sternum give a bird the power it needs to take off and fly through the air.

Food and Flight

A person who does not eat very much is often said to "eat like a bird." This is not an accurate description, however, because a bird actually eats a large amount of food for its size. This food gives the bird the energy it needs to keep its body temperature constant. It also gives the bird the energy it needs to fly. A hummingbird like the one shown in **Figure 15.3** eats nectar about every ten minutes. It eats about twice its body weight in nectar every day!

Figure 15.3 The tiny hummingbird needs to eat a very large amount of food to give it the energy to fly. The wings of a hummingbird beat about 80 times per second.

A bird has no teeth. It uses its beak to catch its food. Birds eat a wide variety of plants and animals. Birds feed on nectar, seeds, insects, worms, fish, and other birds. Each type of bird has a beak that is well suited to the type of food that bird eats. The hummingbird shown in Figure 15.3 has a long, thin beak. The hummingbird uses that beak to reach deep into a flower and sip its nectar.

CONNECTION: Art

Italian artist Leonardo da Vinci (1452–1519) was fascinated with birds and their ability to fly. He used his knowledge of bird wings to draw designs of various flying machines. Da Vinci's design for a human glider looks very similar to a modern-day hang glider.

Origins of Birds

After their success in the ocean and on land, vertebrates moved to the air. Several hundred million years ago, the air was filled with insects that ancient birds ate for food. The air also kept the ancient birds safe from predators in the ocean or on the land. Fossils show that modern birds may have developed from small, two-legged dinosaurs called theropods. **Figure 15.4** shows the relationship between the theropod and modern bird species that may have developed from it.

The fossil record suggests that the earliest bird was *Archaeopteryx*. *Archaeopteryx* lived about 150 million years ago. It had both reptilelike and birdlike features. It was about the size of a crow, and it had feathers and wings. It was not able to fly, however, and it ran to catch its prey. Scientists believe that *Archaeopteryx* was not a direct ancestor of modern birds and represents a smaller branch of the bird family tree.

As You Read

Review the information in the chart in your Science Notebook. Share your information with a partner. Add any other facts that will help you organize what you learn about birds.

What is molting? Preening?

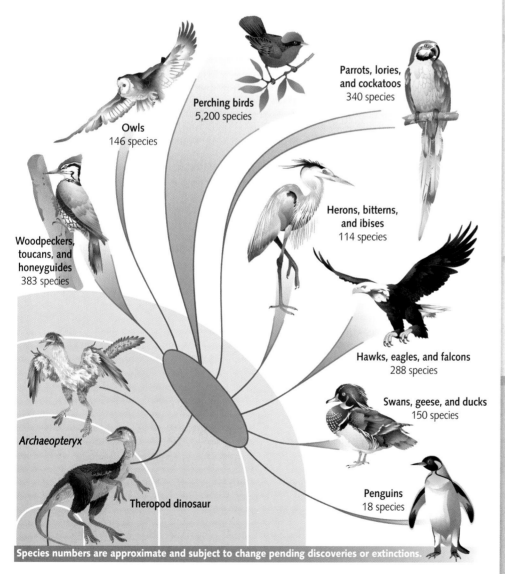

Owls
146 species

Perching birds
5,200 species

Parrots, lories, and cockatoos
340 species

Woodpeckers, toucans, and honeyguides
383 species

Herons, bitterns, and ibises
114 species

Hawks, eagles, and falcons
288 species

Archaeopteryx

Swans, geese, and ducks
150 species

Theropod dinosaur

Penguins
18 species

Species numbers are approximate and subject to change pending discoveries or extinctions.

Figure 15.4 Many scientists believe that birds may have developed from an ancient reptile called a theropod. The skeletons of birds and theropods are similar.

Extend It!

Find out more about *Archaeopteryx*. How was it similar to a dinosaur? What did *Archaeopteryx* eat? What dinosaurs lived during the same time? Write your findings in your Science Notebook.

CONNECTION: Environmental Science

Birds are important to the environment. They help pollinate plants, and they eat insect pests. You can care for and protect the birds in your neighborhood. Building a birdhouse or setting up a bird feeder can ensure birds will continue to populate the area.

Figure It Out

1. What is the function of a bird's wings?
2. Describe how the ostrich, the penguin, and the kiwi are alike and how they are different.

Figure 15.5 The ostrich *(left)*, penguin *(center)*, and kiwi *(right)* are all examples of birds that are not able to fly.

Types of Birds

There are about 10,000 known species of birds. Birds are often divided into a few large groups. These groups are based on one or two physical characteristics of the birds. The four main groups of birds are flightless birds, perching birds, birds of prey, and water birds.

Flightless Birds   The ostrich, penguin, and kiwi shown in **Figure 15.5** are all examples of birds that are not able to fly. Most flightless birds have small wings. The kiwi has no wings. Many flightless birds are fast runners. This helps them catch their prey and avoid predators. Penguins are excellent swimmers.

Perching Birds   Cardinals, woodpeckers, and canaries are types of perching birds. Many of these birds sing songs when they are on their perches. Perching birds can eat insects, worms, or seeds.

Birds of Prey   Hunting birds are also called birds of prey. Hawks, eagles, and owls are all examples of birds of prey. They have sharp claws and strong, curved beaks. Birds of prey fly very fast. The peregrine falcon has reached speeds of up to 290 kilometers per hour while diving for its prey.

Water Birds   Water birds swim and dive in lakes and ponds. Many water birds, such as ducks and swans, have webbed feet for swimming. Other water birds, such as the flamingos shown in **Figure 15.7**, have long legs for wading. Some have long beaks they use to pull food from the mud.

Figure 15.6 The cardinal *(top)* is a type of perching bird. The pygmy falcon *(bottom)* is a bird of prey. It uses sharp claws and a strong beak to catch food.

Figure 15.7 Flamingos wade through the water in search of food.

After You Read

1. What are three common characteristics of all birds?
2. How is a contour feather different from a down feather?
3. From what animal do some scientists think modern birds developed?
4. Describe the four major types of birds.
5. Using the information you wrote in your Science Notebook, write a well-developed paragraph that summarizes what you have learned about birds.

15.2 Adaptations for Flight

Learning Goals

- Describe how a bird's body is adapted for flight.
- Explain how migration is important to the survival of birds.

New Vocabulary

migration

Before You Read

Look through the lesson, taking note of the Learning Goals, headings, diagram, and photographs. What questions about the information come to mind? Write at least three questions in your Science Notebook that you would like answered by the end of the lesson. Think about information you already know that might help you answer these questions.

Have you ever wondered how birds fly? A bird's entire body is adapted to enable it to fly. These adaptations either reduce a bird's body weight or make a bird more compact. The body of a bird is small. It does not weigh very much. A bird's small size and light weight enable it to glide through the air.

A Bird's Body

The wings of a bird are wide and thin. They are similar to the wings on an airplane in that they are slightly curved on the top and thicker in the front. As air flows around the wings, it travels faster over the top of the wing than beneath the wing. Where the air travels faster, the pressure is reduced. Thus, the pressure beneath the wings is greater than the pressure above the wings. This difference in air pressure provides the lift needed for flight. Once in flight, the bird's wings allow the bird to use air currents to stay aloft.

The skeleton of a bird is adapted for flight. As Figure 15.9 on page 264 illustrates, the bones of a bird's skeleton are hollow. Hollow bones keep the bird's skeleton light in weight. The crosspieces in the bird's bones keep the bones strong. Some of the bones are connected to other bones. This makes the bird's skeleton stronger and more stable for flight.

Although a bird's skeleton is light in weight, its muscles are large and heavy. The chest muscles make up about 25 percent of the bird's body weight. These muscles are used to move the bird's wings.

Figure 15.8 The adaptations of the great blue heron's body enable it to glide over the water in search of food.

As You Read

Look at the questions you wrote in your Science Notebook. Use information learned so far in this lesson to help you answer this question.

How do a bird's wings enable it to fly?

People have always dreamed of being able to fly. When people hang glide or parasail, they experience a sense of flight. Birds do it naturally! Some of the characteristic structures of birds that enable them to fly are summarized in **Figure 15.9**.

Figure It Out

1. How are hollow bones useful to a bird?

2. Describe how birds that do not fly are able to move from one place to another.

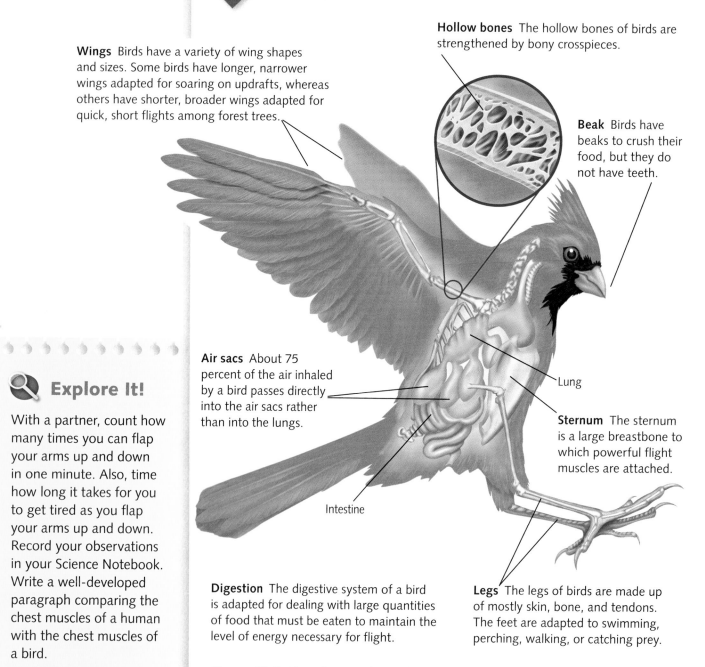

Wings Birds have a variety of wing shapes and sizes. Some birds have longer, narrower wings adapted for soaring on updrafts, whereas others have shorter, broader wings adapted for quick, short flights among forest trees.

Hollow bones The hollow bones of birds are strengthened by bony crosspieces.

Beak Birds have beaks to crush their food, but they do not have teeth.

Air sacs About 75 percent of the air inhaled by a bird passes directly into the air sacs rather than into the lungs.

Lung

Sternum The sternum is a large breastbone to which powerful flight muscles are attached.

Intestine

Digestion The digestive system of a bird is adapted for dealing with large quantities of food that must be eaten to maintain the level of energy necessary for flight.

Legs The legs of birds are made up of mostly skin, bone, and tendons. The feet are adapted to swimming, perching, walking, or catching prey.

Figure 15.9 Many features that enable birds to fly are shown here.

Explore It!

With a partner, count how many times you can flap your arms up and down in one minute. Also, time how long it takes for you to get tired as you flap your arms up and down. Record your observations in your Science Notebook. Write a well-developed paragraph comparing the chest muscles of a human with the chest muscles of a bird.

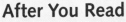

CONNECTION: Physics

Newton's third law of motion states that for every force applied by an object, there is an equal and opposite force applied on that object. A bird's wing can be used to demonstrate this law. When a bird flaps its wings down, the bird's body moves up, and the bird flies.

Bird Migration

Have you ever looked at the sky during the fall months and seen a flock of birds flying toward a warmer climate? Many birds leave their homes before winter to find a warmer place with more food. How do birds find their way to these winter homes?

Migration is the regular movement of a group of animals from one place to another. Birds are not the only animals that migrate, but more birds travel long distances than any other kind of animal. The arctic tern spends its summer in the arctic circle near the north pole. Each year, as winter approaches, this bird flies 35,000 kilometers to travel to Antarctica!

Most birds migrate to find new sources of food in the winter. When they migrate, many birds travel along certain flying routes. Some birds navigate, or determine the route they will take, by observing the Sun and stars. Others navigate by following coastlines, mountain ranges, or wind currents. Scientists think that some birds have magnetic centers in their brains. These magnetic centers act like compasses to help the birds find their way. However, there are many things that scientists have not yet learned about bird migration.

Figure 15.10 Sandhill cranes move to a warmer climate in the winter. A warmer climate will provide more food for the cranes.

After You Read

1. How is a bird's body adapted for flight?
2. What is migration?
3. How do birds find their way during migration?
4. Look at the questions and answers you have recorded in your Science Notebook. Write a well-developed paragraph explaining two bird behaviors: how they fly and how and why they migrate.

Learning Goals

- List the common characteristics of all mammals.
- Compare the three main groups of mammals.
- Describe one physical characteristic of each order of placental mammals.

New Vocabulary

monotreme
marsupial
placental mammal
uterus
placenta
gestation
echolocation
carnivore
herbivore
omnivore

Before You Read

Write the vocabulary terms for this lesson in your Science Notebook. As you read the definitions, write a sentence for each term that includes the definition.

Look closely at the surface of your forearm. What do you see? The hairs sticking up through your skin are not a characteristic of any animal group you have studied so far. Mammals are the only group of animals that have hair. People, dogs, and cats are all types of mammals.

Characteristics of Mammals

All mammals are endothermic vertebrates. At some time during their lives, all mammals have hair or fur. For mammals living in cold climates, this hair or fur helps keep them warm. The musk ox, which lives in Alaska and Northern Canada, has fur that is at least ten centimeters long!

All mammals feed their young with milk produced in their mammary glands. The word *mammal* comes from the word *mammary*. Young mammals depend on their mothers for food. Some young mammals stay with their mothers for a long time.

During the time young mammals stay with their mothers, they learn how to survive on their own. Mammals have the most highly developed brains of all of the animal groups. They also have highly developed senses of sight, hearing, taste, and smell. The brains and senses of mammals help them survive in nature.

Figure 15.11 The long, thick fur of the musk ox *(left)* helps it survive in very cold climates. The Bengal tiger *(right)* uses her mammary glands to nurse her young cubs.

Reproduction in Mammals

Based on the way they reproduce, mammals can be divided into three groups. These three groups are the egg-laying mammals, the pouched mammals, and the placental mammals. The most familiar mammals in the United States are the placental mammals.

Egg-Laying Mammals   Egg-laying mammals, called **monotremes** (MA nuh treems), live in Australia and New Guinea. The eggs have leathery shells. A female incubates her eggs for about ten days. The young nurse by licking milk that oozes from mammary glands onto skin and hair.

The duck-billed platypus and spiny anteater shown in **Figure 15.12** are monotremes. The platypus spends most of its time in the water. It has a broad, flat tail. Its rubbery snout looks like the bill of a duck. The platypus has webbed front feet that help it swim through the water. It also has claws that help it dig through the soil. The spiny anteater has coarse, brown hair. Its back and sides are covered with sharp spines that protect it from predators. The spiny anteater uses its long, sticky tongue to catch insects.

Pouched Mammals   Some mammals have a short period of development inside the mother's body. This is followed by a period of development in a pouch outside the mother's body. These pouched mammals are called **marsupials** (mar SUE pee ulz).

Many marsupials are found in Australia and its surrounding islands. The kangaroo and koala are Australian marsupials. The kangaroo has large, powerful back legs that help it jump large distances. The koala is a very slow-moving animal. It sleeps for almost 19 hours a day. Koalas spend about three hours a day eating tree leaves.

The opossum is the only North American marsupial. The female carries and nurses her babies in her pouch until they are about two to three months old. She carries her babies on her back whenever they leave their den for another one to two months.

Figure 15.12 The duck-billed platypus *(top)* and spiny anteater *(bottom)* are mammals that lay eggs.

Did You Know?

The Australian red kangaroo is the largest marsupial in the world. When a baby kangaroo is born, its mass is about one gram. The baby, called a joey, spends about 235 days in its mother's pouch.

Figure It Out

1. What is the function of a kangaroo's pouch?
2. What is similar about the way these animals reproduce?

Figure 15.13 The kangaroo *(left)* and koala *(center)* are marsupials found in Australia and New Guinea. The opossum *(right)* is the only marsupial found in North America.

Extend It!

Use the library or the Internet to research the gestation period and average number of offspring of different species of mammals. Record your findings in a table in your Science Notebook.

Placental Mammals Most mammals do not lay eggs or have pouches. **Placental mammals** give birth to live young that develop inside the mother's uterus. The **uterus** (YEW tuh rus) is a hollow, muscular organ where the baby develops. A developing mammal gets its food from an organ called the **placenta** (plu SEN tuh). The period of time during which the developing mammal stays in the uterus is called **gestation** (jeh STAY shun). Gestation periods range from 16 days for hamsters to 650 days for elephants.

About 90 percent of all mammals are placental mammals. Scientists classify placental mammals into orders based on their physical characteristics. Some mammals are adapted for walking or running. Others are adapted for swimming. Still others are adapted for flying.

Figure 15.14 The mole *(left)* is an insect-eating mammal. The bat *(center)* is a flying mammal. The beaver *(right)* is a gnawing mammal.

Insect-Eating Mammals Moles and shrews are the only insect-eating mammals that live in North America. They have strong limbs and sharp claws. Because they are active, moles and shrews must eat a large amount of food. A shrew may eat up to 75 percent of its body weight in food every day. Moles feed mainly on worms, and shrews eat insects.

Flying Mammals The bat is the only true flying mammal. A bat's wing is made of thin skin stretched over long finger bones. Bats fly by flapping their wings like birds do. Bats hunt for food at night using echolocation. **Echolocation** is a method some animals use to detect objects. The animal sends out high-pitched sounds that bounce off the object and back to the animal. This helps bats determine how close they are to their prey.

Gnawing Mammals Gnawing mammals that have two pairs of large front teeth are also called rodents. About 40 percent of all mammals are rodents. Mice, rats, squirrels, and beavers are gnawing mammals. The front teeth of a gnawing mammal grow throughout its life. The gnawing mammal chews branches and twigs to wear down its front teeth.

Toothless Mammals Armadillos, sloths, and anteaters are called toothless mammals. They have no front teeth, so they appear to be toothless. Only anteaters have no teeth at all. Armadillos and anteaters use their long, sticky tongues to catch insects. The skin of an armadillo is hard and protects its body like armor. Sloths feed on plants. They have flat teeth that help them grind leaves.

Figure 15.15 The armadillo *(top)* and the anteater *(bottom)* are toothless mammals.

Rodentlike Mammals Rabbits and hares, once classified as rodents, are now placed in a separate order. Rabbits and hares have long teeth for gnawing and long hind legs that help them jump long distances. Rabbits and hares eat plants. Hares are usually bigger than rabbits and have longer ears.

Water-Dwelling Mammals Whales, dolphins, porpoises, and manatees are water-dwelling mammals. Water-dwelling mammals do not have gills like fish. They use their lungs to breathe air, just like people do. A whale takes in air through a hole on the top of its head. Most water-dwelling mammals have teeth and eat fish, which means they are carnivores. A **carnivore** (KAR nuh vor) is an animal that eats other animals. Some whales filter the water to harvest shrimp and other small invertebrates for food. Manatees, shown in **Figure 15.16**, are mainly plant-eaters and can move on land for short periods of time.

Trunk-Nosed Mammals A trunk-nosed mammal has a nose and upper lip that are modified into a trunk. The trunk allows the animal to collect food and water. Only two species of trunk-nosed mammals are alive today. They are the Asian elephant and the African elephant.

Hoofed Mammals Horses, cows, deer, and pigs all have feet with hooves. A hoof is the hard covering on the mammal's toes. Like trunk-nosed mammals, hoofed mammals are herbivores. A **herbivore** (HUR buh vor) is an animal that eats only plants.

Flesh-Eating Mammals Dogs and cats look and act very differently from each other, but they are related. Both dogs and cats are members of the order Carnivora. These mammals have strong jaws and long, sharp teeth for tearing flesh. Lions, tigers, and wolves are also carnivores.

Primates Monkeys, apes, and humans are primates. Except for gorillas, baboons, and humans, most primates live in trees. They have long arms with well-developed hands. Like humans, most primates have hands with four fingers and an opposable thumb. The ability to place the thumb opposite the fingers allows primates to grasp, hold, and use tools. Some primates are herbivores. Others, such as humans, eat plants and animals. Any animal that eats both plants and animals is called an **omnivore** (AHM nih vor).

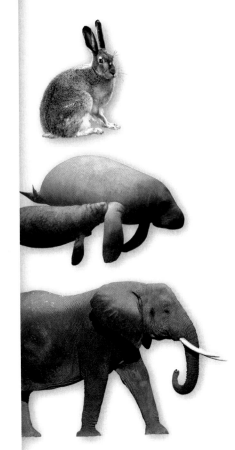

Figure 15.16 The large-eared jackrabbit *(top)* is a rodentlike mammal. The manatee and her calf *(center)* are water-dwelling mammals. The African elephant *(bottom)* is a trunk-nosed mammal.

As You Read

Look at the sentences you wrote for the vocabulary terms. Combine those sentences into a paragraph summarizing the different groups of mammals.

Figure 15.17 The paint horse *(left)* is a hoofed mammal. The cheetah *(center)* is a flesh-eating carnivore. The chimpanzee *(right)* is a primate.

The Origin of Mammals

Many scientists believe that mammals developed from ancient reptiles, just as birds did. The earliest mammals may have lived during the time of the dinosaurs. Fossils show that early mammals may have been about the size of a rat. They looked like rats, too. **Figure 15.18** shows *Eomaia*, the oldest placental mammal fossil discovered. *Eomaia* fossils may be 125 million years old.

Placental mammals may have developed from a reptile called a therapsid (ther AP sid). Therapsids had characteristics of both reptiles and mammals. Figure 15.18 shows an artist's representation of a therapsid.

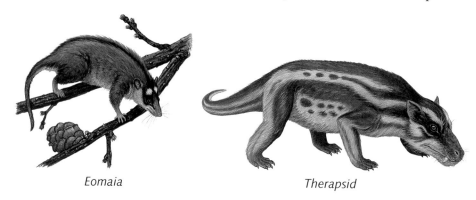

Eomaia   *Therapsid*

Figure 15.18 *Eomaia* is the oldest placental mammal fossil discovered. Based on its body structure, scientists believe it may have lived in trees. Therapsids may have been the ancestors of mammals. Therapsids had jaw bones and middle-ear bones like those of reptiles. Like mammals, they had straight legs held close to their bodies.

When dinosaurs became extinct, mammals began to develop quickly. They increased dramatically in number and developed a variety of shapes, sizes, and characteristics. Without the presence of the dinosaurs, and with climate changes and the appearance of flowering plants, mammals had new areas in which to live and new food sources upon which to feed. After a while, mammals were found in almost every habitat on Earth.

Explain It!

Think of a mammal that you know a lot about, such as a dog or a cat. Then think of a kind of reptile. In your Science Notebook, draw a picture of each type of animal. Write a description that tells how the two animals you drew are alike and how they are different.

CONNECTION: Environmental Science

Manatees are large, water-dwelling mammals that live off the coast of Florida. Manatees are on the threatened species list. Motorboats put manatees in danger. The manatees sometimes collide with boats or get cut by their propellers. Scientists have suggested creating more manatee refuges where boats are not allowed to travel.

After You Read

1. What characteristics do all mammals have in common?
2. How is the development of a pouched mammal different from the development of a placental mammal?
3. How are carnivores, herbivores, and omnivores different from each other?
4. Expand the paragraph you have written in your Science Notebook by identifying and describing the different orders of placental mammals. Then, add a sentence explaining the origin of mammals. What conditions helped the mammals increase in number and diversity?

15.4 Animal Behavior

Learning Goals

- Differentiate between a stimulus and a response.
- Compare innate behavior and learned behavior.
- Describe the social behaviors that occur in a beehive.

New Vocabulary

behavior
stimulus
response
innate behavior
instinct
learned behavior

Before You Read

In your Science Notebook, create a lesson outline. Use the lesson title as the outline title. Label the headings with the Roman numerals *I* through *IV*. Use the letters *A*, *B*, and *C* under each heading to record information you want to remember.

Have you ever noticed that when a dog sees a familiar person, the dog wags his or her tail, playfully jumps up, or barks? A cat may rub up against the person's leg and quietly purr. Such actions are considered animal behaviors.

Behavior is the way an animal responds to its environment. A **stimulus** (STIHM yuh lus) is a signal that causes an animal to react in a certain way. The animal's reaction to a stimulus is called a **response**. The sound of a familiar voice is a stimulus that causes a pet to behave the way that it does. There are different types of animal behavior.

Innate Behavior

A behavior that an animal is born with is called an **innate** (ihn AYT) **behavior**. Innate behaviors do not have to be learned. Cats, for example, do not need to be taught how to use a litter box. An **instinct** is a pattern of behaviors that an animal is born with. A newborn kangaroo uses instinct to crawl into its mother's pouch. It then attaches itself to one of her nipples to get milk.

Most birds build their nests without having to be taught. A species of gull, the kittiwake, has a nest on the side of a steep cliff. The baby kittiwakes innately know not to move away from the nest, because they will fall. Spiders are able to spin complicated webs the first time they spin.

Figure 15.19 Kittiwake gulls build their nests on the side of a cliff. The baby birds innately know not to move from the nest.

Learned Behaviors

Some animal behavior is learned. A **learned behavior** is a behavior that develops through experience. Because of their short life spans, arthropods learn very little. Fishes, amphibians, reptiles, birds, and mammals demonstrate learned behaviors. Because of their complex brains, mammals show more learned behaviors than other groups.

Perhaps you remember when you learned to ride a bicycle. You probably fell many times before you learned how to balance on the bike. Practicing helped you learn how to stay balanced. Newborn lion cubs have claws that help them catch prey, but they do not know how to hunt. Their parents must repeatedly show them the correct way to hunt.

Figure 15.20 Lion cubs must be taught how to hunt by their parents.

As You Read

Review the lesson outline in your Science Notebook with a partner. Make necessary changes.

What is the difference between innate behavior and learned behavior?

Social Behavior

Many animals live by themselves. Other animals live in groups. Each animal in the group has a special role. The role may be to gather food or to protect group members. It may be to help raise the young animals.

Some fishes swim in schools. Some insects live in large hives. Hoofed mammals, such as sheep or bison, form herds. Living in a group helps many animals survive. The group provides food and protection. A group of musk oxen form a circle when they are threatened by a wolf. They place the young musk oxen in the middle of the circle, where they are protected. A wolf often gives up rather than try to defeat the whole group.

Figure It Out

1. Why is it helpful for animals to form groups?
2. Infer why a group of musk oxen provides more protection for young animals than a group of fish does.

Figure 15.21 The school of fish *(left)* includes many individuals that stay together. The musk oxen *(right)* circle together to protect their young from a wolf.

Figure 15.22 This beehive is home to a queen bee, worker bees, and drones. Each type of bee has a different job in the bee society.

Some animals, including ants, termites, and honeybees, live in groups called societies. In a honeybee society, the queen bee is larger than the other bees. Her main function is to lay eggs. The queen also controls the activities in the hive.

The worker bees are females that do not lay eggs. The worker bees build the hive and maintain it. They also protect the hive from predators. Worker bees make honey from flower nectar. There can be many thousands of worker bees in a honeybee colony.

Drones are male bees that mate with the queen bee. They die soon after they mate. There are several hundred drones in each honeybee colony.

The beehive is made of compartments called cells. Some cells hold bee eggs that hatch into bee larvae. Some cells contain honey made by the worker bees. This honey is used to feed all of the bees in the hive.

In a wolf pack, there is a complicated system of societal importance. There is a top male and a top female in the pack, based on strength. These wolves can come and go from the pack as they please. They are usually the only pair of wolves that produce a litter of pups. All the members of the wolf pack help raise the wolf pups. Wolves also hunt as a pack.

Communication

If you have ever heard one dog barking at another dog, you have heard animal communication. Communication is the process by which animals share information. Mammals communicate in many ways.

Dogs and wolves mark their territory by placing their scent in an area. This chemical sign tells other animals to stay away. When skunks raise their tails, other animals know that the scent the skunks are about to spray is very strong and offensive. This drives many predators away.

Animals communicate through sound, too. Humpback whales sing songs to communicate to other humpbacks. Many sounds are made by male animals to attract females during mating season. Female crickets are attracted to the chirping of males. Animals also make sounds when they defend themselves. They may growl, hiss, or snarl to make themselves appear more frightening.

Animals can change the shapes of their bodies to communicate, as well. A cat can arch its back and bristle its tail to make itself look larger and stronger. A cobra snake holds its body upright and spreads out the skin below its head to scare away predators. Even an elephant will avoid a cobra when the cobra's body is in this position.

After You Read

1. How is an innate behavior different from a learned behavior?

2. What is a stimulus? What is a response?

3. Describe the members of a honeybee colony.

4. What are three ways in which animals communicate with each other?

5. Using the outline you have created in your Science Notebook, write three sentences that describe the behavior of animals.

Figure 15.23 The striped skunk *(top)* sprays predators with its strong scent. The humpback whale *(center)* communicates with other humpbacks by singing. The cobra *(bottom)* holds itself upright and spreads out the skin below its head to scare off predators.

KEY CONCEPTS

15.1 Birds

- Birds are endothermic vertebrates with feathers.
- Birds have contour feathers that protect them. They have down feathers that keep them warm.
- Molting is the process by which birds' damaged feathers are replaced. By preening, birds apply an oil to their feathers.
- Birds that can fly have powerful muscles attached to their sternums, or breastbones.
- The four main types of birds are flightless birds, perching birds, birds of prey, and water birds.

15.2 Adaptations for Flight

- Birds have small, lightweight bodies with hollow bones.
- Bird wings are wide and thin and slightly curved on the top. This allows birds to soar through the air.
- Migration is the movement of animals from one place to another.

15.3 Mammals

- Mammals are endothermic vertebrates with hair or fur. They feed their young from mammary glands.
- Based on the way they reproduce, the three main groups of mammals are the egg-laying mammals, the pouched mammals, and the placental mammals.
- Placental mammals give birth to live young that develop in the mother's uterus. The period of time the young stay in the uterus is called gestation.
- A carnivore is an animal that eats only other animals; a herbivore is an animal that eats only plants; an omnivore is an animal that eats both plants and other animals.

15.4 Animal Behavior

- Behavior is the way an animal responds to its environment. An animal is born with innate behavior. Learned behavior develops through experience.
- A stimulus is a signal that causes an animal to react in a certain way. The animal's reaction to a stimulus is called a response.
- Animals that live in groups show social behaviors. Each animal in a group has a specific role.
- Animals communicate with each other through scent, sound, or body positions.

VOCABULARY REVIEW

Write each term in a complete sentence, or write a paragraph relating several terms.

15.1
feather, p. 259
contour feather, p. 259
down feather, p. 259
preening, p. 259
sternum, p. 260

15.2
migration, p. 265

15.3
monotreme, p. 267
marsupial, p. 267
placental mammal, p. 268
uterus, p. 268
placenta, p. 268
gestation, p. 268
echolocation, p. 268
carnivore, p. 269
herbivore, p. 269
omnivore, p. 269

15.4
behavior, p. 271
stimulus, p. 271
response, p. 271
innate behavior, p. 271
instinct, p. 271
learned behavior, p. 271

PREPARE FOR CHAPTER TEST

To prepare for the chapter test, create a question from each Learning Goal. Use the information in your Science Notebook to answer each question. Then use these answers to write a well-developed essay about the chapter. Use the Key Concept on the first page of this chapter as your topic sentence.

MASTERING CONCEPTS

True or False
If the statement is true, write "true." If it is false, change the underlined word or words to make the statement true.

1. <u>Feathers</u> cover the body of a bird and are made of dead cells.

2. <u>Hibernation</u> is the regular movement of a group of animals from one place to another.

3. <u>Marsupials</u> are mammals that lay eggs.

4. <u>Carnivores</u> are animals that eat only plants.

5. <u>Behavior</u> is the way an animal responds to its environment.

6. A behavior that an animal is born with is called a(n) <u>learned behavior</u>.

Short Answer
Answer each of the following in a sentence or brief paragraph.

7. What are three characteristics common to all birds?

8. Why do birds molt?

9. What adaptations does a bird of prey have for catching its food?

10. What are three characteristics common to all mammals?

11. Describe three forms of animal communication.

Critical Thinking
Use what you have learned in this chapter to answer each of the following.

12. **Hypothesize** Why is preening important to a bird?

13. **Relate Facts** Which group of mammals is most similar to birds? Explain your answer.

14. **Apply Concepts** Why do whales come to the surface of the water many times during the day?

Standardized Test Question
Choose the letter of the response that correctly answers the question.

15. The illustrations show a frog taking a bad-tasting insect into its mouth and then spitting it out. The frog never eats this type of insect again. What type of behavior is the frog showing?

    A. innate behavior

    B. instinctive behavior

    C. learned behavior

    D. social behavior

An Endangered Animal Makes a Comeback

ALLIGATORS, the largest reptiles in North America, have powerful jaws full of sharp teeth. They can measure more than four meters in length and have a mass of about 450 kg. The most impressive fact about these animals is that they will eat just about anything.

The American alligator is native to freshwater wetlands from the Carolinas to Texas. In the past, as settlers fearful of alligators moved into these areas, they killed the reptiles by the thousands. By the early 1900s, alligators were scarce at the edges of their range, but they still were not in danger of dying out.

The status of alligators changed after World War I, when alligator skins became fashionable for making belts and shoes. The government also contributed to the demise of alligators. As part of major government construction projects, many swamps were drained, thus destroying alligator habitats. By the 1960s, the alligator population was extremely low. The alligator, which had survived on Earth for more than 200 million years, was now in danger of becoming extinct!

Biologists knew that these animals are an important part of wetland ecosystems and that their extinction would have serious effects on the plant and animal life of these ecosystems. In 1967, the U.S. government gave alligators protection as an endangered species, which made hunting them illegal. As a result of this action, the alligator population began to recover. By 1975, there were enough alligators to remove them from the endangered species list.

American alligators, now listed as a threatened species, currently number about three million. After almost disappearing, they have made an amazing comeback.

Research and Report

Choose an animal. Then create a chart that lists its major characteristics. For example, include where it lives, what it eats, and any threats to its survival. Set up the chart in the way that best displays the information. Include a sketch or photocopy of the animal at the top of the chart.

CAREER CONNECTION ANIMAL SHELTER WORKER

DO YOU LIKE ANIMALS? Do you become concerned when you see an injured or abused animal? Is your first instinct to help? If so, you might enjoy being an animal shelter worker. There is probably at least one animal shelter in or near your community. Animal shelters are places where people take care of lost, homeless, or injured animals.

Animal shelter workers spend their days working hard at a number of tasks. They feed the animals and keep both the animals and their enclosures clean. They watch the animals closely to look for signs of illness. If an animal has a problem, they may assist a veterinarian in treating the animal. Shelter workers also groom the animals—for example, cutting toenails or clipping fur. The shelter workers have fun with the animals, too. They play with the animals and help them get exercise by walking or running with them. Animal shelter workers also deal with the public. They answer questions about the animals and screen people who might want to adopt one.

People who want to become animal shelter workers do not need college degrees. They must be comfortable around animals and have good communication skills. Fortunately, they can learn most of what they need to know on the job. Although this is not a high-paying career, most shelter workers say that helping the animals in their care is the most important reward.

Desert Marvel

WHAT ANIMAL comes to mind when you think of survival in the driest places on Earth? Many people think of the camel. From its fine-haired coat that insulates against heat to its ability to go for long periods without food or water, the camel is made for the desert.

Camels do well in deserts because they don't need much water and can conserve what water they have. A camel can go from four to seven days without drinking. If the weather is cool and the camel isn't very active, it can survive for ten months by getting all of its water from eating plants. Although most animals lose lots of water through sweating when the weather is hot, that is not the case for camels. A camel does not sweat until its body temperature rises fairly high, to about 41°C (106°F). This allows the camel to conserve water. But when camels *do* sweat, they can lose more water without suffering than many other animals can. A 20-percent decrease in body weight due to water loss is fatal to most animals. A 12-percent loss would kill a human being. Camels can survive a 40-percent decrease in body weight due to water loss.

A camel's body has other adaptations for desert environments, as well. Thick eyebrows shield its eyes from bright sunlight. A double row of long eyelashes and an inner eyelid help keep sand out of its eyes. A camel can squeeze its nostrils and lips shut against blowing sand. Its feet spread out like pads so it can walk over soft sand without sinking. Pads of tough material

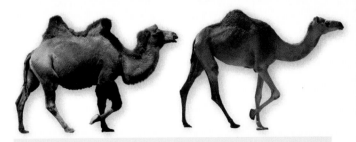

The Arabian camel, or dromedary, has one hump. The smaller Bactrian camel has two.

on its knees and chest protect a camel from hot sand when it kneels or lies down.

Zoos are the only places camels are found in the United States. But that wasn't always the case. In the mid-1800s, the U.S. government bought several dozen camels for use in transporting goods across the deserts of the Southwest. The camels carried heavier loads than other pack animals. They also lived on desert plants, such as cactuses, that mules and horses would not eat. Although the camels did their job well, most people found them noisy, smelly, and bad-tempered. The experiment ended quickly.

Today, camels are found mostly in the deserts of northern Africa, the Middle East, and Asia. They transport people and goods and work as farm animals. People also make camel hair into cloth. In some places, camels provide milk and meat. These odd-looking animals whose adaptations make them successful at surviving and reproducing in an extreme environment have proved to be a desert lifeline.

Solving the Mystery of the Monarchs

PEOPLE HAVE STUDIED monarch butterflies for more than 100 years. In the 1930s, biologists discovered that these insects migrate each spring and fall. However, scientists had no idea *where* the monarchs went.

Then, in the mid-1970s, hikers stumbled upon the monarchs' winter home in the mountains west of Mexico City. Millions of monarch butterflies covered every square centimeter of the trees in the mountains' forest.

Biologists determined that every yearly cycle of migration involves three or four generations. One group of monarchs leaves the northern United States and southern Canada in late summer. This group reaches Mexico in November and stays through March. As the weather warms, these monarchs head north. They stop in Texas and Louisiana to mate and lay eggs, after which they die. Their offspring continue the journey north. This second generation reaches the area around the midwestern United States. There, adults have more offspring and die. A third generation then crosses the Appalachian Mountains to the East Coast and remains there for the rest of the summer. It is this generation that heads south again to Mexico—to a mountain valley these butterflies have never seen.

Unit 6

Ecology

Living Things and Their Environment

KEY CONCEPT Each organism interacts with the living and nonliving things around it.

Mars is colder than the arctic, one of the coldest places on Earth. Its coldest places are so frigid that carbon dioxide freezes out of the atmosphere. Mars is also incredibly dry. All of its water is frozen, and it never rains there. The air is so thin that breathing it would make your lungs bleed. You would probably suffocate first, because there is no oxygen in the atmosphere. The Martian environment is harsh, unforgiving, and deadly to humans.

Scientists are planning a manned mission to Mars. If the mission is to succeed, its members must bring along everything people might need to survive. To do that, they must first understand how living things, such as the polar bear, interact with their surroundings on Earth.

Think About Environments

Imagine you are going to Mars and must pack everything you will need.

- In your Science Notebook, list at least six things you need to survive. Think about what your body needs to function and what you could not live without.
- Label items as things you can find on Mars and things you can find on Earth.

www.sclinks.org
Ecosystems **Code: WGB16**

Learning Goals

- Identify the biotic and abiotic parts of an environment.

- List the levels of organization in an environment.

- Understand how organisms divide up resources in a habitat.

New Vocabulary

environment
biotic
abiotic
ecology
population
community
ecosystem
biosphere
habitat
niche

Before You Read

Create a concept map in your Science Notebook. Start by drawing six circles. Write the word *Plant* in two of the circles, *Herbivore* in another two, and *Predator* in the last two. Predict how these organisms might interact if they lived in the same place. Record your predictions in your Science Notebook.

Look around. Everything you see and feel is a part of your environment. An **environment** (ihn VI run munt) is the place where an organism lives. Your environment includes living things such as plants, animals, people, and even the microbes living inside you. Living things make up the **biotic** (bi AH tihk) parts of the environment. Your environment also includes nonliving things such as your desk chair, the air you breathe, and the sunlight you see each day. Nonliving things make up the **abiotic** (a bi AH tihk) parts of the environment.

Organisms and Their Environments

An organism gets everything it needs to live from its environment. For example, a maple tree gets the energy it needs from sunlight and pulls the water and minerals it needs out of the soil. Living things rely on the abiotic parts of their environment.

Organisms that live in the same environment can affect one another. For example, prairie dogs are rodents that live on the Great Plains in the central part of the United States. They eat grass, and they are always on the lookout for ferrets that want to eat them. If a drought kills the grass, the prairie dogs will starve. The ferrets will also starve unless they can find other animals to eat. All of these organisms are interconnected. Something that affects one part of an environment can affect all of the organisms that live there.

Interactions among organisms and their environments can be very complex. The branch of biology that examines these interactions is called **ecology** (ih KAH luh jee). Scientists who study ecology are called ecologists.

Figure 16.1 Every environment has both living and nonliving parts. How do other living things affect your environment? What about nonliving things?

Figure 16.2 The endangered black-footed ferret depends on prairie dogs for food. If drought or human hunting kills enough prairie dogs, many ferrets will starve.

Levels in Environments

Ecology is an incredibly diverse subject. An ecologist might study a honeybee collecting nectar for its hive, a group of birds competing for food in a field, the interaction of plants and animals in Yellowstone Park, or the way in which sunlight affects the amount of food available at different levels in an ocean. In each case, the ecologist studies relationships between organisms and their environment. For study, ecologists organize these relationships into three levels: populations, communities, and ecosystems.

A **population** of organisms is a group of individuals from the same species living in a specific area at the same time. The cotton-top tamarins shown in **Figure 16.3** represent a population. An ecologist studying a population may look at the ways in which the organisms compete for resources such as food, water, and mates. Or the ecologist might examine how the organisms cooperate to avoid predators or raise young.

A biological **community** is made up of all the interacting populations of species in an environment. An ecologist studying a community may look at the relationships between predators and prey, the ways in which different species compete for the same kinds of food, or the effects of parasites on an infested population.

An **ecosystem** (EE koh sihs tum) is made up of both the community of organisms in an area and their abiotic surroundings. An ecologist studying an ecosystem might look at the mechanisms that move nutrients such as carbon, nitrogen, or phosphorus out of nonliving reservoirs and into living organisms.

All of the ecosystems on Earth are part of the **biosphere** (BI uh sfihr). The biosphere is the part of Earth that contains living things.

Figure 16.3 Members of a population of cotton-top tamarins compete for food but also warn each other about predators. The plants and insects eaten by tamarins are part of the biological community in a South American rain forest. So are the birds and snakes that eat tamarins. The amount of rain and sunlight the forest receives affects which foods are available for tamarins in their ecosystem.

As You Read

Imagine that each of the organisms in your concept map is a unique population of living things. Label each of the six circles with the word *Population*. Then, draw one large circle around all of your population circles.

Which level does the largest circle represent? Explain your answer.

 Explore It!

With a small group of your classmates, use string to mark out a 1-m² space on the grass outside your classroom. If there is no grassy area for you to study, mark out an area on the sidewalk or playground. Count how many different species you can find in your square of space.

Did You Know?

Earth's atmosphere once contained no oxygen. The first photosynthetic bacteria used carbon dioxide and sunlight to make organic compounds and released oxygen as a waste product. Over time, these bacteria changed the atmosphere!

Figure 16.4 The arctic tundra and a rotting log are examples of habitats that vary greatly.

Figure It Out

1. Which of these birds is a tufted titmouse?
2. The caterpillar and adult monarch butterfly are eating on different kinds of plants. How is this fact helpful to this species of butterfly?

Living in an Ecosystem

If you put a penguin in the Amazonian rain forest or a tropical hummingbird in the Antarctic, neither bird would survive for very long. Each bird is adapted to life in a particular place, or **habitat** (HA buh tat). Habitats can be large, like a forest, or small, like a puddle. Some organisms can even be habitats for others. You are the habitat for bacteria that live on your skin and in your intestines. Even the most barren habitats are home to multiple species that must share a limited amount of resources.

Populations of different species avoid fighting over resources by using them in different ways, as **Figure 16.5** shows. Woodpeckers, nuthatches, and tufted titmice, for example, are birds that share the same habitat. They all hunt insects on trees, but they do not compete for the same insects. Tufted titmice eat insects that live on the surface of bark and leaves. Nuthatches look for insects that live under tree bark. Woodpeckers drill into the tree to find insects hiding deep in the wood. These birds can share a tree because they are adapted for eating different insects in different ways.

To avoid competition, woodpeckers *(left)*, tufted timice *(center)*, and nuthatches *(right)* each have a beak adapted for eating different types of insects that live in different parts of trees.

To avoid competition, adult and juvenile butterflies eat different food. A monarch caterpillar *(left)* eats leaves, and an adult butterfly *(right)* sips nectar from flowers.

Figure 16.5 Organisms are adapted to different niches to avoid competition.

Ecologists call the unique strategy a species has for survival its **niche** (NICH). If you think of a species' habitat as its home address, its niche is its occupation. Species hunt, eat, hide, or reproduce in different ways to avoid occupying the same niche as, or competing with, other organisms.

After You Read

1. Review your concept map of a biological community. What elements could you add to make it represent an entire ecosystem?
2. Imagine a community in which rabbits eat grass and bobcats eat rabbits. Predict what would happen to the rabbit population if the bobcat population grew larger. Then, predict what would happen to the bobcat population if a fungus killed the grass. Explain your predictions.

Food and Energy in the Environment

16.2

- Describe how organisms move energy and matter through ecosystems.

- Identify the kinds of organisms found at each trophic level of an ecosystem.

- Explain how energy is lost at each step in a food web.

Before You Read

In your Science Notebook, write the first eight vocabulary terms in this lesson. Leave some space below each term. Then, write a definition for each term in your own words as you read about it in the lesson.

New Vocabulary

producer
autotroph
consumer
heterotroph
herbivore
carnivore
omnivore
decomposer
food chain
trophic level
food web
ecological pyramid
biomass

When you eat a hamburger, in a sense you eat sunlight. The lettuce, tomato, onion, pickle, and roll are all made from plants, and plants build their cells with energy from the Sun. The hamburger patty comes from a cow, which uses the solar energy stored in plant cells to build its own tissues. After you eat the hamburger, your body breaks down the meat and vegetables and uses the stored energy.

Energy—along with carbon, nitrogen, and other elements—moves through the environment in food. Plants build sugars with carbon from carbon dioxide in the air, and they build proteins with nitrogen from the soil. Animals build their bodies with elements from other organisms.

Movement of Energy and Matter

Living things that can capture matter and energy from abiotic sources are called **producers**. Producers, such as plants and green algae, use photosynthesis to turn the Sun's energy, water, and carbon dioxide in the air into complex organic compounds. Other producers, such as the bacteria living near deep-sea vents, can capture energy from other materials. Producers are also called **autotrophs** (AW tuh trohfs), or "self-feeders," because they make organic compounds from inorganic compounds and do not need to eat food.

Most organisms have to eat other organisms to obtain the energy they need to survive. Organisms that eat other organisms are called **consumers**, or **heterotrophs** (HE tuh roh trohfs). **Herbivores** (HUR buh vorz) are consumers that eat only autotrophs. **Carnivores** (KAR nuh vorz) are consumers that eat only other heterotrophs. Consumers that eat both autotrophs and heterotrophs are called **omnivores** (AHM nuh vorz).

When an organism dies, small organisms such as fungi, bacteria, and protozoans break down the complex organic molecules in its body for food. These heterotrophs are called **decomposers** (dee kum POH zurz). Decomposers are important parts of every ecosystem. They release the elements stored in dead tissues into the environment, where they can be reused.

Figure 16.6 Energy and matter flow through an environment as organisms eat one another.

Explain It!

Lions are carnivores that cooperate to hunt large animals such as zebras and wildebeest. In your Science Notebook, explain why lions also depend on the grass in their community.

Ecologists make models to study how energy and matter flow through an ecosystem. The simplest model of the flow of matter and energy in an ecosystem is called a **food chain**. A food chain can be constructed by writing the names of the organisms in an ecosystem according to the role that each plays and drawing arrows between them. The arrows show that one organism is eaten by the next. One food chain from the desert community in **Figure 16.7** could be shown as:

$$\text{seeds} \rightarrow \text{kangaroo rats} \rightarrow \text{rattlesnakes}$$

Each step in a food chain is called a **trophic** (TROH fihk) **level**. In this food chain, seeds make up the first trophic level. The next trophic level contains the kangaroo rats that eat the seeds. Organisms that eat plants are called first-order heterotrophs. In this example, the final trophic level contains the rattlesnakes that eat the kangaroo rats. Organisms that eat first-order heterotrophs are called second-order heterotrophs.

A food chain can illustrate and help explain one feeding pathway, but it does not show the complex feeding relationships in a real ecosystem. Biological communities have many species at each trophic level. Organisms may also eat more than one kind of food. These relationships can be drawn as a series of food chains, or they can be connected to form a food web. A **food web** is a model that shows all the possible feeding relationships at each trophic level. The food web in Figure 16.7 summarizes the feeding relationships in a desert community.

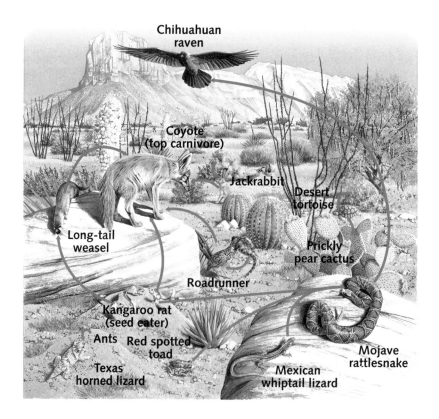

Figure 16.7 Arrows show how energy and matter flow through this desert ecosystem. Every food web starts with producers. Decomposers break down dead tissues at every level.

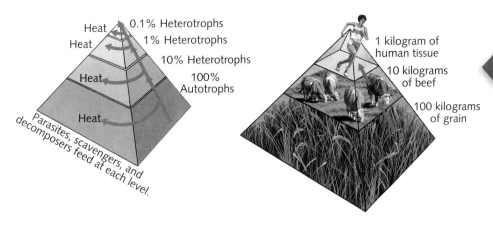

Figure 16.8 An energy pyramid shows that only ten percent of the energy stored in organisms can move to the next level *(left)*. A pyramid of biomass *(right)* shows that each higher level has only ten percent of the mass as the level below it.

Figure It Out

1. How many kilograms of beef must a human eat to make one kilogram of tissue?

2. Use these pyramids to explain why large, fierce animals are rare in ecosystems.

Ecological Pyramids

Food chains and food webs show ecologists "who eats whom" in a community. Another type of model, an **ecological pyramid**, helps ecologists see the relationships among trophic levels. Each layer of an ecological pyramid, as shown in **Figure 16.8**, represents a different trophic level. Autotrophs make up the base of the pyramid, and higher trophic levels are stacked above them.

One type of ecological pyramid shows how energy moves through an ecosystem. Energy is lost as it moves through heterotrophs. Each trophic level contains less energy than the level below it does. This happens because a heterotroph uses most of the energy it gets from food as fuel for life processes. Only about ten percent of the energy a heterotroph takes in is stored in new body tissues. Most of the energy is lost as heat.

Another way to show the stored energy in each trophic level is with a pyramid of biomass, also shown in Figure 16.8. **Biomass** (BI oh mas) is the total mass of living tissue in a trophic level.

An ecological pyramid can also show how many organisms are in each trophic level. Typically, population sizes get smaller in higher trophic levels. There are more autotrophs than herbivores, and there are more herbivores than carnivores. In some cases, however, one organism can feed many smaller organisms. For example, the upside-down pyramid in **Figure 16.9** represents a cow infested with thousands of parasitic roundworms.

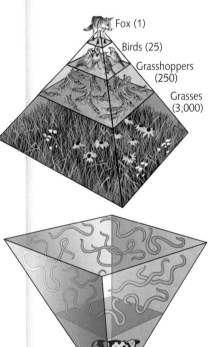

Figure 16.9 A pyramid of numbers *(top)* shows that each level feeds a smaller number of consumers than the level below it. The pyramid can also be inverted *(bottom)*, showing that a large number of parasites feed off of a single organism.

After You Read

1. Explain what would happen if an ecosystem had no decomposers.

2. In your Science Notebook, draw a simple food web of the community in a typical park.

3. Review the list of vocabulary terms you wrote in your Science Notebook. Use the terms to label the species at each trophic level of your food web.

Learning Goals

- Describe how water moves between Earth and the atmosphere.
- Explain how organisms move carbon between Earth and the atmosphere.
- Identify how nitrogen moves between abiotic and biotic reservoirs.

New Vocabulary

biogeochemical cycle
evaporation
condensation
precipitation
transpiration
eutrophication

Before You Read

Read the title, headings, and Learning Goals for this lesson, and look at the figures. Predict what you think you will learn in this lesson. Write your prediction in two or three sentences in your Science Notebook.

Have you ever recycled a plastic bottle? After you send the bottle to the recycling center, it is melted down and reused in a new product. The plastic may become part of a backpack, a pair of sneakers, or playground equipment, but it is still the same plastic that made up the original bottle.

This process is similar to the movement of matter through an ecosystem. Water molecules and elements such as nitrogen and carbon constantly move among the ocean, the atmosphere, and terrestrial ecosystems. Energy is lost as it passes through a community, but matter is recycled and reused over and over again. The recycling systems of an ecosystem are called **biogeochemical** (BI oh jee oh KEM ih kul) **cycles**.

Recycling Water

The water cycle starts with evaporation. **Evaporation** (ih va puh RAY shun) is the process by which heat from the Sun turns liquid water on Earth's surface into water vapor. The water vapor moves into the atmosphere, where it cools and turns into liquid droplets that form around dust particles in the air. The process by which water vapor cools and becomes a liquid is called **condensation** (kahn den SAY shun). The liquid droplets produce clouds, and as more water vapor condenses, more water droplets form. When the drops get heavy enough, they fall back to Earth as **precipitation** (prih sih puh TAY shun), which takes the form of rain, snow, sleet, or hail. Some water falls into lakes and oceans. Water that falls on land may flow to the ocean in streams and rivers or soak deep into the ground to form groundwater.

Water also moves through organisms. Plants pull water out of the ground through their roots and lose it by evaporation from their leaves during **transpiration** (trans puh RAY shun). Animals drink water and lose it when they urinate, sweat, and breathe.

Figure 16.10 Water constantly moves between Earth and the atmosphere.

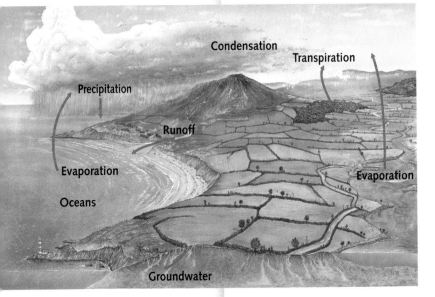

Condensation
Transpiration
Precipitation
Runoff
Evaporation
Evaporation
Oceans
Groundwater

Recycling Carbon

Carbon is the main building block of molecules in living things. Every carbon atom in a living thing was, at some point in the past, part of a carbon dioxide molecule in the air.

Autotrophs use carbon dioxide from the air to make the complex organic molecules found in their tissues. Heterotrophs that eat autotrophs break down the autotroph's molecules. Some of the molecules are broken down for energy. The carbon in these molecules then returns to the air as carbon dioxide. Heterotrophs use other molecules to form their tissues. When a heterotroph dies and decays, some of the carbon in its tissues returns to the air as carbon dioxide. The carbon dioxide given off by all living things is reused by autotrophs.

As You Read

Compare the water cycle with the carbon cycle. In your Science Notebook, explain one way in which the two cycles are the same and one way in which they are different.

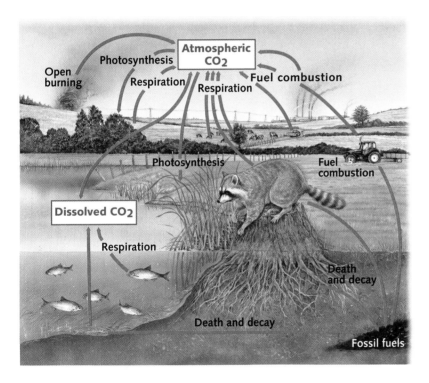

Figure 16.11 Carbon atoms move between the carbon dioxide in the air and the more complex organic molecules inside living things.

CONNECTION: Atmospheric Science

When organisms use carbon to build tissues, they slow down how fast carbon moves through the carbon cycle. The carbon can be locked in an organism's tissues for its lifetime. These molecules form a carbon reservoir, or a pool of stored carbon.

After an organism dies, its stored carbon will be released if the dead organism is eaten. If the dead organism gets buried, however, the carbon in its tissues enters another carbon reservoir, one that over millions of years forms fossil fuels such as coal and oil.

Burning fossil fuels releases carbon into the atmosphere as carbon dioxide. Carbon dioxide is a greenhouse gas. It traps heat that is radiated from Earth and reflects it back to the planet. The increased amount of carbon dioxide in the atmosphere may be at least partially responsible for global warming.

Extend It!

To model the water cycle, put a pot of water on a stove to boil. When the water boils, hold a glass bowl filled with ice over the steam. Be careful to avoid direct contact with the steam.

Where do beads of water appear? Where do evaporation and condensation occur in this model?

Recycling Nitrogen

Nitrogen makes up most of the air you breathe. Yet, like most organisms, your body cannot use it to build its nitrogen-rich proteins and nucleotides. Most of the nitrogen in organisms comes from bacteria that convert atmospheric nitrogen into usable forms.

Plants absorb nitrogen compounds made by bacteria and use the nitrogen to build proteins and other molecules. When herbivores eat plants, they reuse the nitrogen in their own tissues. The nitrogen is reused every time one animal eats another. At each step, some nitrogen leaves an animal in urine. Animal wastes return nitrogen compounds to the soil, where plants and bacteria reuse them. Dead and decomposing organisms also leave nitrogen compounds in the soil, and soil bacteria convert these compounds into nitrogen gas, which returns to the air.

Fertilizers with nitrogen make plants grow taller. However, nitrogen also dissolves in water and can be washed away. If extra nitrogen enters aquatic (water) ecosystems, algae grow uncontrollably. Eventually, exploding algae populations use up the oxygen in the water, and other organisms suffocate. This kind of nutrient pollution is called **eutrophication** (YOO troh fih kay shun).

Did You Know?

Venus's-flytraps, pitcher plants, and sundews are plants that live in acidic, nitrogen-poor soils. Instead of getting nitrogen from the soil, they get it by catching insects and digesting them.

Figure It Out

1. Name both ways in which nitrogen returns to the air.

2. Can you use urine as a fertilizer? Explain why or why not.

Figure 16.12 Nitrogen moves between the air and living things through life processes and complex reactions inside bacteria. The red arrows represent the movement of nitrogen compounds.

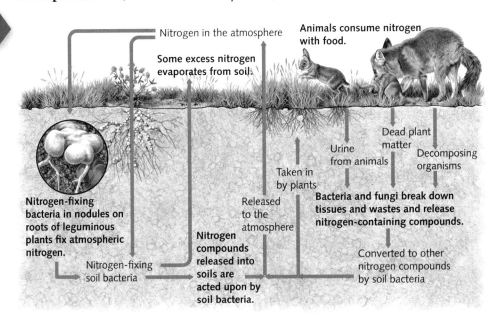

Nitrogen in the atmosphere

Some excess nitrogen evaporates from soil.

Animals consume nitrogen with food.

Nitrogen-fixing bacteria in nodules on roots of leguminous plants fix atmospheric nitrogen.

Nitrogen-fixing soil bacteria

Nitrogen compounds released into soils are acted upon by soil bacteria.

Released to the atmosphere

Taken in by plants

Urine from animals

Dead plant matter

Decomposing organisms

Bacteria and fungi break down tissues and wastes and release nitrogen-containing compounds.

Converted to other nitrogen compounds by soil bacteria

After You Read

1. What drives the water cycle? How does your answer compare with the predictions in your Science Notebook?

2. Where is most of the abiotic nitrogen found?

3. In your Science Notebook, design an experiment to test the idea that plants can only get nitrogen from the soil.

16.4 Changes in the Environment

Before You Read

Draw a T-chart in your Science Notebook. Label one side *Primary Succession* and the other side *Secondary Succession*. In your own words, define the terms *primary* and *secondary*.

On May 18, 1980, an earthquake under Mount St. Helens started a tremendous volcanic eruption. The north face of the mountain slid away in a huge avalanche, releasing a blast of superheated, rock-filled gas that ripped up the trees in its path. By the afternoon, slower, hotter flows of gas and rock had destroyed the trees and killed all living organisms in the soil. Mature forests were turned into ash-covered wasteland.

Since then, hardy plants have reappeared in the ash field. The plants attract herbivores that drop seeds from other plants in their dung. More than 28 years after the eruption, the forest is beginning to regrow.

Figure 16.13 Before May 18, 1980, Mount St. Helens was covered with mature forest *(left)*. After the eruption, the same spot was a barren ash field *(center)*. Now, wildflowers and shrubs have reestablished themselves *(right)*.

Ecosystems change over time. Sudden disturbances such as volcanoes, floods, or fires can affect which species will thrive in an environment. Living things also affect their ecosystems. One set of organisms living in an area can change the area's physical conditions in a way that favors the invasion of a new species. Over time, the sets of species that make up a community change. The steady progression of species invasion and replacement in an ecosystem over time is called **succession** (suk SE shun).

Primary Succession

Some places on Earth have no life. In some, land is being built by erupting volcanoes. In others, the soil has been sterilized by volcanoes or scraped away by glaciers. Living things will eventually invade all of these places. Ecologists call that colonization process **primary succession**.

Figure 16.14 Primary succession occurs when living things invade lifeless places. The first organisms to arrive build resources that other species use later.

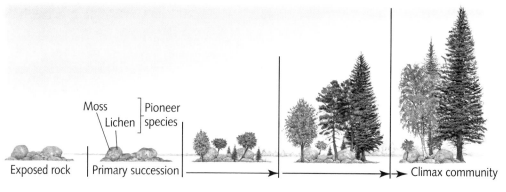

Figure 16.15 Primary succession on land starts when pioneer species colonize exposed rock. As the rock breaks down into soil, species such as ferns and insects move into the area. Later, trees, shrubs, and larger animals replace the smaller species.

The first organisms to move into a barren environment are called **pioneer species**. Pioneer species, such as mosses and lichens, can thrive in places that supply very few nutrients. When they die, their bodies add organic material to the environment and begin to build up soil. Pioneer species change bare rock into an ecosystem that can support more organisms.

A pioneer community attracts other organisms. Plants attract herbivores, herbivores attract carnivores, and the dead bodies attract decomposers. Over time, the community becomes more complex. As more species move in, the community's **biodiversity**, or variety of life, increases. Eventually, species that are better at competing for nutrients push out the pioneer species. The ecosystem then contains a **climax community**, or a stable community that undergoes little or no change in species. Environmental changes still occur, but the species that make up the community change slowly, if at all.

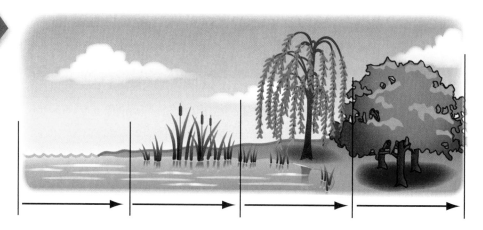

Figure 16.16 Primary succession can also occur in shallow lakes. Plants such as reeds and rushes grow at the edge of a lake and trap soil around their roots, forming new land. Then, plants such as willows grow in the wet soil at the edge of the lake and remove water with their roots. As the ground gets firmer, the willows are replaced by dry-ground climax species such as oaks.

Secondary Succession

Unexpected events can radically change biological communities. Fire can sweep through a forest, burning everything in its path. Floods can carry away land and everything living on it. People can clear woods to create open fields. An avalanche can destroy every tree in its path. Disease can kill a species of tree in an area, affecting every animal that uses that kind of tree for food. Over time, however, organisms move back into the area.

The recolonization of an area by living things after a disturbance is called **secondary succession**. Secondary succession generally takes less time than primary succession does. This is because the ecosystem already contains soil. The soil stores organic material and nutrients. It also contains seeds from older generations of plants. Weedy pioneer species sprout quickly in the cleared areas. Over time, the community of organisms in the area changes. Pioneer species are replaced with a climax community.

Figure 16.17 Natural events such as fires started by lightning can disturb environments and restart succession. This fireweed is a pioneer in secondary succession.

Figure 16.18 Human activities can also restart succession in an ecosystem. Cutting down a forest obviously disturbs an ecosystem, but so do dams on a river and forest fires caused by carelessness.

Some communities depend on regular disturbances. Without the occasional fire to kill tree seedlings, grasslands like the American prairies would eventually become forests. Recent research in the Grand Canyon has shown that the stream communities there depend on occasional floods to build sandbars. Floods also clean silt out of the streambeds where native fish breed. Climax communities are dynamic places that rebound after they are disturbed.

After You Read

1. In the T-chart in your Science Notebook, summarize what you have learned about secondary succession. What is the most significant difference between primary and secondary succession?

2. Are dandelions a pioneer species? Defend your idea.

3. Does secondary succession begin after an old tree falls in a forest? Explain.

Extend It!

Obtain photographs of an area that was disturbed at least 30 years ago. Compare the photos with recent photographs of the same area, or visit the area. In your Science Notebook, record how the ecosystem has changed.

CONNECTION: Wildlife

Wolves were reintroduced to Yellowstone National Park in 1995. Through their hunting, wolves cut the elk population in half. With fewer elk, more cottonwood and aspen trees grew. The trees fed beaver and provided homes for birds. By controlling elk numbers, the wolves have made the forest more diverse. As a result, wolves increased the biodiversity of the area.

KEY CONCEPTS

16.1 A Place to Call Home

- Environments have biotic (living) and abiotic (nonliving) parts.
- Environments are organized in levels: populations, communities, and ecosystems.
- Organisms live in particular places and have strategies for survival. The place in which an organism lives is its habitat. Its strategy for survival is its niche.

16.2 Food and Energy in the Environment

- As organisms eat one another, they move energy and matter through ecosystems.
- Models such as food chains and food webs show how energy and matter move through ecosystems.
- Ecosystems are organized in trophic levels that are defined by what organisms eat.
- An ecological pyramid shows the relationships that exist among trophic levels.
- Energy is lost at each link of a food web.

16.3 Cycles in the Environment

- Biogeochemical cycles move nutrients and water between abiotic reservoirs and living organisms.
- The water cycle is powered by sunlight.
- Autotrophs remove carbon from the atmosphere and incorporate it into edible tissues.
- Nitrogen-fixing bacteria remove nitrogen from the atmosphere and make it available to living organisms.

16.4 Changes in the Environment

- Ecosystems change over time. The steady change in the species that make up a community is called succession. Over time, succession leads to a climax community.
- Pioneer species can thrive under harsh conditions. They change an ecosystem and make it suitable for other organisms.
- If the climax community is disturbed, succession begins again.

VOCABULARY REVIEW

Write each term in a complete sentence, or write a paragraph relating several terms.

16.1
environment, p. 280
biotic, p. 280
abiotic, p. 280
ecology, p. 280
population, p. 281
community, p. 281
ecosystem, p. 281
biosphere, p. 281
habitat, p. 282
niche, p. 282

16.2
producer, p. 283
autotroph, p. 283
consumer, p. 283
heterotroph, p. 283
herbivore, p. 283
carnivore, p. 283
omnivore, p. 283
decomposer, p. 283
food chain, p. 284
trophic level, p. 284
food web, p. 284
ecological pyramid, p. 285
biomass, p. 285

16.3
biogeochemical cycle, p. 286
evaporation, p. 286
condensation, p. 286
precipitation, p. 286
transpiration, p. 286
eutrophication, p. 288

16.4
succession, p. 289
primary succession, p. 289
pioneer species, p. 290
biodiversity, p. 290
climax community, p. 290
secondary succession, p. 291

True or False

If the statement is true, write "true." If it is false, change the underlined word or words to make the statement true.

1. A <u>niche</u> is the place where an organism lives.

2. An organism that does not need to eat food is called a <u>heterotroph</u>.

3. The <u>nitrogen</u> cycle depends on soil bacteria.

4. <u>Secondary succession</u> occurs when a plowed but unplanted field becomes a forest.

5. Without <u>carnivores</u>, ecosystems would be filled with tissues from dead organisms.

6. Energy is <u>gained</u> at each link of a food web.

Short Answer

Answer each of the following in a sentence or brief paragraph.

7. Name three different animal populations that live in your area.

8. A fire destroys 50 square kilometers of mature forest. Summarize how the forest will regrow.

9. One spring, the newspaper reports that a flock of Canada geese has settled at the pond in the town park and is raising goslings. That summer, the newspaper reports that the pond is covered with algae and smells like dead fish. Analyze how these two stories could be related.

10. Discuss why there are more rabbits than wolves living in a forest ecosystem.

11. List a sequence of organisms in a food chain with three trophic levels.

PREPARE FOR CHAPTER TEST

To prepare for the chapter test, create a question from each Learning Goal. Use the information in your Science Notebook to answer each question. Then use these answers to write a well-developed essay about the chapter. Use the Key Concept on the first page of this chapter as your topic sentence.

Critical Thinking

Use what you have learned in this chapter to answer each of the following.

12. **Relate** How is photosynthesis related to the carbon cycle?

13. **Give Examples** Humans are omnivores. Give one example of a food humans eat that makes them first-order heterotrophs. Then give another example of a food humans eat that makes them second-order heterotrophs.

14. **Infer** Two closely related species of lice are associated with human beings. Both species drink blood, but head lice are found only in the hair or scalp, and body lice are found only in clothing. Compare the habitat and niche of each species, and infer why they live in these different places.

Standardized Test Question

Choose the letter of the response that correctly answers the question.

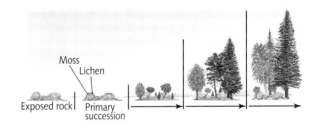

15. The diagram shows the same area over the course of several hundred years. Which term best describes the final picture?

   A. disturbed ecosystem

   B. climax community

   C. abiotic community

   D. pioneer species

Test-Taking Tip

If more than one choice for a multiple-choice question seems correct, ask yourself if each choice completely answers the question. If a choice is only partially true, it is probably not the correct answer.

Chapter 17

Earth's Biomes

KEY CONCEPT Around the world, similar communities develop in places with similar climates.

The tundra is a cold, dry, and windy place found in the far north. Short, cool summers and long, cold winters characterize the climate of the tundra. In the summer, the tundra features marshy soil, cool days, and a thick mat of plants that feed caribou, grizzly bears, and many other animals. Tundra also can be found high up on mountains far from the poles. The tundra is one example of a biome. How does the tundra compare with the place where you live?

Think About Describing a Biome

Earth has an amazing variety of landscapes, plant life, and animal life. If you could visit any natural place in the world, where would it be? Would you classify your selected location as a rain forest, coral reef, grassland, or another type of ecosystem?

- Find your travel spot on a world map or a globe. How far is it from the equator?

- In your Science Notebook, describe the plant and animal life, weather, and land or water features of the natural place you would like to visit.

www.scilinks.org
Biomes **Code: WGB17**

17.1 The Influence of Climate

Learning Goals

- Explain how climatic factors determine the distribution of biomes around the world.

- Describe the rain shadow effect.

New Vocabulary

climate
biome
latitude
tropical region
polar region
temperate region
rain shadow effect

Before You Read

Scientists often make observations and then try to trace each observation back to its cause. In your Science Notebook, describe an example of a cause and its effect from your everyday life. Then look for examples of cause and effect as you complete this lesson.

Around the world, similar communities develop in places with similar climates. **Climate** is the typical pattern of weather that is observed over a long period of time in an area. A place with water-conserving plants and very little rain is a desert, whether it is in Arizona, China, or Saudi Arabia. Climate is determined by such factors as temperature, precipitation, latitude, altitude, nearness to water, and land features. These factors, in turn, influence the communities that develop. Large areas with similar climax communities are called **biomes** (BI ohmz). A climax community is a stable community that undergoes little or no change in species.

There are two main types of biomes: land biomes and water biomes. The seven major land biomes are shown on the map in **Figure 17.1**.

Figure It Out

1. Which biomes are found in Africa near the Tropic of Cancer?

2. According to the map, which biome is typically found just south of the tundra?

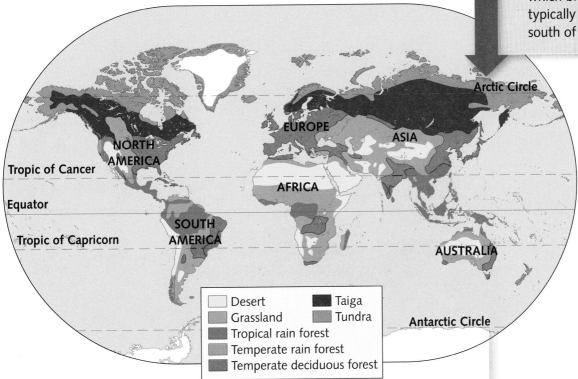

Figure 17.1 Based on the climax communities that form, the land surface of Earth can be divided into seven major biomes.

Temperature and Precipitation

Temperature is a major factor in the development of a community. Some plants have adapted to snow so that it slides off them, and they are able to survive harsh winters. Other plants would be killed by a single frost. The amount of precipitation an area receives also influences the development of a community. Precipitation is water that falls to Earth in the form of rain, snow, sleet, or hail. An oak tree would die quickly from the lack of water in a desert. In a rain forest, a desert cactus would die from too much water. The patterns of temperature and precipitation throughout the year are the most important features of climate.

Latitude and Climate

The **latitude** (LA tuh tewd) of a place describes its distance from the equator. Close to the equator, in the **tropical region**, the Sun's rays strike Earth's surface directly, as shown in **Figure 17.2**. The tropics are hot, with little difference among the seasons. The **polar regions**, near the north and south poles, get less direct sunlight. Polar regions have extremely cold winters and cool summers. During summer at the poles, it is light outside throughout both day and night. But during winter, many days pass without a sunrise. Between the tropical region and the polar regions are **temperate regions**. Temperate regions have four distinct seasons. Conditions in those regions are less extreme than in polar or tropical climates. Look back at Figure 17.1 to see which biomes are found in each region.

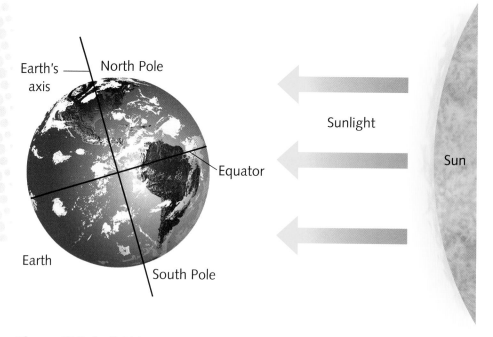

Figure 17.2 Sunlight is more intense near the equator than it is at the north and south poles because the Sun's rays strike the equator more directly. The difference in how the Sun's rays strike various latitudes is due to Earth's curved surface and its tilt on its axis of rotation.

Land and Water Features

Altitude is the distance above or below sea level. As you climb up a hill or mountain, your altitude increases. Increasing altitude has the same effect on climate as increasing latitude. The air tends to get colder and drier at higher altitudes. Plants and other organisms that are adapted to these conditions become more common. In fact, a climber may pass through several biomes while going up a single mountain.

Figure 17.3 Climbing a mountain can involve passing through taiga and tundra communities before reaching a cold desert at the top.

Large bodies of water can also affect climate. Land near an ocean tends to have a milder and wetter climate than areas farther inland do. Even large lakes can affect climate. For example, wet air from the Great Lakes causes deep snow to fall on nearby land. This is called lake-effect snow.

A mountain range close to the coast of a continent also affects the climate nearby. Due to the **rain shadow effect**, the area between the ocean and mountains is wet and mild, while the area beyond the mountains is hot and dry. **Figure 17.4** shows how the mountains along the west coast of the United States create a rain shadow.

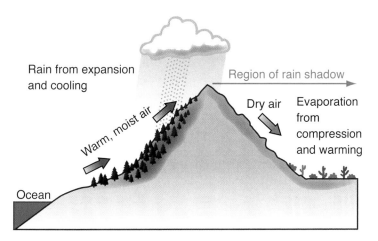

Figure 17.4 Mountains near a coast can create a rain shadow beyond the mountains. Warm, wet ocean air rises over the mountains. As it cools, it releases its moisture as rain. The air that arrives on the other side of the mountains is hot and dry.

After You Read

1. According to the notes in your Science Notebook, what effect does latitude have on climate? Do you think longitude impacts climate? Why or why not? What effect does nearness to a large body of water have on climate?

2. How can the existence of pockets of tundra near the equator be explained?

3. Describe the challenges a polar plant would face if it were moved close to the equator.

Explore It!

Bend two thin slices from a cucumber, and record your observations about their texture. Pour 100 mL of water into each of two shallow bowls. Stir 6 g of table salt into one bowl. Place one cucumber slice in each bowl. Note whether each slice sinks or floats in the water.

After 30 minutes, remove both slices and record your observations about their texture. How are the two slices different? Based on your observations, predict what would happen to a freshwater plant if it were moved to a saltwater environment. Record your answers in your Science Notebook.

Learning Goals

- Describe the characteristics of the seven major land biomes.
- Identify the relationship between specific climatic conditions and the biomes that develop.
- Explain how human activities affect the different biomes.

New Vocabulary

tundra
taiga
temperate rain forest
deciduous forest
grassland
tropical rain forest
desert

Recall Vocabulary

biomass (p. 285)
succession (p. 289)

Figure 17.5 In summer, the tundra is home to many birds, mammals, and insects. The musk ox is one of the few species that remain active through the long, hard winter.

Before You Read

Prepare a table in your Science Notebook with four columns and seven rows. In the first column, record the name of each biome. Use the second and third columns to record precipitation and temperature data. Use the last column to record other important facts you learn about each biome. Preview the lesson and the Learning Goals, and think about where each piece of information will fit into your table.

Tundra

Imagine traveling from the north pole through Canada toward the equator. At the pole, it is so cold and dry that no plants can survive. As you travel south, the temperature gets warmer and low plants emerge. This is the tundra. The **tundra** is a cold, dry, treeless plain. The average daily temperature is –12°C. There is a short summer growing season, but only 15 to 25 cm of precipitation falls each year. Permanently frozen soil, called permafrost, lies beneath the surface and prevents trees from taking root. In the summer, the tundra is filled with flowering plants, lichens, insects, birds, and grazing mammals. During the long, cold winters, most plants and animals become inactive or travel to warmer lands.

Taiga

As you proceed south from the tundra, tall trees appear and the summer days grow milder. This is the **taiga** (TI guh), an evergreen forest that covers more area on Earth than any other biome does. Winter is still long and cold, with temperatures falling as low as –50°C, but during the short summer, temperatures can reach 20°C. Precipitation, from 35 to 100 cm per year, falls mainly as snow. Like the tundra, the taiga is full of birds, mammals, and insects throughout the short summer. Animals that stay active in the winter adapt by growing thick coats and living in burrows to keep warm. The coniferous trees have wax-covered needles that conserve water and shed snow.

Figure 17.6 The taiga is dominated by a few species of coniferous trees. Other species may fill in gaps caused by forest fires. Many animals of the taiga migrate south or hibernate during the winter. Others grow warm coats and blend in, like this snowshoe hare.

Temperate Rain Forest

South of the taiga, several different biomes are possible. Along the rainy Pacific coast lies the largest temperate rain forest in the world. A **temperate rain forest** is a cool, wet, evergreen forest that receives 200 to 400 cm of rain per year. The nearby ocean keeps temperatures, which average 9 to 12°C, mild. Temperate rain forests form between oceans and coastal mountains, where the rain shadow effect causes clouds to drop most of their moisture.

The forest is made up mostly of a few species of hemlock and spruce trees, some of which grow to enormous sizes. Elk, bears, squirrels, and songbirds all thrive in the mild climate. The temperate rain forest has the most biomass of any biome. Biomass is a measure of the density of living and once-living material in an area.

Figure 17.7 Temperate rain forests contain mostly spruce and hemlock trees. These giants can grow for hundreds of years and measure 15 m around at the base. Elk browse the undergrowth and keep the forest floor clear. All but a few small areas of temperate rain forest in North America have been cut down.

Deciduous Forest

Deciduous forests cover much of the eastern United States. **Deciduous** (dih SIH juh wus) **forests** are dominated by hardwood trees that drop their leaves in the cold part of the year. Leaves help trees make food, but they lose water in the winter. Ice can also build up on leafy branches, causing them to break. Deciduous forests receive 70 to 150 cm of precipitation per year. The year is divided into four distinct seasons. Summer temperatures may reach 30°C, while winter temperatures can dip below freezing. Deer, squirrels, turtles, and songbirds can all be found on a walk through a deciduous forest.

The soil of a deciduous forest is rich from the leaves that fall there and decay. European settlers cleared forests for farming and lumber. Some of the forests have since recovered through succession.

As You Read

As you read about each land biome, fill in the table in your Science Notebook.

Which biome has long, cold winters and many coniferous trees?

Figure 17.8 Oak, maple, birch, and elm trees are a few of the hardwood species found in a deciduous forest. Animal life includes birds in the treetops and reptiles, such as this box turtle, on the ground.

1. The climate diagram shows that the Zambian savannah has distinct wet and dry seasons. In which months is there no rainfall at all?

2. What do you notice about temperature during the wet seasons in this area?

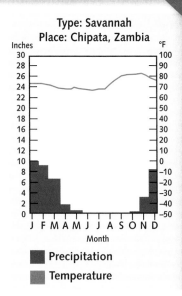

Type: Savannah
Place: Chipata, Zambia

■ Precipitation
■ Temperature

Figure 17.10 A climate diagram shows the average precipitation and temperature range in one area through the year.

Figure 17.9 Grasslands can look like oceans of grass. Many large mammals live there by grazing on the plants that grow from the rich soil. Much of the grasslands in North America have been turned into fields for farming or rangeland for livestock.

Grassland

Grasslands form where there is not enough precipitation to support trees. **Grasslands** are communities in which grasses are the most important plants. In some cases, grasses stretch as far as the eye can see. Trees are found only along the banks of streams and rivers. Grasslands receive 25 to 75 cm of precipitation per year. There is often a long dry season in the summer. The grasslands, or prairies, of North America have cold winters and hot summers. Temperatures range from below freezing to 35°C. Grasslands are found on every continent except Antarctica. In Africa, grasslands are called savannahs; in South America, they are called pampas (PAHM pus); and in Asia, they are called steppes (STEPS).

Many animals live on grasslands, from large grazing mammals such as antelopes to insects such as grasshoppers. Grasses can reach more than 3 m in height. The layers of dead grass that fall on grasslands enrich the soil. Fire is common in grasslands, and both plants and animals are adapted to it. Grasses grow back quickly from their roots after a fire. Animals hide in deep burrows, fly away from harm, or take shelter along a stream.

⊙ CONNECTION: Economics

Economists often try to place a value on natural areas. First, the economists calculate the dollar price of the land. Next, they consider the services provided by the land and the enjoyment people get from visiting the land. Wetlands, for example, provide valuable flood control. What value would you place on a park or other natural place you like to visit?

Tropical Rain Forest

Tropical rain forests are lush forests found near the equator, where it is warm and rainy year-round. The forests receive 200 to 600 cm of precipitation per year, and the temperature usually stays close to 25°C. These climate conditions are ideal for many plants, insects, and birds. The greatest competition in the tropical rain forest is for sunlight and nutrients from the soil. Heavy rainfall washes away much of the organic matter on the ground before it has a chance to build up. The poor soil of the tropical rain forest does not make the land good for farming when it is cleared. Nevertheless, over 100,000 square kilometers of tropical rain forest are cleared each year for human use.

Emergents These giant trees are much higher than the average canopy tree. Birds, such as the macaw, and insects are found here.

Canopy The canopy includes the upper parts of the trees. It's full of life—insects, birds, reptiles, and mammals.

Understory This dark, cool environment is under the canopy leaves but above the ground. Many insects, reptiles, and amphibians live in the understory.

Forest Floor The forest floor is home to many insects, and the largest mammals in the rain forest generally live here.

Figure 17.11 The tropical rain forest contains the most species of any biome. Scientists estimate that 50 to 90 percent of Earth's species live within these rain forests. Each species has a zone in the forest where it lives or spends most of its time.

Figure 17.12 The desert is the driest biome. Desert plants and animals, such as these lizards, are adapted for surviving the lack of water and extreme temperatures.

Desert

A desert forms where there is not enough rainfall to support grasslands. **Deserts** are the driest places on Earth, with less than 25 cm of precipitation per year. In many cases, deserts form because of the rain shadow effect. Moisture from ocean breezes condenses and falls on the sides of mountains. Beyond the mountains, a desert forms. Any rain that does fall on the desert evaporates or runs off the bare soil. Temperatures vary greatly because the dry air does not block the Sun's rays or trap heat overnight. Desert temperatures may rise above 40°C during the day and fall below freezing at night.

Organisms that live in the desert are adapted to temperature extremes and lack of water. Plants such as the cactus store water in their fleshy stems. They have large, shallow root systems that collect rainfall when it occurs. Some cactus species can live for several years off water stored from one rainfall. Other types of plants, including mesquites, have long taproots that reach underground water. Desert plants produce many seeds after a rainfall. These seeds provide food for insects, birds, and small mammals to eat throughout the year. Foxes, coyotes, lizards, and owls are common predators in the desert.

Warm deserts occupy about one-fifth of Earth's land area. Some human activities can increase the size of deserts or make new deserts. Such activities include changing river paths, pumping groundwater, and cutting trees in sensitive areas. Some scientists consider Earth's barren polar regions to be cold deserts. Together, cold deserts and warm deserts occupy about one-third of Earth's land surface.

Explain It!

What is the biome where you live? Find your area on the biome map in Figure 17.1. Use the Internet, books, and other resources to learn what your area was like before its settlement by Europeans. Draw a sketch and write a paragraph to describe a typical natural scene from that time.

After You Read

1. Compare and contrast tropical and temperate rain forests.

2. Which biomes are most affected by the rain shadow effect? Explain how they are affected.

3. Review the table in your Science Notebook. In a well-developed paragraph, discuss which biomes get the least amount of precipitation.

Water Biomes

Before You Read

In your Science Notebook, draw a water depth scale as a vertical line. Label the top of the line *Shallow Water* and the bottom *Deep Water*. Preview the lesson to look for features that change as water becomes deeper. Add any features you find in this lesson to the proper depth on your line.

About 75 percent of Earth's surface is covered by water. The water may be deep or shallow, fresh or salty, moving or still. Each of these factors affects the kinds of organisms that live in the water and the biome that develops.

The Freshwater Biome

Freshwater is water that contains very little salt. Wetlands, ponds, lakes, streams, and rivers all typically contain freshwater. Only about one percent of the water on Earth is usable freshwater. The rest is salty or is frozen in icebergs and the polar ice caps.

Wetlands   Areas where the soil is wet enough for aquatic plants to grow are called **wetlands**. The water in wetlands is very shallow, and the soil may dry out for part of the year. Layers of fallen plants make the soil rich in nutrients. Cattails, water lilies, and cypress trees grow thickly in wetlands. Snails, frogs, birds, and many insects make their homes in wetlands, as well.

Ponds and Lakes   Ponds and lakes form where water pools in a low place in the ground. A shallow body of water with plants growing all the way to the middle is called a pond. A larger body of water is called a lake. In the shallow upper layer of a lake, the water is warm and full of nutrients. Sunlight supports the growth of plants. The plants create plenty of oxygen for animals such as fish, frogs, and snails. Sunlight does not penetrate the cold, deeper regions of a lake. There are fewer nutrients in the deep water, and plants cannot grow there. Decomposers consume the oxygen in deep water, preventing most animals from living in the depths. The layers of lake water usually mix in spring and autumn.

Wetland Losses in the United States
1780s to 1980s

% loss of wetlands
- 0–20
- 20–40
- 40–60
- 60–80
- 80–100

Figure It Out

1. Which regions of the country have lost the highest percentage of their wetlands?

2. Compare this map with the biome map in Figure 17.1. In which biome are the states with the greatest loss of wetlands located?

Figure 17.13 Across the United States, wetlands have been drained to gain usable land and to control mosquitoes. Draining wetlands creates good cropland, but it may increase flooding problems downstream.

As You Read

Look for information about how temperature, oxygen level, light, and nutrients are affected by water depth. Add pertinent information to your water depth scale.

Does shallow water contain more or less oxygen than deep water does? Explain your answer.

Rivers and Streams Rivers and streams contain moving water. The movement of the water mixes in oxygen from the air. Sunlight reaches the bottom of shallow rivers and streams. Slow-moving rivers have muddy bottoms with plants growing in them. Swift streams have rocky bottoms. The organisms in a fast-moving stream are adapted for clinging to the rocks so they will not be washed downstream. Many fish, insects, and other invertebrates live in rivers and streams. Birds, otters, and alligators visit the streams to eat the creatures living there.

Water gradually travels downhill until it reaches the ocean. The area where the freshwater of a river mixes with the salty water of the sea is called an **estuary** (ES chuh wer ee). Rivers bring many nutrients to the estuary. Estuaries provide rich habitats for grasses, young sea animals, and birds.

Figure 17.14 The freshwater biome includes wetlands *(left)*, ponds and lakes *(center)*, and rivers and streams *(right)*.

Extend It!

Plankton form the base of the ocean food chain. Research plankton and find out the difference between phytoplankton and zooplankton. Record this information in your Science Notebook. Also explain the importance of plankton and support your explanation with an illustration of a marine food chain or food web.

The Marine Biome

The oceans and seas of the world make up the **marine**, or saltwater, biome. The marine biome covers most of Earth's surface. Like freshwater lakes, oceans can be divided into shallow and deep layers. Sunlight penetrates about 200 m into the water, creating a warm layer where plants can grow. The plants produce oxygen, allowing many sea creatures to live. The base of most ocean food chains is made up of **plankton**, tiny algae, bacteria, and animals that float in the water. Even the blue whale, the largest animal on Earth, survives by eating plankton. The deep water of the ocean is cold, dark, and largely lifeless except for areas around undersea volcanoes. Autotrophic bacteria that do not use sunlight as their energy source form the base of the food chains in these areas.

Coral reefs form in the shallow, warm water of the tropics. **Coral reefs** are the accumulated skeletons of tiny coral animals. Algae and mollusks that secrete calcium-containing shells help cement colonies of different types of corals together into a reef. The complicated shapes of the reefs provide habitats for many species of invertebrates and fish. Birds and sea mammals depend on the reef animals for food. Coral reefs are sensitive ecosystems that can be damaged easily by ships, pollution, and other human activities.

Seashores are the edge between land and marine biomes. At high tide, the seashore is wet and salty. At low tide, the seashore dries out. The organisms that live on the seashore are adapted to these changing conditions. Near the seashore, forests of giant seaweed called kelp support sea urchins, sea stars, and sea otters.

⊕ CONNECTION: Earth Science

The discovery of thriving ecosystems deep in the ocean was a surprise to both Earth scientists and biologists. Sunlight penetrates only about 200 m into the ocean. As a result, scientists expected to find only decomposers in the depths.

The first photographs from thousands of meters below the surface of the ocean were taken in the 1970s. The pictures showed colorful shrimps, worms, and layers of bacteria. Researchers discovered that some bacteria can live off the hydrogen sulfide and other chemicals released by underwater volcanoes. These bacteria form the basis of a new ecosystem never before seen by humans.

Undersea communities are exciting for many reasons. First, they may represent the earliest life on Earth. In addition, scientists have found that some deep-sea creatures contain chemicals that may be useful in medicine and industry. Scientists also recognize that there is more to study and learn about the oceans of the world. Finally, by studying these communities, scientists hope to learn if and how life can survive on planets other than Earth.

Figure 17.15 Shallow, well-lit ocean water supports many species of fish, including tuna *(top)*. Coral reefs *(center)* provide homes to colorful fish and invertebrates. Plants and animals of the seashore *(bottom)* are adapted to changing water, temperature, and salt levels.

After You Read

1. Compare the amount of freshwater to the amount of salt water on Earth.

2. According to the depth scale in your Science Notebook, where would you expect to find plants growing in water?

3. Create a sample food chain for a freshwater ecosystem.

Summary

KEY CONCEPTS

17.1 The Influence of Climate

- Large areas with similar climax communities are called biomes.
- Climate determines which biome develops in an area.
- Major factors influencing climate include temperature, precipitation, latitude, altitude, land features, and water features.
- In the rain shadow effect, the area between the ocean and mountains is wet and mild, while the area beyond the mountains is hot and dry.

17.2 Land Biomes

- Earth's land surface can be divided into seven major biomes: tundra, taiga, temperate rain forest, deciduous forest, grassland, tropical rain forest, and desert.
- Each biome has characteristic plant life and animal life.
- Each biome has characteristic ranges of temperature and precipitation.
- Organisms are adapted to the conditions of their biome.
- Human activities have made major changes in some biomes.

17.3 Water Biomes

- Most of Earth's surface is covered by the freshwater biome and marine biome.
- The freshwater biome includes wetlands, ponds and lakes, and rivers and streams.
- An estuary is the area where the freshwater of a river mixes with the salt water of the sea.
- The marine biome includes oceans and seas, coral reefs, and seashores.
- Depth and movement have important effects on water conditions.

VOCABULARY REVIEW

Write each term in a complete sentence, or write a paragraph relating several terms.

17.1
climate, p. 295
biome, p. 295
latitude, p. 296
tropical region, p. 296
polar region, p. 296
temperate region, p. 296
rain shadow effect, p. 297

17.2
tundra, p. 298
taiga, p. 298
temperate rain forest, p. 299
deciduous forest, p. 299
grassland, p. 300
tropical rain forest, p. 301
desert, p. 302

17.3
freshwater, p. 303
wetland, p. 303
estuary, p. 304
marine, p. 304
plankton, p. 304
coral reef, p. 305
seashore, p. 305

PREPARE FOR CHAPTER TEST

To prepare for the chapter test, create a question from each Learning Goal. Use the information in your Science Notebook to answer each question. Then use these answers to write a well-developed essay about the chapter. Use the Key Concept on the first page of this chapter as your topic sentence.

True or False

If the statement is true, write "true." If it is false, change the underlined word or words to make the statement true.

1. <u>Deciduous forest</u> trees lose their leaves during winter.

2. Most of the water on Earth is <u>marine</u> water.

3. The weather pattern of an area is called its <u>latitude</u>.

4. Plants grow all the way to the center of <u>lakes</u>.

5. <u>Tropical rain forests</u> and <u>deserts</u> have very rich soil.

6. Freshwater and marine water mix in a(n) <u>estuary</u>.

Short Answer

Answer each of the following in a sentence or brief paragraph.

7. Explain what prevents forests from growing in the tundra.

8. List four differences between a deep lake and a fast-moving stream.

9. Describe the challenges a tree from a tropical rain forest would face in a grassland.

10. Diagram a possible food chain in a marine biome.

11. Describe the seasons in a tropical region, a temperate region, and a polar region.

12. Explain the factors that allow a tropical rain forest to be home to more species than any other biome.

13. Identify the biomes in which the temperature never falls below freezing.

Critical Thinking

Use what you have learned in this chapter to answer each of the following.

14. **Evaluate** Choose one biome that has been largely removed by humans and another that has not been so severely affected. Evaluate why the biomes have been treated differently by humans.

15. **Relate** How do the discoveries of undersea vents and deep underground life relate to the search for life on other planets?

16. **Analyze** Choose an organism that lives near you. Analyze how this organism is adapted to the climate in which you live.

Standardized Test Question

Choose the letter of the response that correctly answers the question.

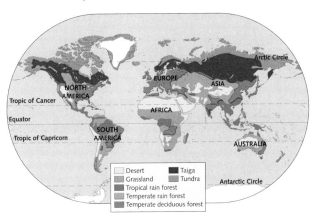

17. According to the map, grasslands are found in _____ regions.

   A. temperate and polar

   B. polar and tropical

   C. polar, temperate, and tropical

   D. temperate and tropical

Test-Taking Tip

If you finish before time is up, check your answers. Make sure you answered each part of every question and did not skip any parts.

Chapter 18

Interactions Among Living Things

KEY CONCEPT Population sizes and individual survival depend on interactions among organisms, their environment, and humans.

The Serengeti Plain in Africa is popular with tourists. Travelers from around the world come to see herds of wildebeest, zebras, and elephants.

The Serengeti is home to many individual animals, but it is striking that some species are much more common than others. Why are there so many zebras but only a few lions? What happens to the animals in times of drought? Why are scientists concerned that elephants will die out in the wild? This chapter is about what determines population sizes and how different species, including humans, interact with one another.

Think About Population Sizes

There are no herds of zebras in your neighborhood, but there is one group of animals that lives and feeds near humans almost everywhere: birds.

- Find a place with wild birds. In your Science Notebook, list the names of the birds you observe, or make a sketch or write a description of each one.

- For each species, note the largest group you see at one time. Why don't you see the same number of each kind of bird? How do you think that living so close to humans affects the birds? Record your thoughts.

www.scilinks.org
Environment **Code: WGB18**

18.1 Population Growth and Size

Learning Goals

- Describe population growth without limiting factors.
- Identify the different reproductive styles and their effects on population growth.
- Discuss the limiting factors for a population's growth and size.
- Explain how a population grows to its carrying capacity.
- Understand how natural events can affect populations.

New Vocabulary

factor
reproductive cycle
limiting factor
density
competition
range
carrying capacity

Recall Vocabulary

population (p. 281)
habitat (p. 282)
ecosystem (p. 281)

Before You Read

Make a T-chart in your Science Notebook. Label the columns *Factors that Increase Population* and *Factors that Decrease Population*. Preview the lesson, and predict what you will write in each column.

Have you ever noticed how the number of each kind of tree in one area tends to remain constant, while the number of weeds changes throughout the year? If so, you have made observations about populations of living things. A population is a group of organisms of the same species living in a specific area. The size of a population depends on several **factors**, or elements influencing a result. Population ecologists study the factors that lead a population to grow, shrink, or reach a steady level.

Population Growth

The early English settlers in North America were homesick. They missed the climate, animals, and plants of their homeland. While they could not change the climate, they did send for many of their favorite animal and plant species from England. The dandelion was one plant they introduced to the continent. From a few seeds, the species grew and thrived. Native Americans, westward-traveling pioneers, and the wind took its seeds across the country. Soon, the plant was growing from coast to coast.

The population growth of the dandelion is typical for a new species in an area with many resources. At first, the population grows slowly. Each plant, however, produces many seeds in its lifetime. The population grows faster and faster. The speed at which a population can grow is set by the species' **reproductive cycle**, or rate and timing of reproduction.

Some species, such as whales, sharks, and humans, produce a few large offspring after a long period of growth. Others mature quickly and produce many young. In addition to dandelions, many insects and small rodents follow this pattern. Of the two patterns, organisms that mature early will produce the larger population over a given number of years. This is because after only a short time, there will be many breeding members of the population.

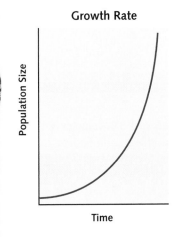

Figure 18.1 Just after it arrived in North America, the dandelion showed uncontrolled population growth. A line graph with a J-shaped curve represents this pattern. What would happen if this rate of growth continued for many years?

Growth Rate

Population Size

Time

Figure 18.2 The orange shading on this map shows the range of the barn owl. What limiting factors could determine the edge of the range in each direction?

As You Read

Complete the T-chart in your Science Notebook. Write each factor that affects population size and growth in the correct column.

Would the presence of a lot of waste products tend to increase or decrease population growth?

Figure 18.3 Building purple martin houses has helped increase the population of this desirable species. What limiting factors are decreased by taking this measure?

Limiting Factors

No population can continue to grow uncontrolled for long. Why not? Imagine a population of houseflies, starting with 100 eggs, half male and half female. If the 50 females each produced 2,000 offspring in 50 days, there would be 100,000 flies in the second generation. Fifty days later, there would be 100,000,000 flies. Within ten generations, the population would have a mass equal to that of planet Earth!

Long before the fly population could reach such a size, individual flies would run short of food and places to lay eggs. They would be subjected to conditions that would limit their rate of growth. Any factor that reduces the growth rate of a population is called a **limiting factor**.

Food, Water, and Nutrients Every organism must take in food or sunlight for energy. In addition, each organism needs water and a supply of vitamins and minerals to build its body. As a population grows in **density** (DEN suh tee), or number of individuals per unit of area, the nutrients available to it may become harder to find.

Living Space and Nesting Sites Some organisms prefer to live closer together than others, but almost all need some personal space or they show signs of stress. Wolverines are an extreme example of spacing, with a population density of one individual per 100 square kilometers. At the other extreme are herbs such as garlic mustard, with millions of individuals per square kilometer.

Plants must have access to light and room for their roots to travel through the soil. Certain animals must have places in which to construct their nests. Many species of birds, for example, have plenty of space in which to live but are limited by the number of available nesting sites, such as holes in large, dead trees.

Wastes and Disease As a population's density increases, its waste products can build up to unhealthy levels. The wastes can poison the organisms directly, or they can spread diseases. Earlier in history, waste and disease were often factors that limited the population growth of humans in cities. Fortunately, humans have developed systems for removing and treating wastes. Without this limiting factor, the populations of cities are able to grow in a healthy way.

Whenever a population's resources are limited, competition will arise. **Competition** is the struggle among living things for the food or space they need in order to survive and produce offspring. There are some places where the limiting factors are so severe that no members of a species can live. The edges of these areas define the **range** of the population, or where it can be found.

Carrying Capacity

The number of organisms of one species that an environment can support for a long time is called the environment's **carrying capacity**. In most cases, populations start small and grow until they are larger than the carrying capacity of their habitat. Limiting factors cause the population to drop below the carrying capacity for a period of time before growing once again. The whole cycle is shown in **Figure 18.4**.

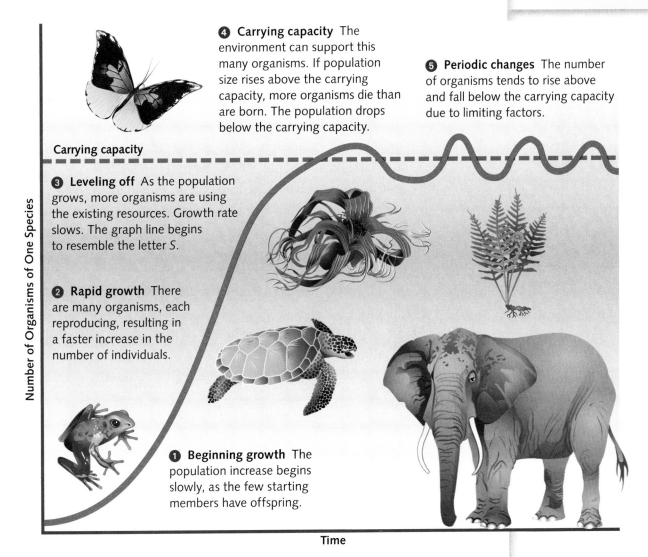

4 Carrying capacity The environment can support this many organisms. If population size rises above the carrying capacity, more organisms die than are born. The population drops below the carrying capacity.

5 Periodic changes The number of organisms tends to rise above and fall below the carrying capacity due to limiting factors.

Carrying capacity

3 Leveling off As the population grows, more organisms are using the existing resources. Growth rate slows. The graph line begins to resemble the letter *S*.

2 Rapid growth There are many organisms, each reproducing, resulting in a faster increase in the number of individuals.

1 Beginning growth The population increase begins slowly, as the few starting members have offspring.

Number of Organisms of One Species

Time

Figure 18.4 A population in a habitat with few limiting factors grows larger until it reaches the habitat's carrying capacity. Populations grow and change periodically whether they are plant or animal, whether on land or in the ocean.

Natural Events

Imagine a stable ecosystem in which most populations are near their carrying capacities. Suddenly, a nearby volcano erupts, or a hurricane floods the area, or a swarm of grasshoppers passes through. In each case, a natural event has changed the rules of the environment for a time, and the populations will respond.

Some natural events destroy almost all of the populations in the area. However, what is a disaster to some populations can be good luck for others. The grasshoppers may eat every blade of grass, allowing clover to sprout and find sunlight. While rabbits may starve without plants to eat, birds, foxes, and mice will feast for a time on insects. Populations may rise or fall dramatically, but in time, most will recover to their earlier levels.

The time it takes a population to return to its carrying capacity depends on the organisms' reproductive style. Aphids, tiny insects that are born pregnant, can repopulate an area in a matter of weeks. Sharks, however, produce only a few young each year, starting when they are several years old. It may take decades for a shark population to recover.

Population Equation

The number of births and deaths does not always give enough information to calculate the size of a population. Individuals in a population sometimes move to a new community, as well. The size of a population can be calculated using the following equation.

> **This Year's Population** = **Last Year's Population + Births + Individuals Moving In – Deaths – Individuals Moving Out**

All of the factors you have just read about are limiting factors that affect at least one part of this equation.

Figure It Out

1. In which year did the population of black grouse drop suddenly?
2. If the population is half male, predict the carrying capacity for grouse in this area.

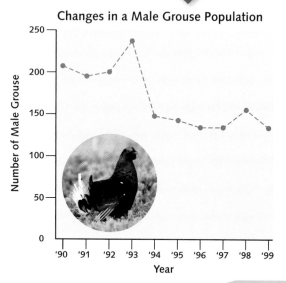

Changes in a Male Grouse Population

Number of Male Grouse (y-axis: 0, 50, 100, 150, 200, 250)

Year (x-axis: '90 '91 '92 '93 '94 '95 '96 '97 '98 '99)

Figure 18.5 This black grouse population dropped significantly in the 1990s.

After You Read

1. What pattern does a population follow as it grows from a few individuals in a new environment?
2. According to the chart you made in your Science Notebook, what are three factors that tend to decrease a population's growth rate?
3. Last year's population of a species of bird in a certain location was 1,842. During the year, there were 425 births and 72 deaths. A total of 38 individuals moved out of the area, and 83 individuals moved into the area. Calculate this year's population of the bird species.

18.2 Relationships Among Populations

Before You Read

Create a concept map for the word *Relationships* in your Science Notebook. Use the headings, subheadings, Learning Goals, and vocabulary terms in the lesson to construct your map. Include definitions of the terms, examples of each type of relationship, and other important details.

No place on Earth is home to only one species of living organism. Everywhere you look, there are communities made up of many different populations that share the same resources. At first, a forest may appear to be made up mainly of trees. Upon closer examination, you may notice caterpillars eating the tree leaves, fungi growing on the trees, squirrels living in the trees, and bees carrying pollen from one tree to another.

Symbiosis

A close relationship between individuals of two or more species is called **symbiosis** (sihm bee OH sus, plural: symbioses). The word *symbiosis* comes from Greek, in which *syn-* means "together" and *-bios* means "life." The partners in symbiosis typically have effects on each other's lives. They also affect each other's **fitness**, which is the ability of a living thing to survive and reproduce. Symbioses are classified according to which species benefits from the relationship. There are three kinds of symbiosis: mutualism, commensalism, and parasitism.

Mutualism A symbiotic relationship in which both species benefit is called **mutualism** (MYEW chuh wuh lih zum). There are many examples of mutualism involving species from every kingdom of life. The human body is involved in mutualism with the bacteria in the digestive tract. These bacteria get nutrients from food that is eaten, and they get a safe place to live within the body. At the same time, they fight harmful bacteria and produce vitamin K and other chemicals the body cannot make on its own. The proof of mutualism is that both organisms suffer when they are apart from each other.

Most plants form a type of mutualistic relationship with fungi that is called a mycorrhiza (mi koh RI zuh). Fungi have hairlike structures called hyphae (HI fee), which take up water and nutrients. The fungi live off sugars and other food they collect from plant roots. In turn, the fungi use their networks of hyphae to absorb water and minerals that are shared with the plant. Plants that are grown without mycorrhizal fungi are more sensitive to droughts and may not get enough phosphorus from the soil. Mycorrhizal fungi cannot survive away from their plant partners.

Figure 18.6 Relationships between honeybees and flowering plants are mutualism because both species benefit. A bee gets nectar and pollen to eat and also spreads pollen from one flower to another. Separately, the bees would starve and the flowering plants might be unable to make seeds.

Figure 18.7 The remora is a fish that is commensal with sharks. Remoras attach to sharks with a sucker, catching a ride and scraps of food from the sharks' meals. The sharks are neither harmed nor helped.

As You Read

Fill in the concept map in your Science Notebook. For each type of relationship, give a definition, an example, and one or two details. Also indicate which partner benefits from the relationship.

What is the name for an organism fed upon by a parasite?

Commensalism A symbiotic relationship in which one organism benefits and the other is not affected is called **commensalism** (kuh MEN suh lih zum). One common type of commensalism occurs between jellyfish and young fish. A young fish benefits from the relationship because it is protected from predators by the jellyfish's stinging tentacles. The jellyfish is neither helped nor hurt by the presence of the fish. **Figure 18.7** shows another example of commensalism.

When two species live closely together, they almost always affect each other's fitness. Over time, most of the examples of commensalism have been reclassified as other types of symbiosis.

Parasitism **Parasitism** (PER uh suh tih zum) is a symbiotic relationship in which one organism, the parasite, feeds off the living body of the other organism. The parasite benefits, while the **host**, or the organism off which the parasite lives, is partially eaten or harmed. The host thus loses fitness. Some parasites, including fleas, mosquitoes, and ticks, live on the outside of the host. Other parasites, such as tapeworms, can live only inside the host's body.

Parasitism is common in nature. All viruses are parasites, as are many bacteria and fungi. Almost every free-living species on Earth is host to at least one parasite. Parasites have evolved in such a way that while they hurt the host, they usually do not kill it. If the host were to die, the parasite would likely also die unless it could quickly find another host.

Many diseases are caused by parasites. Malaria, which infects over 350 million people per year, is caused by a protist called *Plasmodium*. *Plasmodium* uses a mosquito as its first host. When the mosquito bites a person, it injects *Plasmodium*-containing saliva into a vein. The parasite lives and reproduces inside the person, eating red blood cells. When another mosquito bites an infected person, the *Plasmodium* can be passed on to a new host.

Competition

When two species struggle for the same limited resources, they are in competition. Each is hurt by the presence of the other. Plants compete for sunlight, water, and minerals, while animals may compete for food and nesting places. The need to compete successfully for resources is one of the most important causes of evolution by natural selection.

Figure 18.8 Zebra mussels are a species of mollusk from Russia that was accidentally brought to the American Great Lakes on ships. They reproduce very quickly and compete with native mussels for food and places to live. As zebra mussels spread to new waterways, scientists expect native populations to decrease.

Predation

When one creature kills and eats another creature, it is called **predation** (prih DAY shun). Predation involves two individuals—a predator and its prey. The predator benefits by gaining food, while the prey that is eaten can no longer survive and reproduce. Unlike parasitism, which lasts a long time without killing the host, predation is a one-time relationship. Predator and prey species can affect each other's carrying capacity in a community.

Often an ecosystem contains chains or networks of predators and prey. For example, grass is eaten by snails, which are eaten by birds, which are eaten by snakes. If the top predator in a community is removed, all of the other species in the network will be affected, for better or worse. In this example, the removal of snakes will have an effect on the populations of birds, snails, and grass.

Humans have killed off wolves throughout most of the United States. This was done to prevent wolves from eating livestock and to reduce competition between human hunters and wolves. One unexpected result was the great increase in the number and range of coyotes. It turns out that coyotes were competing with wolves for small animal prey. Also, wolves killed and ate many young coyotes. So as the population of wolves decreased, the population of coyotes increased. Elk, rabbit, mouse, and deer populations also grew rapidly after the removal of wolves. These increases led to overgrazing and destruction of crops and plant communities. Without the wolf population, people have to shoot, trap, and poison many of its former prey themselves.

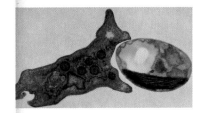

Figure 18.9 Predators come in all sizes. This amoeba, which is a protist, is preying on an algal cell.

Interspecific Relationships		
Relationship	**Organism 1**	**Organism 2**
mutualism	+	+
commensalism	+ commensal	0 host
parasitism	+ parasite	– host
competition	–	–
predation	+ predator	– prey

Key: + helped   – hurt   0 not affected

Figure 18.10 This table shows a summary of the possible relationships between two species.

Figure It Out

1. Which two relationships affect both species in the same way?

2. What does the term *interspecific* mean?

Explore It!

Choose a large animal or plant to be your study subject. Carefully observe all of the interactions your subject has with the plants, animals, and other organisms around it. Record your observations in your Science Notebook. Classify each relationship as mutualism, parasitism, commensalism, competition, or predation.

After You Read

1. Explain what would happen to the population sizes of a parasite and its host if they were separated from each other.

2. According to the concept map in your Science Notebook, what are some benefits an organism could gain from symbiosis?

3. Imagine whitetail deer becoming extinct in the United States. How would this affect ticks, woody plants, mountain lions, and mule deer?

Before You Read

In your Science Notebook, write the headings in this lesson. Leave enough space to write a few lines under each heading. As you read the lesson, look for examples of each type of adaptation.

So far, scientists have found and named more than 2.5 million species. They estimate that another 2 to 50 million species remain undiscovered. Each of these species interacts with many others through symbiosis, predation, and competition. Why don't the most skilled predators simply eat up all of their prey? Why doesn't one species of tree take over all of the forests of the world? The answers lie in the ability of species to adapt, or change in ways that improve their survival and reproductive success. Natural selection favors the individuals that are best adapted to their environment and the other organisms in it.

Adaptation to Symbiosis

Two species whose lives are closely connected tend to become adapted to each other. Parasites get better at finding good hosts and feeding on them without causing their deaths. Host species adapt by poisoning parasites or attacking them with their immune systems. Commensal species adapt by seeking partners that will provide the full benefit of the relationship. If small fish are more likely to survive when they are protected by a jellyfish's tentacles, the fish species will likely adapt to recognize and attach themselves to jellyfish.

Mutualism results in some of the most dramatic adaptations. Pollinators, such as birds and insects, and the flowers they pollinate are a good example. Plants produce flowers with the shape, color, smell, and timing that their pollinators prefer. Some flowers even evolved to look much like females of the insect pollinator's species. Mutualism can develop from other relationships, such as the one between predators and prey. Many fruits, for example, contain seeds that are carried by fruit eaters to new places where they can grow.

Sometimes, one partner in symbiosis performs certain life functions for both species. Lichens (LI kunz), for example, are formed by fungi and their bacterial or algal partners. The algae or bacteria produce all of the food for lichens through photosynthesis. The fungi absorb water and minerals and shelter the algae or bacteria. Lichens can grow in habitats that neither species could survive in alone.

Figure It Out

1. What relationship exists between the hummingbird and the plants that feed it?

2. Predict what would happen to lobster claw plants with a mutation that caused them to have white flowers.

Figure 18.11 Hummingbirds and the flowers they pollinate are both adapted to their symbiotic relationship. These lobster claw flowers are red, hummingbirds' favorite color. The plant even produces two lengths of flowers that match the beak lengths of male and female birds.

Adaptation to Competition

When two species use the same limited resources in the same way, they will compete with each other. The species best adapted to the environment will eventually take over the habitat. The only way for the two species to live together is for each one to have a different adaptation for using the same resource. For example, two species of birds called finches may eat the same kinds of seeds on separate islands. Where the finches are found together, one species has a beak for cracking seeds that is smaller than the other species' beak. This adaptation reduces competition enough for both species to succeed.

Another adaptation to competition involves making the environment unpleasant for competitors. Trees such as black walnuts release from their roots chemicals that prevent the growth of other plants. This reduces competition for sunlight and nutrients. In another case, bluebirds and wild mice are both limited by the number of nest holes in their habitat. Humans have built tens of thousands of nest boxes to increase the population of bluebirds, but many of them have been taken over by mice. Years after a mouse has used a box, the smell will still keep bluebirds away.

Adaptation to Predation

Predators and prey both adapt to their relationships. Prey species adapt by becoming harder to find and less tasty to predators. They may also adopt a reproductive strategy of having many offspring at once. For example, all of the maple trees in a population produce a heavy crop of seeds at the same time every few years. There are too many seeds for predators to eat them all. By the time the predator populations have grown larger, the young trees are too big to be eaten.

Predators also adapt to increase their fitness. They can become better at catching and killing their prey. Behaviors such as speed, teamwork, and quiet movement help lions kill animals larger than themselves. In addition, many predators can survive for a long time without catching any prey. Spiders can wait months between meals, and wolverines store meat under the snow.

Figure 18.12 The katydid on this leaf uses camouflage, or colors and patterns that are hard to see. The fitness of the katydid is greater because a predator is less likely to find it.

As You Read

Record in your Science Notebook examples of adaptations that organisms make to each type of relationship.

What is an adaptation an organism has that reduces competition for food?

Extend It!

Choose one species and research its relationships and adaptations within its community. Record answers to the following questions in your Science Notebook.

- Does your species form any mutualisms or commensalisms?

- What are your species' most important parasites? Is your species ever a parasite?

- What are your organism's predators and prey? Which other species compete for the same resources?

- What adaptations does your species have that relate to the other organisms in its environment? Which other organism affects the fitness of your species the most?

PEOPLE IN SCIENCE Lynn Margulis 1938–

In 1966, biologist Lynn Margulis suggested an entirely new way that a prey species can adapt: it can become a working part of the predator! What Margulis meant was that some of the organelles within complex cells started out as simple bacteria that were eaten by other bacteria. The prey bacteria survived inside their hosts and even reproduced themselves. In some cases, the prey bacteria made a useful product, such as sugar, through photosynthesis. Meanwhile, the prey bacteria were protected from other predators and from oxygen, which could poison them. The symbiotic cell had greater fitness than the simpler cells around it.

Margulis's ideas about cell predation turning into mutualism are called the endosymbiotic theory. Margulis predicted that organelles formed in this way would have double membranes and DNA separate from the rest of the cell. In fact, the chloroplasts that carry out photosynthesis and the mitochondria that make energy for cells have both of these predicted characteristics.

Although it took many years for Margulis to convince other scientists, her endosymbiotic theory is today the accepted model for the origin of eukaryotic cells.

Adaptation to Natural Events

Natural events can have important effects on populations. Organisms that can adapt to fires, droughts, and floods will gain in fitness. One adaptation developed by some species is forming seeds or eggs that will be ready to take advantage of a natural event. Plant seeds can survive underground for many years until a fire or fallen tree opens a space for the young trees to grow. Jack pine seeds will not germinate, or start to sprout, until a fire has scorched them.

Some species adapt to natural events by leaving the affected area for a time. Other organisms, such as mosquitoes, take advantage of floods to breed rapidly before puddles have time to dry. All of these adaptations save animals and help them reproduce. They also shorten the time it takes for a community to recover from a natural disturbance.

Figure 18.13 Most generations of these aphids are wingless. If food becomes scarce, the aphids produce young with wings that can fly to a better habitat. This adaptation helps the population survive in an unpredictable environment.

After You Read

1. What evidence indicates that a change in a species is an adaptation?

2. According to the notes in your Science Notebook, what limiting factor is reduced when birds develop different sizes of beaks?

3. Imagine a plant living in the rain forest, where it must compete for sunlight and limited nutrients in the soil. There are many grazing animals at ground level, as well as insect parasites. The forest floods every ten years. What adaptations might increase the plant's fitness in this environment?

18.4 People and the Environment

Learning Goals

- Describe human population growth over the last 1,000 years.
- Explain the ways in which humans affect the environment.
- Identify steps that can be taken to conserve wild populations and habitats.

New Vocabulary

technology
natural resource
renewable resource
nonrenewable resource
sustainable use
habitat destruction
introduced species
pollution
global warming

Before You Read

Preview the lesson. Read the headings and the Learning Goals, and look at the pictures. Think about what you expect to learn from this lesson. Write the headings in your Science Notebook, leaving space for a few sentences under each heading.

It's easy to imagine how modern humans can significantly affect the environment. They can use chainsaws to cut down trees, dams to flood valleys, and guns to kill the fiercest animals. In fact, human impact on the environment began many thousands of years ago. At the end of the last ice age, there were llamas, camels, giant sloths, mammoths, and five kinds of wild horses in North America. Today, there are none. All of these species died out soon after the first humans arrived on the continent.

Human Population Growth

All populations are limited by factors that set a carrying capacity, or maximum steady population size. Look at **Figure 18.14**, which shows human population growth over the last 1,000 years. How has the human population continued to increase in size without facing limiting factors? Humans use **technology**, or the ability to control nature, to grow more food, to provide materials, and to treat diseases. So far, humans have not reached their carrying capacity on Earth.

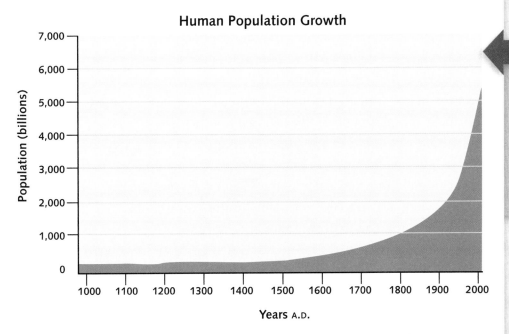

Figure 18.14 The human population has grown faster and faster over the last 1,000 years.

Figure It Out

1. Around what year did the human population reach one billion?

2. What are some factors that could eventually limit the human population?

As You Read

Summarize the key points of each section under the correct heading in your Science Notebook. Be sure to add specific details and facts.

How does habitat destruction affect the well-being of both wild populations and humans?

As a result of using technology, humans impact the natural world more than any other species does. Without limits on their behavior, humans can damage the environment in many ways.

Resource Use   As the human population grows, it requires more and more resources to keep its members healthy and comfortable. Each new family needs a place to live, food, water, clothing, medicine, and tools. It takes a lot of wood, plastic, grain, cotton, and metal to meet the needs of just the United States!

Natural resources include water, trees, sunlight, coal, animals, wind, and all other useful products of the natural world. Some of these resources are **renewable resources**, or resources that can be used many times if managed properly. Water, trees, air, and sunlight are examples of renewable resources. Other resources, such as coal and crude oil, take millions of years to form. They are **nonrenewable resources** because once they are used, they are gone forever. Responsible planners ensure that they make **sustainable use** of natural resources, taking only as much as can be replaced by nature. Hunting, cutting trees, and pumping water for human use can all be sustainable activities if humans limit themselves to a level that nature can replenish.

Habitat Destruction   No matter how carefully people plan, land that is used to grow crops or to build houses is no longer suitable for most of the organisms that once lived there. **Habitat destruction** occurs when a habitat is removed and replaced with some other type of habitat. As a result, the organisms living at the site must move or be destroyed. Examples of habitat destruction include draining wetlands, clearing land for farming, building housing developments, and strip-mining.

Habitat destruction is the most important reason that species are threatened with extinction today. Habitat destruction can hurt humans, too. Scientists believe that the terrible destruction caused by Hurricane Katrina in 2005 resulted from the removal of wetlands that would normally take up much of the regional floodwaters.

Killing of Sensitive Species   Throughout history, humans have hunted for several reasons. They obtained food and necessary materials from animals and eliminated competition for crops and prey. Many species were driven to extinction by hunting thousands of years ago. As the human population grew, more and more species of animals were threatened with extinction from too much hunting. Laws were passed to ban the killing of sensitive species and to limit hunting. Nevertheless, poaching, or illegal hunting, continues to threaten many populations.

Some sensitive species of animals and plants play important roles in the ecosystems to which they belong. Whenever one species is removed, countless others are affected.

Introduced Species   Wherever humans go, they take with them animals, plants, and microorganisms. If these species take hold in a new land, they are called **introduced species**. Introduced species, especially diseases, can have serious effects on an ecosystem and on the human population. The smallpox virus, which Spanish explorers brought to North America, killed 25 percent of the Aztecs and 60 to 90 percent of the Inca empire. American chestnut trees, which used to make up a quarter of eastern forests, have been almost completely killed off by an introduced fungus from Asia.

While the introductions of some species are accidental, others are intentional. People sometimes bring in natural enemies of introduced species in order to control them. Unfortunately, only about one in five biological-control efforts is successful. Others, such as the introduction of cane toads to Australia, end up causing even more damage.

Pollution and Atmospheric Change   Over the last few centuries, humans have relied heavily on technology to feed and provide for their growing population. All of this machinery and production requires a lot of energy. In just the last 50 years, world use of fossil fuels—coal, oil, and natural gas—has increased by four times.

Two side effects of today's heavy use of fossil fuels are pollution and global warming. **Pollution** is the release of harmful substances into the environment. Sources of pollution include fertilizer in runoff from fields, smoke from power plants, and oil spills. Fertilizers can cause major changes in water ecosystems. Air pollution can result in human sickness and acid rain that kills trees and pond life. Oil spills kill animals by poisoning them.

Global warming is the rise in the average temperature of Earth's air and oceans. During the twentieth century, the planet warmed by about 0.6°C. While this may not sound like a big change, it was enough to speed up melting of polar ice caps and affect weather all over the world. Hurricanes, tornadoes, flooding, and droughts have become more common. Animals such as polar bears that depend on the arctic ice are likely to become extinct. Some species, especially parasites, will be able to expand their ranges as Earth warms, but scientists predict that the total number of species in the world will decrease.

Figure 18.16 Fertilizer in runoff from farms ends up in the Gulf of Mexico and causes algal blooms (green and light blue areas of the map). As the algae die, decomposers use up all the oxygen in the water, killing most of the sea life in an area, now the size of New Jersey, called the Dead Zone (red area of the map).

Figure 18.15 A common frog is sitting on the head of a South American cane toad in this photo. The very large cane toad was introduced to Australia in the hope that it would destroy sugarcane beetles. Instead, the cane toad eats important pollinators and poisons native animals that try to eat it.

Figure 18.17 Pollution such as this smog over Los Angeles damaged human health and the environment, leading to the passage of the Clean Air Act.

Choices for Conservation

There are many ways in which humans can change and damage the natural world. There are also steps people can take to preserve it.

Protective Laws and Treaties   Governments often make laws to protect species or ecosystems when it is clear that they are in danger of being destroyed by humans. The laws might protect only one or two species, like the Eagle Protection Act of 1940. They might protect whole ecosystems, as in the Wetlands Conservation Act of 1989. In each case, the laws make it illegal to kill or damage the protected species and ecosystems or to develop property in a way that would threaten them.

Some threats are too big for any one country to handle alone. In such cases, many countries will sign a treaty, or international agreement, to prevent further ecological damage. A total ban on the import and export of endangered species products was put in place with the Convention on the International Trade in Endangered Species (CITES). CITES has worked effectively to increase the elephant populations in some countries.

Protective laws are often controversial, meaning that not everyone approves of them. People in the United States value freedom and dislike laws that restrict their actions. Many protective laws are examples of efforts to balance the needs of humans with the needs of the environment.

Antipollution Measures   Some environmental laws try to protect the environment as a whole, rather than protect certain species or sensitive habitats. In the nineteenth and early twentieth centuries, there were few laws controlling factory smokestacks and pollution from mines and farms. When it was clear that the health of people and wildlife suffered from the pollution, the Clean Air Act, the Clean Water Act, and related acts were passed. These laws require businesses, farms, and governments to clean up any water or factory waste before it enters the environment.

Preserves   Sometimes, the best way to protect a species or a habitat is to set aside land as a nature preserve, or protected area. Preserves may belong to local, state, or national governments, private organizations, or individuals. Nature preserves are managed with the environment in mind, but they usually allow visitors to hike on marked paths. Many preserves have the motto, "Take only pictures; leave only footprints."

Personal Choices   You can do a lot to help preserve the natural world. Keep in mind that every purchase you make and every product you use has an impact. Take the catalogs your family gets in the mail as an example. The average American receives more than 60 catalogs in the mail each year. More than nine billion kilograms of catalogs are mailed yearly in the United States. Catalogs are usually made directly from lumber, often from the old-growth forests of Canada. What can you do to help reduce the effects on the environment? You can cancel catalogs you don't read, recycle used catalogs, or call or write to companies urging them to use recycled paper.

Ordinary citizens can improve human effects on the natural world by insisting that it is the right thing to do. Millions of concerned people boycotted, or refused to buy, tuna caught in fishnets that also drowned dolphins. The result of this campaign is clear: More than 90 percent of the tuna now sold in the world is caught using dolphin-safe nets.

The bald eagle is a large bird of prey with a wingspan of 2 to 2.5 m. It lives near water in every U.S. state except Hawaii, nests in tall trees, and mates for life. Bald eagles hunt for fish and eat dead animals. Without humans, limiting factors for eagle populations include less food in the winter, diseases, and predation on the young by owls and raccoons. **FIgure 18.18** summarizes the human threats to bald eagles and what has been done to preserve the species.

Figure It Out

1. In which year did the eagle population rise past 5,000 nesting pairs?

2. When the population reaches its carrying capacity, will it be above or below the level of 1782? Explain your prediction.

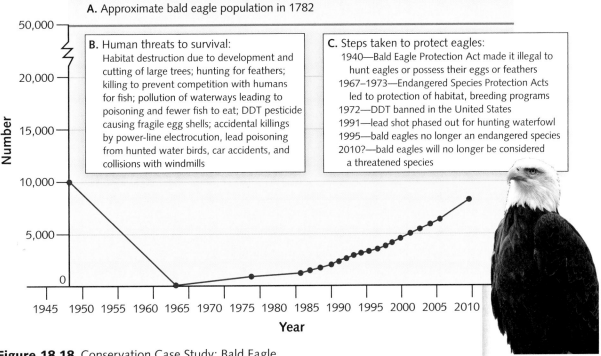

Nesting Pairs of Bald Eagles in the Continental United States

A. Approximate bald eagle population in 1782

B. Human threats to survival:
Habitat destruction due to development and cutting of large trees; hunting for feathers; killing to prevent competition with humans for fish; pollution of waterways leading to poisoning and fewer fish to eat; DDT pesticide causing fragile egg shells; accidental killings by power-line electrocution, lead poisoning from hunted water birds, car accidents, and collisions with windmills

C. Steps taken to protect eagles:
1940—Bald Eagle Protection Act made it illegal to hunt eagles or possess their eggs or feathers
1967–1973—Endangered Species Protection Acts led to protection of habitat, breeding programs
1972—DDT banned in the United States
1991—lead shot phased out for hunting waterfowl
1995—bald eagles no longer an endangered species
2010?—bald eagles will no longer be considered a threatened species

Figure 18.18 Conservation Case Study: Bald Eagle

After You Read

1. Explain why the human population has not yet reached its carrying capacity.

2. What are five actions people can take to save species and ecosystems?

3. Describe two effects that humans have on the environment that can be repaired and two others that are nearly impossible to undo.

KEY CONCEPTS

18.1 Population Growth and Size

- A population's growth rate is determined by its reproductive style, limiting factors, and natural events.
- Each population has a carrying capacity, or maximum number of individuals that its environment can sustain.

18.2 Relationships Among Populations

- Communities are made up of interacting populations that may help or hurt each other.
- A close relationship between individuals of two or more species is called symbiosis.
- Different species compete with each other for limited resources, causing all to evolve.
- Ecologists design experiments to determine what relationships exist between species, now and in the past.

18.3 Adaptations: Challenges and Opportunities

- Adaptations to relationships with other species increase the fitness of organisms.
- Two species with closely connected lives tend to become adapted to each other.
- Two species that use the same resources can survive in the same area only if they are adapted for using the resources in different ways.

18.4 People and the Environment

- Humans have raised Earth's carrying capacity for the human population through the years by making use of technology.
- Humans can cause enormous changes to the environment by altering habitats, killing other organisms, introducing new species, and releasing pollution.
- There are many steps that governments and people can take to prevent or repair environmental damage.
- Each person's choices about how to use resources affect the environment.

VOCABULARY REVIEW

Write each term in a complete sentence, or write a paragraph relating several terms.

18.1
factor, p. 309
reproductive cycle, p. 309
limiting factor, p. 310
density, p. 310
competition, p. 310
range, p. 310
carrying capacity, p. 311

18.2
symbiosis, p. 313
fitness, p. 313
mutualism, p. 313
commensalism, p. 314
parasitism, p. 314
host, p. 314
predation, p. 315

18.4
technology, p. 319
natural resource, p. 320
renewable resource, p. 320
nonrenewable resource, p. 320
sustainable use, p. 320
habitat destruction, p. 320
introduced species, p. 321
pollution, p. 321
global warming, p. 321

PREPARE FOR CHAPTER TEST

To prepare for the chapter test, create a question from each Learning Goal. Use the information in your Science Notebook to answer each question. Then use these answers to write a well-developed essay about the chapter. Use the Key Concept on the first page of this chapter as your topic sentence.

MASTERING CONCEPTS

True or False
If the statement is true, write "true." If it is false, change the underlined word or words to make the statement true.

1. In <u>parasitism</u>, both species in the relationship benefit.
2. To avoid <u>predation</u>, some species can run fast and have camouflage.
3. <u>Commensalism</u> occurs when people bring plants to a new environment.
4. The rate and timing of a species' reproduction is called its <u>carrying capacity</u>.
5. Humans have not yet reached their carrying capacity due to their use of <u>technology</u>.
6. Two species struggling for the same limited resource is called <u>predation</u>.

Short Answer
Answer each of the following in a sentence or brief paragraph.

7. What are the differences between a parasite and a predator?
8. Describe the limiting factors that affect trees in a forest.
9. Why is habitat destruction the most important reason that species are threatened with extinction?
10. What determines the range of a population?
11. What are some steps that your school could take to make its use of resources more sustainable?
12. Why have most examples of commensalism been reclassified over the years?

Critical Thinking
Use what you have learned in this chapter to answer each of the following.

13. **Analyze** In what ways are snakes adapted to their predatory lifestyle?
14. **Predict** Goats are introduced to an island that previously had no large mammals. Predict the effect on native plants, mice, fleas, and eagles.

15. **Diagram** A fire reduces the populations of two species of mice to very low numbers. One species breeds rapidly at a young age, while the other has only a few offspring at a much later age. Draw a diagram showing how each population grows to its carrying capacity.
16. **Devise a Plan** Pollution, being hunted for meat and oil, and noise in the ocean have nearly caused the blue whale to become extinct. Devise a plan to help the blue whale population recover.

Standardized Test Question
Choose the letter of the response that correctly answers the question.

17. The relationship between this mosquito and the human it is biting is best described as

    _____.

    A. predation
    B. commensalism
    C. parasitism
    D. mutualism

Test-Taking Tip

Don't get stuck on a difficult question. Instead, make a small mark next to the question. Remember to go back and answer the question later. Other parts of the test may give you a clue that will help you answer the question.

Going to Extremes

MOST LIVING THINGS on Earth exist within strict boundaries—for example, in places that are neither too hot nor too cold, neither too damp nor too dry. But, extremophiles are different. They live comfortably in environments that would kill most other forms of life. They live in boiling springs, bone-dry deserts, dark caves, and even the vents of volcanoes.

Scientists have explored all sorts of extreme environments on Earth in search of these organisms, which are mostly microorganisms. They have found many types. Some thrive in temperatures over 80°C (176°F). Others prefer freezing cold temperatures. There are even those that can withstand environments as acidic as the inside of a car battery.

Amazing ecosystems of extremophiles exist on the ocean floor. They live near seafloor cracks,

These extremophiles live in darkness on the seafloor.

or vents, in Earth's crust. These vents spew out boiling hot water full of chemicals such as hydrogen sulfide. Until scientists discovered these life-forms, they thought green plants or algae formed the first trophic level of just about every food chain. The extremophile ecosystems on the ocean floor, however, never have access to sunlight. The unusual bacteria at the base of these deep-sea food chains use chemicals that pour out of hot-water vents for the energy they need to make organic compounds. These organisms live in constant darkness, crushing pressure, and scalding temperatures. They include unusual species of fishes, shrimps, crabs, and worms nearly 1 m long.

Thermophilic (heat-loving) bacteria thrive in this hot spring at Yellowstone National Park.

Extremophiles challenge what we have always thought about life on Earth. Scientists now know that life can exist in environments on Earth that were once thought to be incapable of supporting life. Many scientists think that it is important to look more closely at the hostile environments on other planets. Life could be lurking there, too!

CAREER CONNECTION ARBORNAUT

RAIN FORESTS are the most diverse ecosystems on Earth, but they are very difficult to study. In a rain forest, most of the plant and animal life is high up in the forest canopy. The upper parts of the canopy can reach 30 m (100 ft) above the ground. If you're a biologist who studies rain forests, there is just one thing to do: climb into the trees.

Biologists who do their work at the tops of rain forests are called canopy biologists, or arbornauts. For these scientists, getting to work is an amazing process. They use all sorts of strange contraptions—towers, cranes, rope systems,

platforms, and even devices that look like small ski lifts. Once they are up among the branches, they are in danger of falling or being caught in sudden storms. But, the bird's-eye view of the plants and animals of the rain-forest ecosystem makes all of the dangers worthwhile.

How do you become an arbornaut? Most arbornauts go to school for years to earn a doctorate degree in biology. Then, they specialize in studying treetop environments. It also helps to have no fear of heights!

Disappearing Ice, Disappearing Bears

TROUBLING THINGS are happening in the arctic. Temperatures are rising, having increased an average of 7 to 9°C over the past century. As a result, the thick arctic ice cap is thinning and breaking up. Satellite images show that it has been shrinking at a rate of about 9 percent each decade in recent years.

For most scientists, the cause is clear: global warming. Earth's average temperature is slowly rising, largely due to human actions. Burning huge amounts of fuels such as oil, gas, and coal releases carbon dioxide and other gases into the atmosphere. Carbon dioxide is a natural greenhouse gas—a gas that holds in Earth's heat. As carbon dioxide builds up in the air, the heating effect is increased. In the arctic, the heating effect is magnified.

Polar bears are one species that has been feeling the effects of global warming. Polar bears hunt seals and raise their young on ice floes that drift on arctic waters from late fall through spring. After the ice breaks up in the spring, the bears stay on land and live off their stored fat until the ice returns and they can hunt again. For years, however, the ice has been breaking up earlier and earlier in the spring. This gives the polar bears less time to hunt and build up fat to last them through the summer months. In some cases, female polar bears have become lighter in

Research and Report

Both the arctic and antarctic are cold and bleak but not lifeless. With a partner, research an arctic or antarctic ecosystem. Present your findings in a report that includes a labeled drawing of a typical arctic food chain.

weight and less successful at reproducing. Polar bear young are also less likely to survive.

Although polar bears are good swimmers, melting ice has another tragic effect: more open water between ice floes and between floes and the land. The bears must swim longer distances between floes to hunt. Exhausted bears are drowning in the open water. If polar bears stay on land to hunt for food, they must compete with more aggressive land animals such as the grizzly bear.

As global warming continues, the worldwide population of 20,000 polar bears could drop by more than 30 percent in the next 45 years. Some scientists predict that continued warming could even bring about an ice-free Arctic Ocean by the end of the century. Unfortunately, that would mean a polar bear-free arctic, as well.

New Marine Species Found

SCIENTISTS EXPLORING the Pacific Ocean off Papua, Indonesia, recently found something quite unusual—a patch of ocean so full of life that it is being called "Earth's richest seascape." The area contains about 1,200 species of fish and 600 species of reef-building coral—about 75 percent of the world's total. They have also discovered some strange marine species, including a shrimp that looks like a praying mantis and a shark that can walk along the seafloor on its fins.

Scientists now face the problem of ensuring that this biodiversity is protected. Local people fish these waters using practices that could damage the ecosystem. Fishing teams often stun the fish first with poisons such as cyanide or blasts of dynamite. Huge fleets of factory fishing ships may soon move into the area. These ships drag along the ocean bottom huge nets that sweep in all marine life and destroy marine ecosystems.

The goal of scientists is to encourage sustainable use of this

This shark species walks on the seafloor.

rich area. That means allowing fishing to occur in ways that will conserve fish populations and protect ecosystems. Ten percent of the area is now protected. There is hope that the government of Indonesia will protect a larger part of these seas.

Skin, Skeletal System, and Muscular System

KEY CONCEPT The skin, skeletal system, and muscular system protect the body, give it shape, and aid in motion.

Imagine a complex machine that can play and do chores. This machine thinks and dreams. It reacts to situations and solves problems. It even feels emotions. This machine is your body! The human body enables you to move, work, and enjoy a sunny day. It enables you to think and wonder. It enables you to feel happiness, sadness, fear, and excitement.

The human body is organized into systems that carry out different functions to keep you alive. Three of these systems consist mainly of the skin, the skeleton, and the muscles.

Think About the Human Body

Most people take for granted all the extraordinary things the human body can do.

- Think about the things your body does every day. What do you think are some functions of your skin, muscles, and bones?

- Make a three-column chart in your Science Notebook, labeling one column *Skin*, one column *Bones*, and one column *Muscles*. Use your chart to list what you think your skin, bones, and muscles do. How do they work to keep you healthy? Review your chart and add to your lists as you read this chapter.

www.scilinks.org
Body Systems **Code: WGB19**

Learning Goals

- Describe the levels of organization in the human body.
- Review the basic structure of cells.
- Identify the four types of tissue found in the human body.

New Vocabulary

cell
tissue
epithelial tissue
connective tissue
muscle tissue
nervous tissue
organ
organ system

Before You Read

Create a concept map in your Science Notebook by writing *The Human Body* in a circle. Draw several smaller circles around this circle, and connect each small circle to the center circle with a line. Preview the material in this lesson, and predict what you will learn by filling in the smaller circles of your concept map.

It's the top of the ninth inning. The score is six to five in favor of the home team, which will win if it can hold on to its lead. For the visitors, the tying run is at third base, waiting for his chance to even the score. The go-ahead run is on first base. With one out, a new batter comes to the plate. He swings at the first pitch and sends the ball speeding down the first-base line. The first baseman dives for the ball, catches it, and tags the bag for the second out. He hurls the ball to home plate. The catcher tags the runner. It's a double play, and the game is over. The home team has pulled off another win.

Levels of Organization

To win a baseball game, the players on a team must work together. They must be organized, too. Each player has his own role and his own responsibilities. Just as the players of a baseball team work together, the cells of your body work together to keep you alive and healthy. Like baseball players, your cells are organized. Each has a special role.

A baseball team is made up of 30 or fewer members. Your body is made up of about 100 trillion cells! How does your body keep so many individual cells working together?

Figure 19.1 How is your body like a baseball team? Just as the players on a baseball team work together to win a game, the parts of your body work together to keep you alive and healthy.

Cells

A **cell** is the basic unit of life. All organisms are made up of one or more cells. Cells are too small to be seen with the unaided eye. Cells in the human body are similar to cells in other animals. They are composed of smaller structures that together carry out all the functions of life. **Figure 19.3** shows some of the parts of most human cells.

Figure 19.2 All living things are made up of cells. These human cells have been magnified 1,200× with a microscope.

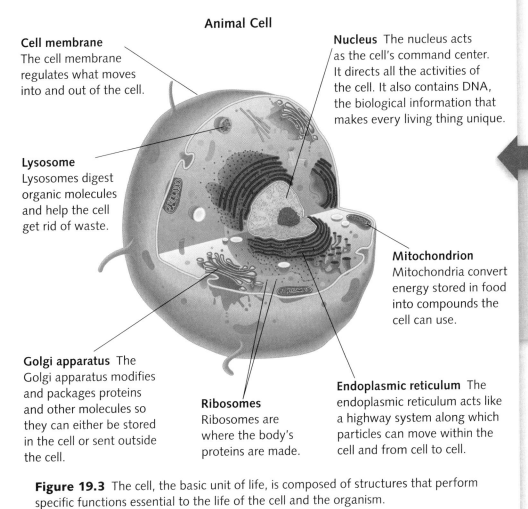

Animal Cell

Cell membrane The cell membrane regulates what moves into and out of the cell.

Nucleus The nucleus acts as the cell's command center. It directs all the activities of the cell. It also contains DNA, the biological information that makes every living thing unique.

Lysosome Lysosomes digest organic molecules and help the cell get rid of waste.

Mitochondrion Mitochondria convert energy stored in food into compounds the cell can use.

Golgi apparatus The Golgi apparatus modifies and packages proteins and other molecules so they can either be stored in the cell or sent outside the cell.

Ribosomes Ribosomes are where the body's proteins are made.

Endoplasmic reticulum The endoplasmic reticulum acts like a highway system along which particles can move within the cell and from cell to cell.

Figure 19.3 The cell, the basic unit of life, is composed of structures that perform specific functions essential to the life of the cell and the organism.

Figure It Out

1. What are the major structures of a human cell?

2. Hypothesize about the number of mitochondria that would be found in a cell that has large energy requirements.

PEOPLE IN SCIENCE Robert Hooke 1635–1703

Robert Hooke is remembered as one of the great scientists of his time. Hooke studied many fields of science, including astronomy, physics, and biology. What Hooke is probably best known for, however, is discovering and naming the cell.

Hooke developed a compound microscope that he used to observe such things as fleas, feathers, and shells. While looking at a thin slice of cork bark, he saw that the bark appeared to be made of many small empty chambers, which he named *cells*. What Hooke saw were the cell walls of dead plant cells. It would be another 200 years before the cell was recognized as the smallest unit of life.

Tissues

The cells in the human body, as well as in most multicellular organisms, are specialized to perform particular functions. Specialized cells group together to form tissues. A **tissue** is a group of cells that work together to perform a specific function. There are four main types of tissue in the human body.

Epithelial Tissue   Gently touch your arms or face. You are touching epithelial tissue! **Epithelial** (eh puh THEE lee ul) **tissue** covers the outside of the body and lines many of the structures inside the body. Your skin is made up of epithelial tissue that you can see and feel. Cells that form this kind of tissue are closely packed. There is little space between them. This gives epithelial tissue its function as a protective barrier for the body. It protects the body from injury, disease, and loss of essential fluids.

Connective Tissue   As its name implies, **connective tissue** holds together other tissues in the body. Tendons, ligaments, bones, and even blood are types of connective tissue. Unlike the cells of epithelial tissue, the cells that form connective tissue are loosely packed.

As You Read

In the outer circles of your concept map, fill in words that describe how the body is organized. Add more circles to the map as you need them. Next to the lines connecting the smaller circles to the large circle, describe the relationships between the connected words. Look to see if some of the smaller circles can be connected to one another.

What are cells made up of? What are tissues made up of?

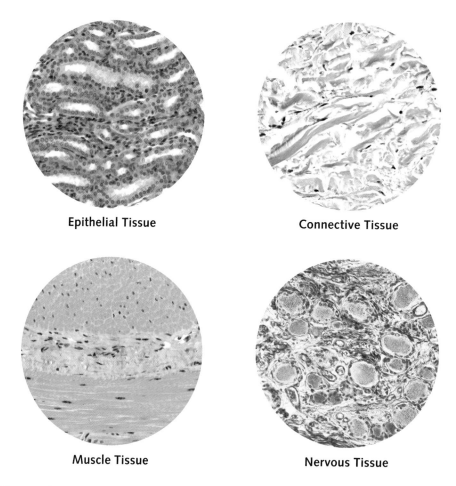

Epithelial Tissue   Connective Tissue

Muscle Tissue   Nervous Tissue

Figure 19.4 Compare the four main types of human tissue. How are they similar in appearance? How do they differ?

Muscle Tissue   Muscle cells group together to form **muscle tissue**. Muscle tissue surrounds most of the bones in the body. It also combines with other tissues to make up many of the body's internal structures, such as the heart, stomach, and lungs. Muscle tissue gives parts of the body, such as the arms and legs, their ability to move. It allows you to do things such as run, play sports, fold laundry, and reach up for things. Muscle tissue in the heart causes the heart to contract, or tighten, and relax. These motions cause blood to be pumped throughout the body. You will learn more about muscle tissue and what it enables the body to do in Lesson 19.4.

Nervous Tissue   Specialized cells called neurons make up **nervous tissue**. Nervous tissue enables the body to sense and respond to things in the environment. You can see, smell, feel, taste, and hear things because of nervous tissue. You can think about and remember things because of nervous tissue. Nervous tissue gathers and transmits information throughout the body.

Organs

Tissues group together to form organs. In Chapter 3, you learned that an **organ** is a group of different types of tissue that work together to perform a specific function. The heart is an organ. So, too, are the stomach, lungs, and eyes. Each of these organs is made up of epithelial tissue, connective tissue, muscle tissue, and nervous tissue working together to perform certain functions. In the stomach, these tissues help the body digest food. In the lungs, they enable the body to take in oxygen and give off carbon dioxide. They work together in the eyes to allow the body to see.

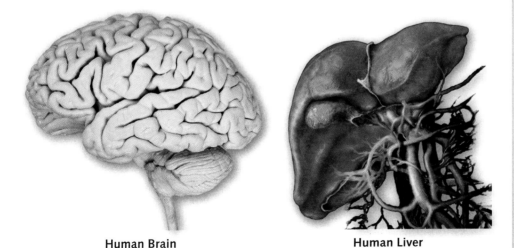

Human Brain   **Human Liver**

Figure 19.5 The brain and liver are two of the body's vital organs. They are made up of tissues that work together to perform specific functions in the body. What other organs can you name?

Organ Systems

Cells are organized into tissues. Tissues are organized into organs. The parts of your body are further organized into organ systems. An **organ system** is a group of organs that work together to carry out one or more body functions.

Figure 19.6 These major organ systems keep you active and healthy.

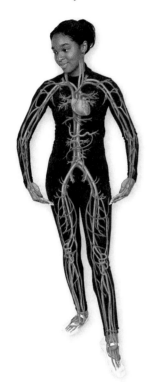

Circulatory system This system pumps blood throughout the body, brings nutrients to the cells, and carries away waste from the cells.

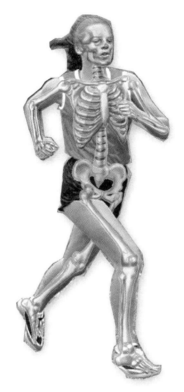

Skeletal system This system supports the body and gives it shape.

Muscular system This system enables the body to move.

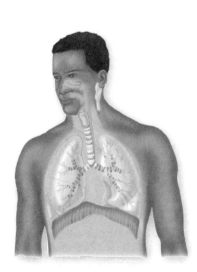

Respiratory system This system is responsible for taking in oxygen from the air and giving off carbon dioxide.

Digestive system This system is responsible for taking in food and breaking it down into usable energy.

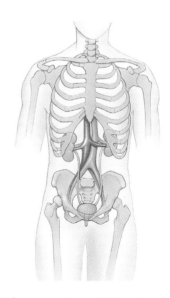

Urinary system This system rids the body of liquid and dissolved wastes and maintains the balance of salts and water in the blood.

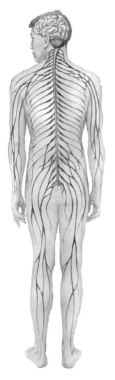

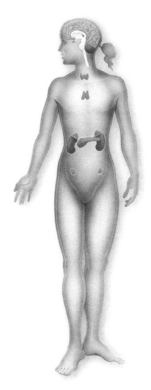

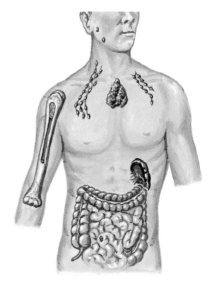

Nervous system This system enables humans to take in and respond to information from the environment.

Endocrine system This system helps regulate the body's functions.

Immune system This system helps the body fight disease.

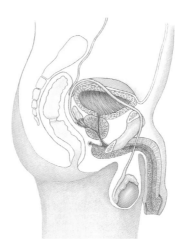

Reproductive system The female *(left)* and male *(right)* systems enable humans to reproduce, or produce offspring.

After You Read

1. Review your concept map. Use it to describe the levels of organization in the human body.

2. Describe the basic structure of a human cell. What does each cell part do?

3. What are the four types of tissue in the human body?

4. What makes up organ systems? In what ways do organ systems help you?

Learning Goals

- Identify the layers of the skin and their functions in the body.
- Describe how nails and hair are part of the body's skin.

New Vocabulary

epidermis
keratin
melanin
dermis
subcutaneous tissue

19.2 Skin

Before You Read

Create a K-W-L-S-H chart in your Science Notebook. Think about the title of this lesson. Then look at the pictures and headings in the lesson. In the column labeled *K*, write what you already know about skin. In the column labeled *W*, write what you would like to learn about skin.

Cool raindrops. Prickly cacti. Smooth snakes. Have you ever wondered what allows you to feel different things? The answer is your skin! Without skin, you could not feel temperature or wind. You could not tell rough objects from smooth ones, or wet objects from dry ones.

The Largest Organ in the Body

Your skin is the largest organ of your body. If you could peel it off and lay it flat, the skin of an average adult would cover an area of about two square meters. That is about the same size as a classroom chalkboard. Skin is more than just a sense organ. It does other important jobs to keep you healthy. It covers your body's organs and protects them from injury. It helps regulate your temperature and keeps you from getting overheated. It helps get rid of harmful wastes and prevents loss of water, blood, and other essential fluids. It even protects your body from disease. The skin has three main layers: the epidermis, the dermis, and the subcutaneous tissue.

Figure 19.7 Skin gives humans their sense of touch. This sense of touch enables people with impaired vision to read and to feel things that they cannot see with their eyes.

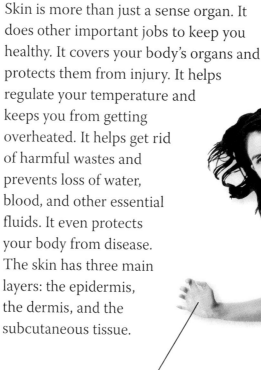

Your skin is thinnest on your eyelids.

Your skin is thickest on the palms of your hands and the soles of your feet.

Figure 19.8 Skin covers the body and protects internal organs from injury and disease. Although it is only a few centimeters thick, your skin is the largest organ of your body.

The Epidermis

The **epidermis** (eh puh DUR mus) is the layer that you see. It is the tough, protective outer layer of the skin. It is made up mostly of cells called keratinocytes (KER eh tehn oh sites), which are flat cells stacked like bricks in a wall. These cells produce keratin. **Keratin** is a protein that helps make the skin tough. Keratinocytes protect deeper cells from damage and from drying out. They also help keep out harmful bacteria and other microorganisms.

As keratinocytes mature, they become filled with keratin. They also get pushed upward as new cells are produced in the bottom layer of the epidermis. Because these cells are now far from a food supply, they soon die. In time, they flake off and are replaced. The body produces an entirely new epidermis about every 28 days through this process.

The epidermis also contains melanocytes (meh LAH neh sites) and Langerhans (LAHNG ur hahnz) cells. The melanocytes produce melanin. **Melanin** is a chemical that is responsible for skin color. The more melanin cells produce, the darker the skin is. Melanocytes also absorb sunlight, thereby protecting other cells from the Sun's harmful rays. As melanocytes absorb sunlight, they produce more melanin. This is what causes skin to become tan after exposure to sunlight. The Langerhans cells play an important role in the immune system. They help protect the body from diseases and infections.

CONNECTION: Health

Too much sunlight can be harmful to your skin. It can keep the skin from functioning properly and can cause skin cancer. It is important to protect your skin by wearing sunscreen even on cloudy days. Sunscreen contains chemicals that keep your skin from absorbing sunlight. Each brand of sunscreen is rated with an SPF, or sun protection factor. The SPF describes how much sunlight is blocked. Sunscreens with higher SPF values block more sunlight. You should wear a sunscreen with an SPF of at least 30 on bright sunny days.

Figure 19.9 The chemical melanin produced in the skin is responsible for variations in skin color. It also causes skin to tan upon exposure to sunlight.

The Dermis

The **dermis** is the thick layer of skin that lies beneath the epidermis. It is made up mostly of nerve endings, blood vessels, and connective tissue. It also contains fibers, oil glands, sweat glands, and hair follicles.

The nerve endings of the dermis give the body its sense of touch. They collect signals from the environment and send them to the brain, where they are processed. The blood vessels of the dermis help control the body's temperature. They expand, or widen, to allow heat to escape through the skin. They contract, or narrow, to keep heat in. The fibers of the dermis contain chemicals that are responsible for giving your skin its elasticity. These chemicals, known as collagen and elastin, allow skin to stretch and bend without tearing.

Like blood vessels, sweat glands play a role in regulating the body's temperature. Sweat glands produce sweat when the body starts getting too warm. As sweat evaporates from the skin, heat is released and the body is cooled. In addition to keeping the body cool, sweat also plays a role in getting rid of poisonous body wastes. Waste products dissolve in sweat, which is made mostly of water, and are released through the skin.

The dermis's oil glands are located near the hair follicles. The glands produce oil that keeps the skin smooth and waterproof and keeps hair from becoming brittle.

As You Read

In the column labeled *L* in your K-W-L-S-H chart, write what you have learned about skin.

What are the three main layers of skin?

Figure 19.10 When the body becomes warm, sweat glands produce sweat, and blood vessels expand as a way of releasing heat. The expansion of the blood vessels causes lots of red blood cells to flow through the skin. This is what makes your skin look flushed when you are hot.

Figure It Out

1. Which structures are found in the dermis layer of the skin?
2. Explain why you think sweat glands are located in the skin rather than deeper in the body.

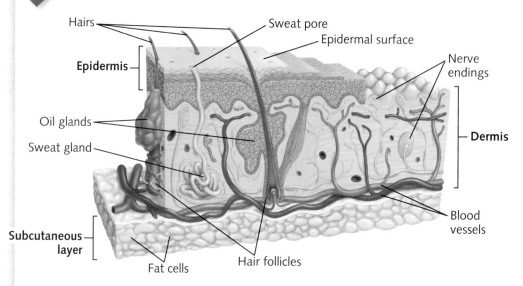

Figure 19.11 The skin is composed of three layers: the epidermis, the dermis, and the subcutaneous tissue.

The Subcutaneous Tissue

The innermost layer of skin is the **subcutaneous** (sub kyoo TAY nee us) **tissue**. This layer is made up mostly of connective tissue and fat cells. In fact, most of the body's fat cells are in this layer of skin. Fat cells act as insulators for the body. They help conserve heat and thus play an important role in maintaining a healthy body temperature.

Nails and Hair

It might surprise you to learn that nails are modified skin. Nails are made mostly of keratin. They grow from nail roots under the skin at the base and sides of the nails. Nails protect the soft tissue at the tips of your toes and fingers. They help you pick up objects, scratch an itch, and untie knots. They could even get you into the Guinness Book of World Records, as **Figure 19.12** shows.

Hair is also part of the skin. Hair covers the entire body except for the palms of the hands, the soles of the feet, and the lips. Hair grows from hair follicles, which are tiny saclike structures in the dermis, and extends out through the epidermis. Like skin color, hair color is determined by melanin. The more melanin cells produce, the darker the hair is.

Like nails, hair is made mostly of keratin. Hair protects the eyes and nose from dust and harmful materials. It also helps keep the body warm. How does it do this? First, humans lose most of their body heat through their heads. A thick head of hair helps keep in some of this heat. Second, muscles attached to the hair shafts contract with cold, fear, or emotion. This causes "goose bumps" to form on the skin and causes hair to stand on end. The lifted hair traps heat and warms the body.

Figure 19.13 An average adult has more than 100,000 hairs on his or her head.

Figure 19.12 Lee Redmond started letting her fingernails grow in 1979. In this photo, 27 years later, her nails are 84 cm (33 in.) long.

🔍 Explore It!

Work with a partner to observe hair and nails under a microscope. Gently pull a strand of hair from your scalp, and place it under the lens of a microscope. Draw what you see. Then, trim the end of one nail. Place the trimming under the lens of the microscope, and draw what you see. How does your nail look compared to your hair?

After You Read

1. Identify and describe the layers that make up skin.

2. What are some ways in which skin helps keep the body healthy?

3. Describe the structure and function of nails and hair.

4. Use your completed K-W-L columns to summarize this lesson in a paragraph in your Science Notebook. Then, write in the *S* column what you would still like to know about skin and how it helps keep you healthy. Complete the chart by writing in the *H* column how you can find this information.

Before You Read

Create a lesson outline in your Science Notebook, using the title, headings, and subheadings of this lesson. Use the outline to summarize what you learn in this lesson.

What do the words *hammer, stirrup,* and *anvil* make you think of? You might say tools, and you would be correct, but think again. They are also the names of bones! In fact, they are the smallest bones in your body. Though they may be tiny—only a few millimeters across—they do a very important job. What do they do? Here is a clue: They are found in your ears. The hammer, stirrup, and anvil help you hear. What else do bones help you do?

Your Bony Body

The 206 bones of the human skeleton make up your **skeletal system**. Together, bones have many functions in the body. They give the body its shape. They support the body and protect delicate internal organs, such as the brain, heart, and lungs. They allow the body to move. They even store minerals and produce blood for the body. Without bones, you would be a disorganized, formless heap of skin and other tissue.

You may think of bones as hard, nonliving material, but they are far from dead matter. Just like the other organs of your body, bones are made up of living tissue. In fact, if bones were not living, you would not grow during childhood, and broken bones would never heal.

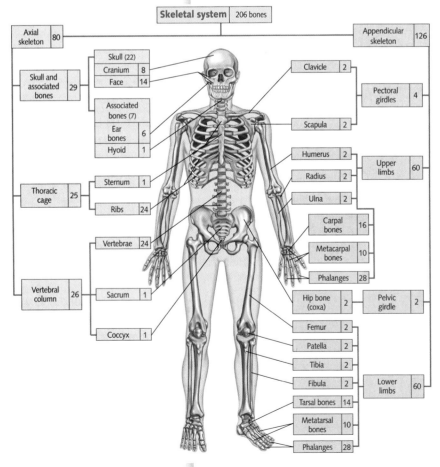

Figure 19.14 The human skeleton is divided into two main parts. The bones labeled on the left support the body and keep you upright. The bones labeled on the right support your limbs—your arms and legs.

Shape and Function   Bones come in many shapes and sizes. Each one is shaped for a specific function. The short, slender bones in your fingers allow your fingers to move and grasp things. The wide, flat bones in your skull protect your brain. Bones are classified as:

- long bones, such as those found in your legs and arms
- short bones, such as those found in your fingers and toes
- flat bones, such as those found in your skull and pelvis
- irregular bones, such as those found in your backbone and ears

Bone Composition   A bone is made up of three layers of connective tissue. The outermost layer of a bone is the **periosteum** (per ee AH stee um). This thin layer of tissue contains blood vessels that bring nutrients and oxygen to the bone. **Compact bone** lies beneath the periosteum. This dense layer of bone consists of blood vessels, nerve cells, and living bone cells called osteocytes. Compact bone is held together by a framework of hard, nonliving minerals such as calcium and phosphorus. **Spongy bone** is found in the ends of long bones and in the middle part of short bones and flat bones. Spongy bone is more porous and lightweight than compact bone, but it is strong and supportive. Cells known as osteoblasts and osteoclasts are found in both compact and spongy bone. Osteoblasts produce bone, while osteoclasts break down bone. Together, these cells are responsible for bone growth and repair.

Two types of **bone marrow**, red and yellow, run through a cavity in the center of many long bones. Red bone marrow produces red blood cells, which transport oxygen throughout the body. It also produces some kinds of white blood cells. White blood cells are part of the immune system. They aid in fighting disease. Yellow bone marrow is made up mostly of fat cells that are a source of stored energy.

Extend It!

The most common kinds of bone injuries are broken bones, which heal in time with proper care. Working with a partner or in a small group, do research to find out about other kinds of bone injury and about bone disease. Prepare a presentation, and share your findings with your classmates.

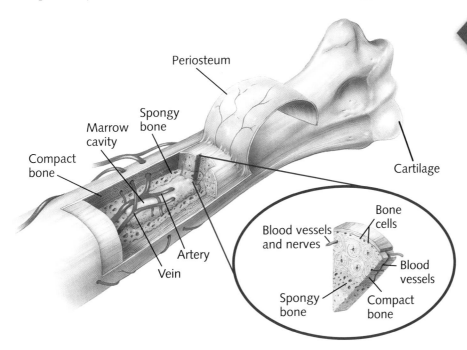

Periosteum

Spongy bone

Marrow cavity

Compact bone

Cartilage

Artery

Vein

Blood vessels and nerves

Bone cells

Blood vessels

Spongy bone

Compact bone

Figure It Out

1. What three layers of tissue make up bone?

2. Describe how bone marrow aids the body.

Figure 19.15 The most obvious feature of a long bone is the long shaft, or center, which contains compact bone. Within the shaft are hollow spaces containing marrow.

Formation of Bones

The formation of bones begins early in a human's life—during the fifth or sixth week of pregnancy. At this early stage, soft, flexible connective tissue called **cartilage** (KAR tuh lihj) begins to form the basis of the skeleton. Within a few weeks of its formation, much of this cartilage is changed into hard bone through a process called ossification. At birth, humans have more than 300 bones. As humans develop, these bones grow longer and fuse together to form the 206 bones of a mature adult. By the time humans reach the age of about 18, their bones are completely ossified and stop growing longer.

Joints

The place where two or more bones meet is called a **joint**. Joints are classified as immovable, slightly movable, or movable based on the amount of movement they allow.

Immovable Joints   Immovable joints allow no movement. The bones that make up your skull meet at immovable joints. Bones that meet at immovable joints are fused together.

Slightly Movable Joints   Slightly movable joints, as their name suggests, allow for a limited amount of movement. Your vertebrae, or backbones, meet at slightly movable joints where adjacent vertebrae are separated by discs that contain cartilage, as shown in **Figure 19.17**.

Movable Joints   Movable joints allow movement in one or more directions. A ball-and-socket joint allows for the widest range of movement. The bones of the thigh and pelvis meet at ball-and-socket joints. A gliding joint allows for some movement in all directions. You can spread your fingers wide apart and can move them up and down because of gliding joints. A hinge joint allows for back-and-forth motion in one direction. The knee is a hinge joint. A pivot joint, such as the one found between the forearm and the upper arm, allows one bone to rotate around another. The four main types of movable joints are shown in **Figure 19.18**.

As You Read

Use your lesson outline to take notes about the skeletal system as you read. Create a list of unanswered questions that you have about this system in the outline, as well.

What are the three types of joints?

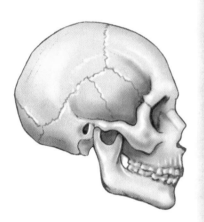

Figure 19.16 The human skull is made up of 29 flat bones that meet at immovable joints.

Did You Know?

Scientists classify animals into two main groups based on whether or not they have backbones. Animals with backbones are called vertebrates. You are a vertebrate. Animals without backbones are called invertebrates. Sponges, snails, and spiders are examples of invertebrates, as are most of Earth's animals.

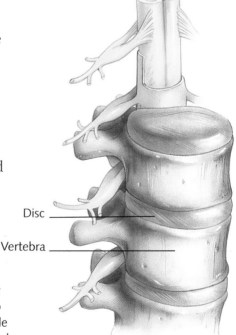

Disc

Vertebra

Figure 19.17 The vertebrae meet at slightly movable joints. The joints are rigid enough to keep your body strong and upright but flexible enough to allow you to twist, bend, and stretch.

Bones that meet at movable joints are held together by thick connective tissues called **ligaments** (LIH guh munts). They are also protected from stress by a layer of cartilage that forms on the surface of the bone around the joint. Synovial (suh NOH vee ul) fluid, a watery liquid, further cushions the joint and aids in smooth movement.

Figure It Out

1. What kind of motion does a hinge joint allow?

2. Compare the four types of movable joints. How are they alike? How do they differ?

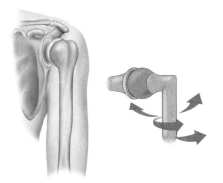

Ball-and-socket joint Ball-and-socket joints allow movement in all directions. The joints of the hips and shoulders are ball-and-socket joints; they allow you to swing your arms and legs around in many directions.

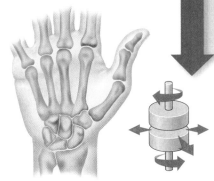

Gliding joint Gliding joints, found in the wrists and ankles, are similar to sliding doors. They allow bones to slide past each other.

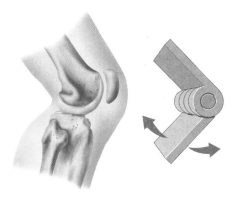

Hinge joint Hinge joints are found in the elbows, knees, fingers, and toes. They allow back-and-forth movement like that of a door hinge.

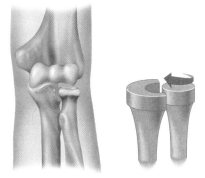

Pivot joint Pivot joints allow bones to twist around each other. One example is in your forearm, between the ulna and the radius. It allows you to twist your forearm around.

Figure 19.18 Joints, or places where two bones meet, are shaped according to how they function in the body.

After You Read

1. What are some of the functions of the skeletal system?

2. Use your lesson outline to describe the composition of bone.

3. What are the three main types of joints?

4. Based on the information in your outline, hypothesize why bones have different shapes.

Learning Goals

- Identify the major functions of the muscular system.
- Describe three types of muscle tissue.
- Explain how skeletal muscles allow movement of the bones.

New Vocabulary

muscular system
skeletal muscle
smooth muscle
cardiac muscle
tendon

Before You Read

Write the title of this lesson in your Science Notebook. Then, look at the pictures and review the headings in the lesson. Record some questions you have about the muscular system.

Weightlifters often impress crowds with their ability to lift and carry tremendous amounts of weight. The strongest women can bench-press more than 225 kg. They can do squats with 338-kg weights.

What would it take for you to have as many muscles as these athletes? Nothing! Your muscles may not be quite as developed or defined, but like these athletes, you were born with the more than 600 muscles that are in your body. Together, your muscles make up your **muscular system**. Big or small, your muscles are responsible for every move you make, from blinking your eyes and breathing to kicking a soccer ball and lifting weights.

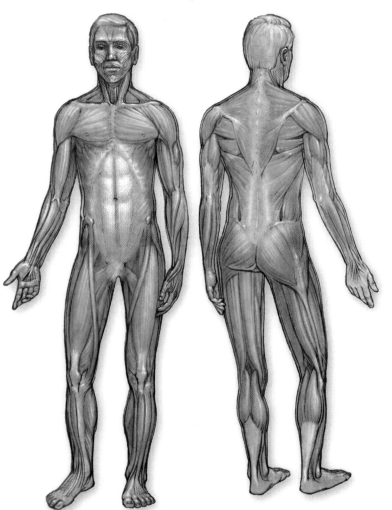

Your Muscular Body

More than 40 percent of the body mass of an average human is muscle. Without this muscle, humans could not move. They could not lift crying babies or pick up groceries. They could not talk, breathe, or eat food. Whether you realize it or not, just about everything your body does to keep you alive involves some kind of movement that is made possible by muscles.

Figure 19.19 shows some of the major muscles of the human body. As you can see, they are found throughout the body. Some, such as those worked on by weightlifters and bodybuilders, are visible beneath the skin. Others are "buried" within the body and help make up vital organs such as the lungs, the heart, and the stomach.

Figure 19.19 The muscles shown here are skeletal muscles, one of the three types of muscles in the human body. Smooth muscles and cardiac muscles make up organs deep within the body and thus cannot be shown in this diagram.

Three Types of Muscle Tissue

Each muscle in the body is made up of muscle tissue. There are three main types of muscle tissue: skeletal muscle, smooth muscle, and cardiac muscle. Each type is specialized to perform a different function in the body.

Skeletal muscle tissue makes up the skeletal muscles. **Skeletal muscles** are the muscles that attach to bones. The biceps and triceps in the upper arms and the quadriceps in the thighs are examples of skeletal muscles. Skeletal muscle tissue is made up of long, thin cells that each contain more than one nucleus. Skeletal muscle is often called striated muscle because it appears to have alternating light and dark stripes or bands.

Skeletal muscles pull on bones to make the body move. They are responsible for all the body's voluntary movement, such as dancing, playing a piano, and in-line skating. Voluntary movement is movement you can consciously control.

Smooth muscle tissue makes up the muscle found in internal organs such as the stomach, intestines, kidney, and liver. Smooth muscle tissue is not striated. It is made up of spindle-shaped cells that each contain only one nucleus.

Smooth muscle tissue is not usually under voluntary control. Instead, it is responsible for involuntary movement. This is movement you cannot consciously control. The movement of food through the digestive system and the movement of blood through blood vessels are examples of involuntary movements performed by smooth muscle tissue.

Cardiac muscle tissue is the muscle tissue found in the heart. Like smooth muscle tissue, cardiac muscle is not under voluntary control. You do not have to think about your heart beating for it to happen. Like skeletal muscle tissue, cardiac muscle is striated. However, its cells usually contain just one nucleus each. The cells that make up cardiac muscle are smaller than those that make up skeletal muscle.

Figure It Out

1. Compare the three types of muscle tissue. How are they similar? How do they differ?

2. Describe what each type of muscle tissue does in the body.

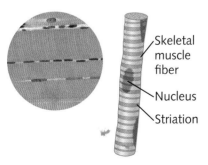

Skeletal muscle tissue Skeletal muscle fibers appear striated, or striped, under a microscope.

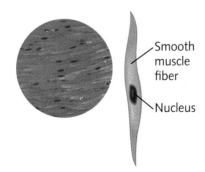

Smooth muscle tissue Smooth muscle fibers appear spindle-shaped when magnified.

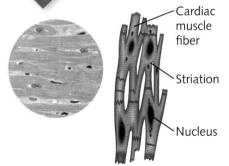

Cardiac muscle tissue Cardiac muscle fibers also appear striated when magnified.

Figure 19.20 The three main types of muscle tissue are skeletal muscle tissue, smooth muscle tissue, and cardiac muscle tissue. Each type of tissue is specialized to perform a different function.

Skeletal Muscle Structure

Skeletal muscle cells, often called muscle fibers, consist mainly of two types of protein strands: actin and myosin. Where these two strands overlap, the muscle fiber looks dark. Where the strands do not overlap, the fiber looks light. The alternating light and dark bands give skeletal muscle tissue its striated appearance. These protein strands are responsible for the ability of muscles to contract and relax. The contracting and relaxing of muscles is what causes movement.

At the end of each muscle fiber is a motor nerve. Motor nerves connect the skeletal muscles to the brain. They enable the voluntary control of the skeletal muscles. Blood vessels running through muscle tissue bring oxygen and food to the cells. They give muscle fibers the energy required to contract and relax to make the body move.

Muscle fibers are grouped together into bundles. Small muscles, such as those found in the fingers, can be made up of just a few bundles. Larger muscles, such as those in the legs, have hundreds of bundles.

As You Read

Look for the answers to the questions you have about the muscular system as you read the text. Record the answers in your Science Notebook.

What are the three types of muscle tissue?

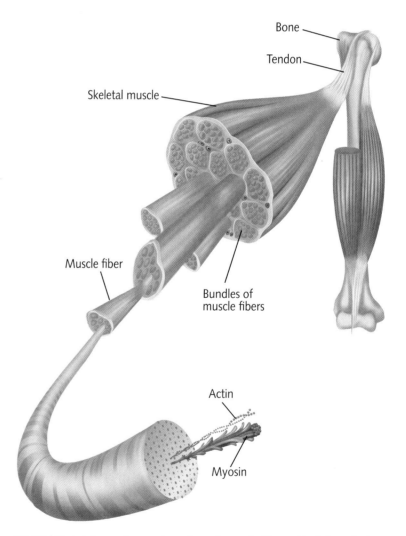

Figure 19.21 Skeletal muscles are bundles of muscle fibers. Each bundle is covered by connective tissue. Each muscle fiber is a single cell that can be as long as 40 cm.

Muscles and Movement

Skeletal muscles are attached to bones by thick connective tissues called **tendons** (TEN dunz). Touch the back of your ankle. The tough, thick, cordlike structure you feel is your Achilles tendon. It is the largest tendon in the body. It attaches the outermost calf muscle to the heel bone. Hundreds of tendons throughout the body attach the skeletal muscles to the bones. They allow the muscles to pull on the bones and move the body.

Skeletal muscles can only pull on bones. For this reason, most skeletal muscles in the body work in pairs to give the body a range of movement. For example, the biceps and triceps attach to the bones of the forearm and allow the arm to move in a variety of ways. When muscle fibers of the biceps are stimulated by nerves, the actin and myosin in the fibers slide past each other, causing the muscle to shorten, or contract. When the muscle contracts, it pulls on the bone to which it is attached. The forearm lifts up.

When muscle fibers of the triceps are stimulated, the actin and myosin in these fibers slide past each other, causing this muscle to contract. As a result, the forearm stretches out. When one muscle in the pair contracts, the other muscle relaxes. Skeletal muscles throughout the body work in this way to allow the body to move.

Explain It!

Skeletal muscles, tendons, and joints work together like levers to move bones. Find out about levers. Then, write a paragraph in your Science Notebook explaining how muscles, tendons, and joints are similar to these simple machines. Use diagrams to enhance your explanation.

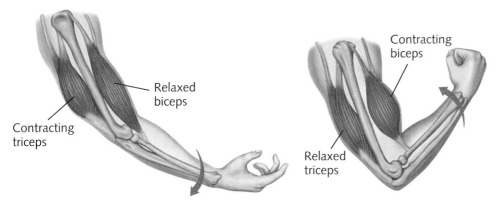

Contracting biceps

Relaxed biceps

Contracting triceps

Relaxed triceps

Figure 19.22 If muscles did not work in pairs, movements of the arm such as these would be impossible. What muscle must contract in order for you to bend your elbow?

After You Read

1. What is the main function of the muscular system?

2. Describe the three types of muscle tissue.

3. Explain how muscles help move bones.

4. Use your lesson outline, including the questions you had about the muscular system and the answers you found, to summarize what you learned in this lesson.

KEY CONCEPTS

19.1 Human Body Organization

- The body is organized into cells, tissues, organs, and organ systems.
- Cells are the basic building blocks of life. Cells are specialized to perform specific functions in the body.
- Specialized cells group together to form tissues. Four main types of tissue in the human body are epithelial tissue, connective tissue, muscle tissue, and nervous tissue.

19.2 Skin

- Skin covers the body and protects internal organs. It helps regulate the body's temperature and helps prevent disease and loss of essential fluids.
- Skin is made up of three main layers of tissue. The epidermis is the outer layer. The dermis lies beneath the epidermis. The subcutaneous layer is the innermost layer.
- Hair and nails are part of the skin.

19.3 The Skeletal System

- The skeletal system gives the body shape, supports and protects the body, enables movement, stores nutrients, and produces blood.
- Bones are composed of compact and spongy bone and are covered by a layer of tissue known as the periosteum. Bone marrow runs through the center of many long bones.
- A joint is the place where two or more bones meet. Joints are classified as immovable, slightly movable, or movable based on the amount of movement they allow. Ligaments hold bones together at the joints.

19.4 The Muscular System

- The muscular system is responsible for the body's motions.
- The three main types of muscle tissue are skeletal, smooth, and cardiac muscle tissue.
- Skeletal muscles are attached to bones by tendons. When a muscle contracts, it pulls on the bone it is attached to, causing the bone to move.

VOCABULARY REVIEW

Write each term in a complete sentence, or write a paragraph relating several terms.

19.1
cell, p. 331
tissue, p. 332
epithelial tissue, p. 332
connective tissue, p. 332
muscle tissue, p. 333
nervous tissue, p. 333
organ, p. 333
organ system, p. 334

19.2
epidermis, p. 337
keratin, p. 337
melanin, p. 337
dermis, p. 338
subcutaneous tissue, p. 339

19.3
skeletal system, p. 340
periosteum, p. 341
compact bone, p. 341
spongy bone, p. 341
bone marrow, p. 341
cartilage, p. 342
joint, p. 342
ligament, p. 343

19.4
muscular system, p. 344
skeletal muscle, p. 345
smooth muscle, p. 345
cardiac muscle, p. 345
tendon, p. 347

PREPARE FOR CHAPTER TEST

To prepare for the chapter test, create a question from each Learning Goal. Use the information in your Science Notebook to answer each question. Then use these answers to write a well-developed essay about the chapter. Use the Key Concept on the first page of this chapter as your topic sentence.

True or False

If the statement is true, write "true." If it is false, change the underlined word or words to make the statement true.

1. A <u>tissue</u> is the basic unit of life.

2. Skin is the <u>smallest</u> organ in the human body.

3. Three types of muscle tissue are <u>smooth</u>, skeletal, and cardiac muscle tissue.

4. The skeletal system aids in the production of <u>blood</u>.

5. Many bones contain <u>yellow</u> bone marrow, which produces red and white blood cells.

6. Organ systems are made up of <u>tissues</u> that work together to perform specific functions in the body.

Short Answer

Answer each of the following in a sentence or brief paragraph.

7. What are the basic functions of the skeletal system?

8. In what ways does hair help keep the body healthy?

9. Why are muscles essential to your survival?

10. What are bones made up of?

11. What is the epidermis? Describe its main functions in the body.

Critical Thinking

Use what you have learned in this chapter to answer each of the following.

12. **Explain** Why do most skeletal muscles work in pairs?

13. **Compare and Contrast** What are the three main types of joints? How are they alike? How do they differ?

14. **Analyze** Why do you think skin is made up of several layers of tissue?

Standardized Test Question

Choose the letter of the response that correctly answers the question.

15. How are skeletal muscles attached to bones?
    - A. by ligaments
    - B. by tendons
    - C. by skin cells
    - D. by smooth muscles

Test-Taking Tip

For multiple-choice questions, use the process of elimination. First eliminate answers you know are incorrect. Then compare your remaining choices to solve the problem or to make an educated guess.

Respiratory and Circulatory Systems

KEY CONCEPT The respiratory and circulatory systems move essential materials around the body.

If the human body were very, very flat, it would need neither a heart nor lungs. Flatworms live without these organs. However, these worms are never more than a couple of millimeters thick, because all of their cells must be close to the outside environment.

Every human cell gets nutrients and discards wastes by diffusion across its cell membrane. Through diffusion alone, however, an oxygen molecule would take about three years to move between a lung and a foot. Most animals need a faster way to move materials. The heart and lungs are pumps that can quickly move essential materials around the body.

Think About Moving Fluids

Your heart and lungs are pumps. Your lungs move air into and out of your body. Your heart pushes blood around your body.

- In your Science Notebook, make a list of three manufactured systems that use pumps to move air or water.

- Think about these pumps. How do they move fluids? Do they push them or pull them? Write what you think next to each of the pumps on your list.

NSTA

SCiLINKS
THE WORLD'S A CLICK AWAY

www.scilinks.org
Respiratory System **Code: WGB20A**
Circulatory System **Code: WGB20B**

20.1 What Is Respiration?

Learning Goals

- Identify the three stages of respiration.
- Explain how breathing moves air into and out of the lungs.
- Distinguish between respiration and cellular respiration.

New Vocabulary

respiration
ventilation
lung
diaphragm
alveolus
cellular respiration

Before You Read

Make a prediction about what you think respiration is. Write your prediction in your Science Notebook, using descriptive words and examples.

Take a deep breath, and then let it out. Your body is respiring. Now, take a deep breath and hold it. Your body is still respiring. How can that be? The answer is in the distinction between respiration and breathing. **Respiration** (res puh RAY shun) is the process that moves oxygen and carbon dioxide between the atmosphere and the cells inside the body. Respiration is more than just breathing.

Oxygen moves into the body in three stages. The first stage is **ventilation** (ven tuh LAY shun). Ventilation moves air over the tissues of the lungs where gas exchange occurs. In the second stage of respiration, oxygen diffuses into the bloodstream from the outside environment. The third stage of respiration occurs when oxygen diffuses from the blood into the cells. Carbon dioxide leaves the cells using the same pathway in reverse.

Ventilation

Humans ventilate their bodies by breathing. When humans breathe, they pull air into a pair of organs inside the chest called **lungs**. The lungs are surrounded by a cage of bones called the ribs. They are separated from the digestive organs by a thin muscle called the **diaphragm** (DI uh fram).

The lungs fill with air when humans inhale (ihn HAYL). Inhalation occurs when muscles in the diaphragm and between the ribs contract, or shorten. As the muscles shorten, the ribs move upward and outward and the diaphragm moves down. These movements make the space inside the chest cavity larger. As the chest expands, the air pressure inside the space decreases. Air gets sucked into the lungs and expands them like balloons until the pressure inside the chest is the same as the pressure in the outside environment.

Air is pushed out of the lungs when humans exhale (eks HAYL). Exhalation occurs when the muscles around the chest cavity relax. The diaphragm and ribs spring back to their original positions and force the air in the lungs out of the body. Inhalation is an active process that uses energy, but exhalation requires no extra energy. The processes of inhalation and exhalation are illustrated in **Figure 20.1** on page 352.

Did You Know?

Other animals ventilate their bodies in ways that are different from the method used by humans. Fish pump water across their gills to pull oxygen from the water. Frogs pump their throats up and down to force air into their lungs.

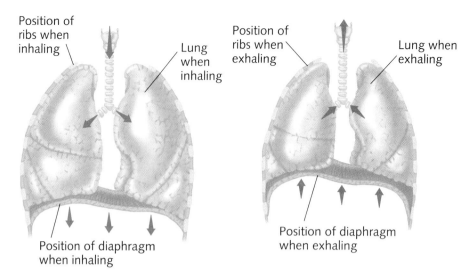

Position of ribs when inhaling

Lung when inhaling

Position of ribs when exhaling

Lung when exhaling

Position of diaphragm when inhaling

Position of diaphragm when exhaling

Figure 20.1 When the muscles in the diaphragm and chest contract, the chest cavity gets larger and fills with air *(left)*. When the muscles in the diaphragm and chest relax, the chest cavity gets smaller and pushes air out of the lungs *(right)*.

Gas Exchange

The oxygen that is inhaled may be inside the lungs, but it is not inside the body. The space inside the lungs is still part of the outside environment. In the second stage of respiration, oxygen enters the body by diffusing across lung tissue and into the bloodstream.

The inside of a lung is divided into many small compartments called **alveoli** (al VEE uh li, singular: alveolus). The alveoli are surrounded by tiny blood vessels called capillaries. Gases move between the alveoli and the capillaries easily, because the walls of both are only one cell thick.

Inside the alveoli, oxygen moves into the bloodstream because there is more oxygen in the air in the alveoli than in the blood. Carbon dioxide leaves the bloodstream because there is more carbon dioxide in the blood than in the air in the alveoli. The blood moves the oxygen to other parts of the body, and the carbon dioxide is exhaled.

The third stage of respiration occurs when the oxygenated blood reaches cells inside the body. Oxygen diffuses out of the blood and across cell membranes to the insides of cells. Carbon dioxide diffuses from the inside of each cell into the blood, where it can be carried to the lungs.

Alveoli

O_2-rich blood

Capillary network

Alveolus

CO_2-rich blood

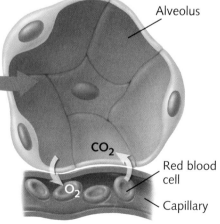

Alveolus

CO_2

Red blood cell

O_2

Capillary

Figure 20.2 A network of capillaries brings CO_2-rich blood to each alveolus and carries away O_2-rich blood *(left)*. Gases move between the outside air and the blood through the thin walls of an alveolus and a capillary *(right)*.

Figure It Out

1. When is the oxygen concentration lowest?
2. Explain why the concentration of nitrogen does not change between the time air is inhaled and the time it is exhaled.

Comparison of Gases in Inhaled and Exhaled Air

Gas	Inhaled Air	Exhaled Air
nitrogen	78.00%	78.00%
oxygen	21.00%	16.54%
carbon dioxide	0.03%	4.49%
other gases	0.97%	0.97%

Figure 20.3 Exhaled air is different from inhaled air. Compare the relative percentages of gases in inhaled and exhaled air.

Respiration Inside Cells

Oxygen can move from the blood into the cells because cells contain relatively little oxygen. The cells in the body constantly use oxygen to break down glucose and store its energy in a complex chemical called ATP. The process of breaking down nutrients for energy is called **cellular respiration**.

One waste product of cellular respiration is carbon dioxide. Because cells are constantly breaking down glucose for energy, the concentration of carbon dioxide inside a cell will steadily increase unless the gas is removed. The respiratory system has two functions. It supplies the body with oxygen needed by the cells for cellular respiration, and it also removes waste carbon dioxide from the body.

After You Read

1. Review the two definitions of respiration that you recorded in your Science Notebook. Then, write a well-developed paragraph explaining what organ-level respiration and cellular respiration have in common.
2. Name the three stages of respiration.
3. Sometimes people suffer an injury that punches a hole through the chest wall. After such an injury, these people cannot inhale. Why?
4. Why do you start feeling uncomfortable after you hold your breath for a minute?

 Explore It!

How much air is inside your lungs?

- Measure 1 L of water. Pour it into a plastic bread bag, and use a permanent marker to draw a line on the bag at the top of the water level. Write *1 L* at the line. Repeat the process to make marks at the 2-, 3-, and 4-L levels. Pour all the water out of the bag.
- Close the opening of the bag around a straw, and seal it with a rubber band. Flatten the bag so there is no air inside it, and then take a deep breath and exhale into the bag through the straw. When you are finished, pinch the straw to keep the air from escaping.
- Carefully flatten the bag from the top until all the air is pushed to the bottom of the bag. How many liters of air did you exhale? Write the answer in your Science Notebook.

Learning Goals

- Describe how air moves through the respiratory system.
- Name the structures that make up the human respiratory system.
- Identify the causes and symptoms of colds, bronchitis, pneumonia, and asthma.

New Vocabulary

respiratory system
pharynx
epiglottis
larynx
trachea
bronchus

Before You Read

Create a sequence chart in your Science Notebook that starts with the word *Nose*. As you read, complete the chart with the names of the structures that air travels through after inhalation. Use arrows to indicate the correct sequence.

Each time a person inhales, the air travels through the respiratory system before it gets to the exchange surfaces in the lungs. The **respiratory system** consists of the lungs and a series of passages.

The Pharynx and Larynx

Air can enter the body through two openings: the nose and the mouth. When a person breathes through the nose, the air enters spaces inside the head that warm and moisten the air. Mucus on the walls of these spaces helps clean and filter the air by trapping dust and pollen.

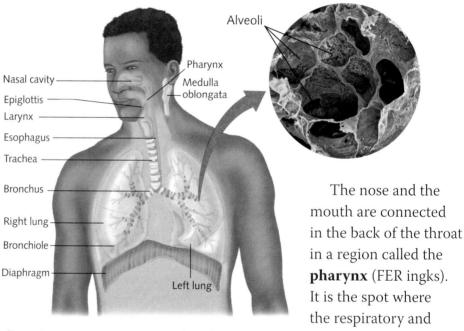

Alveoli
Nasal cavity
Pharynx
Medulla oblongata
Epiglottis
Larynx
Esophagus
Trachea
Bronchus
Right lung
Bronchiole
Diaphragm
Left lung

Figure 20.4 Air travels through a series of passageways before it reaches the alveoli of the lungs.

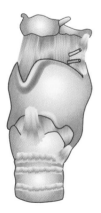

Figure 20.5 The larynx is made of cartilage and muscle and contains folds of tissue called vocal cords.

The nose and the mouth are connected in the back of the throat in a region called the **pharynx** (FER ingks). It is the spot where the respiratory and digestive systems cross. Food and air both travel through the pharynx. Food is usually kept out of the respiratory passage by a flap of cartilage at the lower end of the pharynx called the **epiglottis** (eh puh GLAH tus). When a person breathes, the epiglottis stays upright and lets air flow deeper into the lungs. When a person swallows, it folds over the respiratory passageway and deflects food into the esophagus.

After the pharynx, air moves through the **larynx** (LER ingks), which is also called the voice box because it contains the vocal cords. When air is forced through the vocal cords, they vibrate and produce sound.

The Trachea and Bronchi

After passing through the larynx, air enters a large tube that runs down the front of the neck called the **trachea** (TRAY kee uh). It is reinforced with a series of *C*-shaped pieces of cartilage. The cartilage keeps the trachea from collapsing when a person swallows or turns his or her head. The lower end of the trachea splits into two short tubes called the **bronchi** (BRAHN ki, singular: bronchus). The bronchi carry air into the lungs.

Lungs

Humans have two lungs, one on each side of the chest. Each lung is attached to one of the bronchi. As soon as a bronchus enters the lung, it splits into several smaller tubes. The tubes split again and again, carrying air into even narrower and thinner-walled tubes as the air moves deeper into the lung.

Each tube ends in clusters of tiny sacs called alveoli. Recall that each alveolus is only one cell thick and is covered with capillaries. The inner walls of the alveoli are wet, so oxygen and carbon dioxide can diffuse easily between the blood inside the capillaries and the air inside the alveoli. The insides of the alveoli are also coated with a slippery material called surfactant. Surfactant keeps the walls of the alveoli from sticking together and makes it easier for them to expand during inhalation.

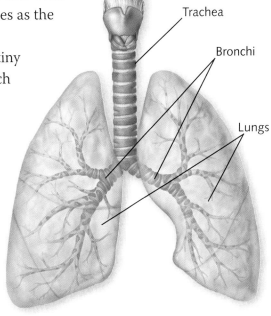

Trachea

Bronchi

Lungs

Figure 20.6 The trachea and bronchi contain rings of cartilage that support the air passageway and hold it open. Clusters of alveoli and capillaries make up the lung tissue.

The cells lining the respiratory passages are covered with cilia that constantly sweep mucus out of the lungs. The mucus traps dust and pollen, and the cilia push it into the throat, where it can be coughed up, sneezed out, or swallowed.

Respiratory Disorders

In a healthy person, air moves through the respiratory system easily. Many diseases, however, attack the respiratory system and make it hard to breathe. Some of these diseases can cause discomfort for a few days, but others can be life threatening.

As You Read

On the same sequence chart, use different-colored arrows to draw the path of an exhalation.

Is the path of an exhalation identical to the path of an inhalation?

 Explain It!

A choking person can die in minutes if his or her airway is not cleared. Abdominal thrusts, also called the Heimlich maneuver, are one way to get an object out of the trachea in an emergency. One person stands behind the choking person and quickly pulls his or her fist up into the victim's diaphragm. Explain why pushing on the diaphragm can dislodge objects that are stuck in the trachea.

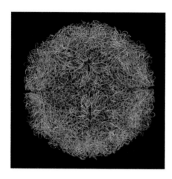

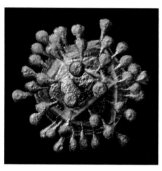

Figure 20.7 Most human colds are caused by rhinoviruses *(top)* or coronaviruses *(bottom)*.

Figure It Out

1. How many more people were diagnosed with asthma in 1993–94 than were diagnosed in 1980?

2. In which age groups did asthma cases increase most?

Everybody catches a cold now and then. When this happens, the throat feels scratchy, the person coughs and sneezes, and it gets hard to breathe through the nose. Viruses that attack the cells inside the nose and pharynx cause colds. These viruses cannot be treated with antibiotics. There are also no vaccines to prevent colds because the viruses that cause colds mutate quickly. Each cold a person catches is really a different disease.

The same viruses that cause colds can also cause a condition called bronchitis (brahn KI tus). Infected bronchi swell up and produce extra mucus, causing coughing and shortness of breath. People who smoke get bronchitis more often than nonsmokers do.

An infection inside the lungs is called pneumonia (noo MOH nyuh). The alveoli swell and fill with fluid, making it difficult to breathe. A person with pneumonia usually has a fever and a cough. Pneumonia can be caused by bacteria, viruses, or fungi.

Asthma (AZ muh) is not caused by an infection. It is a condition that contracts the bronchi and makes it hard for a person to exhale. Asthma attacks are often caused by allergies, but they can also be triggered by exercise or stress. A person having an asthma attack may cough or wheeze. Asthma is treated with medicines that are inhaled to relax the bronchi.

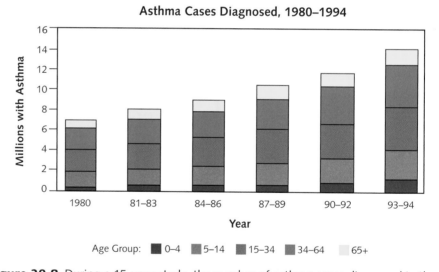

Figure 20.8 During a 15-year study, the number of asthma cases diagnosed in the United States increased.

After You Read

1. Use your sequence chart to draw and label a diagram of the human respiratory system in your Science Notebook. Write the names of the four respiratory conditions you have learned about. Draw an arrow from each name to the part of the respiratory system it affects.

2. Explain one advantage of breathing through the nose instead of through the mouth.

The Body's Transport System

Learning Goals

- Identify the three roles of the cardiovascular system.

- Describe how the cardiovascular system carries out each of these roles.

New Vocabulary

cardiovascular system
cell metabolism

Before You Read

Read the lesson title, the headings, and the Learning Goals, and look at the figures. Predict what you think you will learn in this lesson. Write your prediction in two or three sentences in your Science Notebook.

How do the cells in a human's toes keep from being oxygen-deprived when they are more than a meter away from the lungs? How does the brain get nutrients when it is just as far from the intestines? And why aren't muscle cells swimming in their own waste? All of the body's cells depend on a transport system to move materials such as oxygen, nutrients, and wastes around the body.

The body's transport system is called the **cardiovascular** (kar dee oh VAS kyuh lur) **system**. It includes the heart, blood, and a series of blood vessels long enough to circle Earth's equator four times. The system has three jobs. It moves a variety of molecules around the body. It also regulates the body's fluids and temperature. Lastly, it helps protect the body from disease and infection.

Transportation

Blood carries many different molecules around the body. Blood moves oxygen and carbon dioxide between the lungs and the cells. It absorbs nutrients, salts, vitamins, and water from the food inside the small intestine and brings them to the cells. The blood carries cell wastes to the kidneys for disposal. Blood absorbs drugs such as alcohol and carries them to the liver. Finally, blood carries hormones to the tissues they affect.

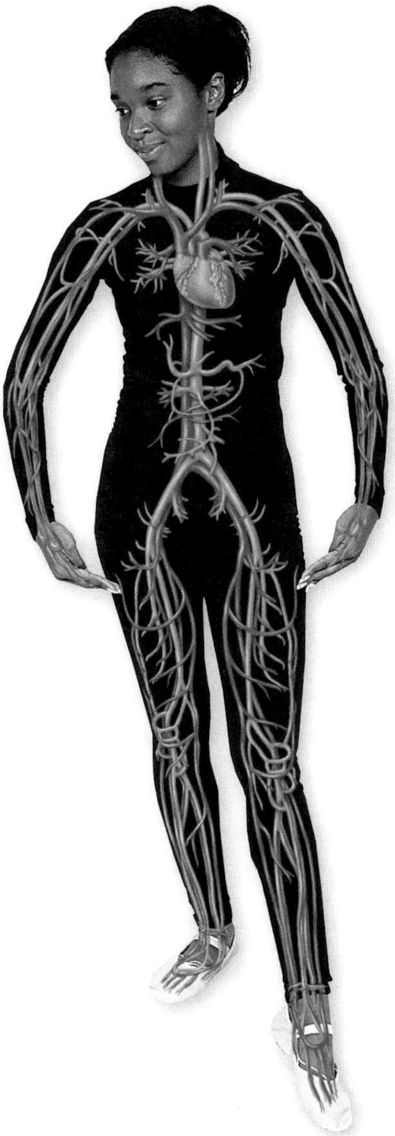

As You Read

Look at the prediction you wrote in your Science Notebook. Compare the role of blood in transporting materials described in this lesson with your prediction. Work with a partner to make necessary corrections to your prediction.

What materials are transported by blood?

Figure 20.9 The cardiovascular system is also called the circulatory system. It moves many materials around the body, suspended in the blood.

Regulation

The cardiovascular system also regulates the condition of the fluids inside the body. As a result of **cell metabolism** (muh TA buh lih zum), which is the sum of all the chemical reactions that occur in a cell, ions move into the spaces outside the cells, and fluids often follow by osmosis. Blood can move excess ions away from the tissues.

Blood also regulates body temperature by moving heat from one part of the body to another. Human cells work best at a temperature of about 37°C (98.6°F), but the temperature of the tissues is always changing. Blood picks up excess heat from warm parts of the body and moves it to cooler parts of the body.

The cardiovascular system can conserve heat when the outside environment is cold, and it can get rid of extra heat when the outside environment is hot. Blood vessels in the skin can squeeze closed to keep warm blood away from the surface of the body and close to internal organs. The same blood vessels can open up and allow blood to reach the skin and give off extra heat into the environment. Countercurrent systems are another way to regulate body temperature. **Figure 20.10** shows an example of how one of these systems works.

Figure It Out

1. Is the blood at point A warmer or cooler than the blood at point B?

2. Illustrate one way to arrange blood vessels to cool a warm part of the body.

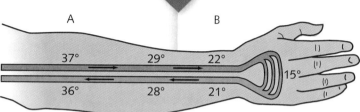

Air temperature = 5°C

Figure 20.10 Blood vessels carrying blood in opposite directions lie next to one another. Warmer blood flows toward the hand. Cooler blood flows back toward the core of the body. Cooler blood pulls heat away from the warmer blood.

Protection

If a person gets a cut, the person bleeds. Harmful organisms may take advantage of the injury and use it to invade the body. Blood can protect against such an invasion. Blood contains cells that attack foreign bacteria and viruses that get into the tissues. It also contains chemicals that make clots, so that there is not too much blood loss from an injury.

After You Read

1. Compare the three roles of the cardiovascular system with the predictions in your Science Notebook. In a well-developed paragraph, explain how the cardiovascular system carries out its three roles.

2. Why do your hands get cold when you are outside on a cold day?

3. Name three things that blood carries to the cells. Why would you not include carbon dioxide in your answer?

20.4 Heart and Blood Vessels

Learning Goals

- Describe how blood moves through the heart.
- Compare the structure and function of arteries, capillaries, and veins.
- Explain what determines blood pressure and heart rate.

New Vocabulary

atrium
ventricle
vena cava
aorta
artery
capillary
vein
blood pressure
pulse

Before You Read

Create a K-W-L-S-H chart in your Science Notebook. Think about the title of this lesson. Then, look at the pictures, headings, and Learning Goals. In the column labeled *K*, write what you already know about the heart and blood vessels. In the column labeled *W*, write what you would like to learn.

The human heart starts beating before birth. It contracts about 70 times every minute, pushing blood through the body with each beat. Over the course of a person's life, the heart will beat about three billion times!

The Heart

The heart is a muscular organ about the size of a fist. It sits inside the rib cage between the lungs. It has four chambers. The two upper chambers are small. Each one is called an **atrium** (AY tree um, plural: atria). The two lower chambers, or **ventricles** (VEN trih kulz), are large.

The heart is a pair of pumps that work together. Oxygen-poor blood from the body enters the right side of the heart. The right atrium and right ventricle pump oxygen-poor blood to the lungs. Waste products are exchanged for oxygen in the lungs. Oxygen-rich blood returns to the left side of the heart from the lungs. The left atrium and left ventricle pump oxygen-rich blood to the rest of the body.

Did You Know?

Ventricles cannot expand on their own after they contract in a heartbeat. The blood that is forced into the ventricles when the atria contract pushes the ventricle muscle into position for the next heartbeat.

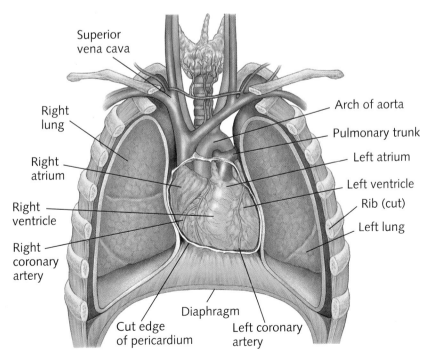

Figure 20.11 Heart contractions move, or "pump," the blood around the body.

Movement of Blood

The heart pushes blood with rhythmic contractions that always follow the same pattern. Both atria contract first, followed by both ventricles. When the atria contract, they force blood into the ventricles. When the ventricles contract, they force blood out of the heart. The two-beat pattern makes the "lub-dub" sound of your heartbeat.

Blood can move in only one direction through the heart. Oxygen-poor blood from the body enters the right atrium through two large veins called the **venae cavae** (vee nee KAY vee, singular: vena cava). When the right atrium contracts, blood is pushed into the right

Figure 20.12 One-way valves like this one keep the blood from moving backward when the heart beats.

ventricle. When the right ventricle contracts, the blood is pushed into blood vessels inside the lungs. After the blood picks up oxygen in the lungs, it returns to the heart and enters the left atrium. Contraction of the left atrium pushes blood into the left ventricle. When the left ventricle contracts, blood is pushed into a large artery called the **aorta** (ay OR tuh). From there, it goes to the rest of the body. One-way valves separate each atrium and ventricle to keep blood from sloshing backward when the heart beats.

Blood Vessels

There are three kinds of blood vessels inside the body. After blood leaves the heart, it flows through each type of vessel in order: first through arteries, then into capillaries, and finally into veins. Blood moves fastest in the largest arteries and slowest in the capillaries.

Arteries (AR tuh reez) carry blood away from the heart. They have thick walls containing layers of stiff connective tissue and smooth muscle. Artery walls are elastic. Every time the heart beats, blood is pushed against the artery walls and makes them expand. The arteries store some of this energy and release it slowly between heartbeats. This makes the blood flow smoothly.

As blood moves away from the heart, the arteries get smaller and narrower. They also split again and again, forming a branching network inside the body. When each artery divides into two or three smaller arteries, the total diameter of the new vessels is actually larger than the diameter of the original vessel. The blood slows down, allowing it to deliver nutrients or pick up wastes as it moves through the arteries and into the capillaries.

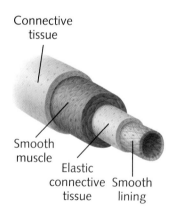

Connective tissue

Smooth muscle

Elastic connective tissue

Smooth lining

Figure 20.13 Arteries have thick walls that can store energy when blood pushes against them.

Capillaries (KAP uh ler eez) are microscopic vessels that connect arteries and veins. They spread through every part of the body. Each capillary is only one cell thick, as **Figure 20.15** shows. Blood moves slowly through the capillaries, so there is more time for nutrients and oxygen to diffuse through the thin walls into the surrounding cells. Blood also picks up wastes from the cells when it is in the capillaries.

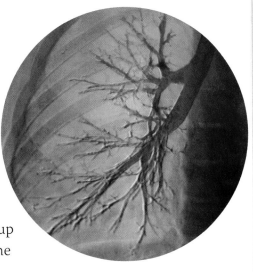

Figure 20.14 As blood moves away from the heart, it travels through smaller and smaller vessels.

Veins (VAYNZ) carry blood back to the heart from the capillaries. Veins have thin walls made of layers of connective tissue and smooth muscle. Veins also contain a series of one-way valves that keep blood flowing toward the heart. When blood flows forward, it pushes the valves open. But if it flows backward, the blood presses against the valves and closes them.

Heartbeats are not strong enough to push the blood inside the veins back to the heart. Instead, contractions of skeletal muscles squeeze the veins and move the blood inside them past the valves.

As You Read

In the column labeled *L* in the chart you made in your Science Notebook, write three or four things you have learned about the heart and blood vessels.

Through which vessels do nutrients and gases move out of the blood and into the tissues?

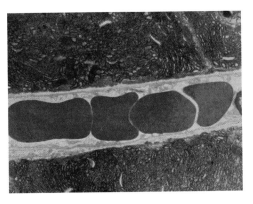
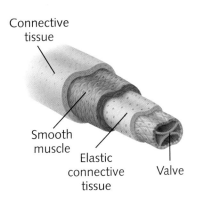

Connective tissue

Smooth muscle

Elastic connective tissue

Valve

Figure 20.15 A capillary *(left)* is so narrow that red blood cells must squeeze through it in single file. Veins *(right)* contain one-way valves that keep blood from moving backward through the cardiovascular system.

Blood Pressure and Heart Rate

When the heart beats, the muscular contractions push oxygen-rich blood through the aorta and to the rest of the body. As blood is forced out of the heart, it pushes against the walls of the arteries and stretches them. When the heart relaxes between beats, the pressure on the artery walls drops. The force that the blood exerts on the artery walls is called **blood pressure**.

1. According to this chart, what are the systolic and diastolic pressures in the aorta?

2. Calculate the drop in blood pressure when the blood reaches the capillaries.

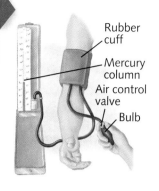

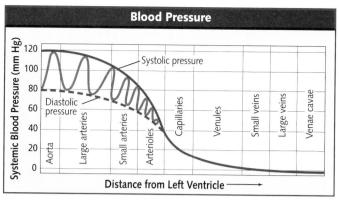

Figure 20.16 Blood pressure rises and falls with each heartbeat. It gets weaker as the blood moves away from the heart.

Explore It!

How does your heart rate change when you exercise?

- Draw a T-chart in your Science Notebook. Title one column *Resting* and the other *Exercising*. Sit at your desk. Have a partner take your pulse for one minute. The number of beats counted in one minute is your heart rate. Write the result in the *Resting* column.

- Jog in place for one minute. Immediately have your partner take your pulse for one minute. Write the result in the *Exercising* column of your chart.

- Find your resting and exercising pulses three or four more times. Then, calculate your average resting and exercising pulses.

Blood moves through the arteries in waves. Blood pressure increases when the ventricles contract and push blood out of the heart. Blood pressure drops when the ventricles relax and refill. You can feel these waves of changing pressure by putting your fingers on your throat or your wrist. These waves create your **pulse**. The pressure waves are the reason blood pressure is measured with two numbers, such as 120 over 80. The first number is the systolic (sihs TAH lihk) pressure. It is the pressure on the arteries when the ventricles contract. The second number is the diastolic (di uh STAH lihk) pressure. It is the pressure on the arteries when the ventricles relax.

Measuring the pulse tells you how quickly the heart is beating. The heart rate is set by a bundle of cells on top of the right atrium called the pacemaker. The pacemaker sends a regular electrical impulse over both atria and makes them contract together. The impulse also makes a second set of cells at the base of the right atrium fire off its own electrical impulse. This impulse makes the ventricles contract. The faster the pacemaker fires, the faster the heart beats. Heart rate increases when a person exercises or is frightened or excited. It slows down when a person is at rest or calm.

After You Read

1. Describe the path of blood through the heart.

2. Compare and contrast the structure of the arteries and the structure of the veins.

3. Use the completed *K*, *W*, and *L* columns of the chart in your Science Notebook to describe the functions of the heart and blood vessels. Complete your chart by indicating in the *S* column what you would still like to learn about the heart and blood vessels and noting in the *H* column how you can find this information.

20.5 Blood

Before You Read

Create a three-column chart in your Science Notebook. Label the first column *Red Blood Cells*, the second column *White Blood Cells*, and the third column *Platelets*. As you read about these three parts of blood, add information to your chart that describes their characteristics.

Learning Goals

- Identify the functions of red blood cells, white blood cells, platelets, and plasma.
- Explain the function of the lymphatic system.
- Identify and describe blood types.

New Vocabulary

hemoglobin
antibody
platelet
plasma
lymphatic system
lymph
antigen

Blood is a liquid tissue that contains several different kinds of cells. Each kind has its own job in the body.

Red Blood Cells

Most of the cells in your blood are red blood cells. Red blood cells carry oxygen. A red blood cell has no nucleus, and each side of the cell is pushed inward to form a disc. This shape makes red blood cells thin and flexible and gives them a cell membrane with a large surface area for oxygen to diffuse across.

The inside of each red blood cell is packed with molecules of a protein called **hemoglobin** (HEE muh gloh bun). Each hemoglobin molecule can bind to four oxygen atoms, increasing the total amount of oxygen the blood can carry. When blood is in the lungs, the hemoglobin fills up with oxygen. When the blood gets to tissues with a lower oxygen concentration, the hemoglobin releases oxygen, which then diffuses into the cells.

Hemoglobin can also help move carbon dioxide. Most carbon dioxide combines with water in the blood plasma to form bicarbonate. Some carbon dioxide, however, binds to hemoglobin molecules, which carry it to the lungs.

White Blood Cells

There are several different types of white blood cells, and these cells make up about one percent of the blood. Each type has a specific role to play in the immune system. They work together to protect the body against disease. Some types engulf and destroy foreign organisms, such as bacteria. Some cause swelling around a wound. Others reduce swelling or produce **antibodies** (AN tih bah deez), which are proteins that detect and bind to foreign proteins.

Did You Know?

Red blood cells, white blood cells, and platelets are all made by stem cells in the bone marrow. Red blood cells live 120 days in the bloodstream before they are destroyed by the spleen, liver, and lymph nodes. Platelets live only ten days before they are destroyed.

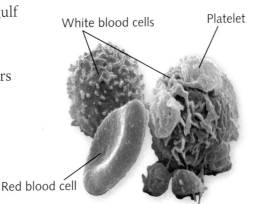

White blood cells

Platelet

Red blood cell

Figure 20.17 White blood cells are much larger than red blood cells. Platelets are smaller than both red and white blood cells and have no nuclei.

Platelets

When a person gets a small cut, he or she bleeds for a little while, but the blood eventually clots. Clotting is controlled by **platelets** (PLAYT luts) in the blood. Platelets are small fragments of cells that stick to damaged tissues. They also stick to one another, forming a plug that seals the wound and stops the bleeding. They then release chemicals that help build a linked network of a protein called fibrin, forming a leathery clot called a scab that covers the wound until it heals.

Plasma

The blood cells and platelets are suspended in the liquid part of the blood, which is called the **plasma** (PLAZ muh). Plasma makes up more than half of the blood's volume. It is mostly water, but it also contains dissolved proteins, minerals, vitamins, and small organic molecules such as amino acids, fatty acids, and glucose. Blood carries waste products such as carbon dioxide and urea dissolved in the plasma.

The Lymphatic System

As blood moves through the capillaries, some fluid from the plasma leaks out of the vessels and gets into the spaces between the cells. This fluid bathes the cells and keeps them moist. If too much fluid stays in the tissue, however, it can cause swelling. Excess fluid normally leaves the tissue through small vessels in the **lymphatic** (lihm FA tihk) **system**.

The fluid that enters the lymphatic capillaries is called **lymph** (LIHMF). The lymphatic capillaries join to form a network of small, thin-walled tubes that eventually drain into veins near the heart. On the way, lymph passes through filtering tissues called lymph nodes, which are found in many parts of the body, including the tonsils. White blood cells produced by the thymus and the spleen enter the fluid in the lymph nodes to trap bacteria and other foreign particles.

As You Read

The chart you made in your Science Notebook describes the cellular parts of blood. Under your chart, define plasma, and explain how plasma becomes lymph.

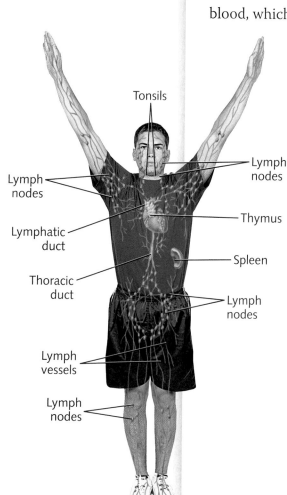

Tonsils

Lymph nodes

Lymph nodes

Lymphatic duct

Thymus

Thoracic duct

Spleen

Lymph nodes

Lymph vessels

Lymph nodes

Figure 20.18 The lymphatic system drains fluid from the tissues and moves it back to the cardiovascular system. Like veins, lymphatic vessels have valves to keep the fluid inside them from flowing backward.

CONNECTION: Chemistry

Each hemoglobin molecule contains four iron (Fe) atoms. When iron binds to oxygen (O), it turns red. This means that blood changes color when it is carrying oxygen. Oxygenated blood in the arteries is bright red. Deoxygenated blood in the veins is a darker red.

Blood Types

For nearly three hundred years, doctors tried to move blood from healthy people into their sick patients. But in most cases, the transfusion of blood would make the patient even sicker. About a hundred years ago, scientists discovered that many people have proteins on the surfaces of their red blood cells that can react with another person's immune system. These proteins are called **antigens** (AN tih junz).

Three important blood antigens are called A, B, and Rh. The combination of proteins on a person's red blood cells determines the person's blood type. Blood types are inherited traits.

Blood type is expressed as a letter and a plus or minus sign. The letters A and B indicate which antigens are on the surfaces of red blood cells. If a red blood cell has A antigens, the person has type A blood. If it has B antigens, the person has type B blood. If it has both antigens, the person is type AB. If it has no antigens, the person is type O. People can also have an antigen called Rh on the surfaces of their red blood cells. People who have the Rh antigen have Rh positive (+) blood. People who lack the Rh antigen have Rh negative (−) blood. If a person has B⁺ blood, the person has both the B and Rh antigens on the red blood cells.

Extend It!

Are some blood types more common than others? Find out the percentage of the human population that has each blood type by doing research in the library and on the Internet. Summarize your findings in a pie chart. If the members of your class can find out their blood types, collect this information and determine the percentage of each type. Do the class percentages match your research findings?

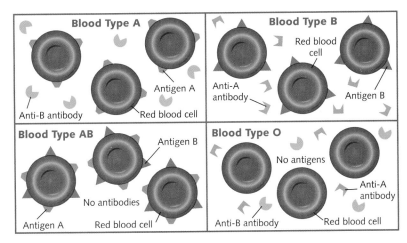

Figure 20.19 Doctors prefer to give patients blood that matches their own blood type, but in an emergency, people can receive any blood that does not contain antibodies that would attack their red blood cells. For example, this chart shows that a person with type A blood could receive type O blood.

Figure It Out

1. What kind of antibody does a person with type B blood have in his or her plasma?

2. Predict which types of blood a type AB person could receive if there were no AB blood at the hospital.

After You Read

1. On one side of the chart you made in your Science Notebook, write the words *Antigen* and *Antibody*. Draw an arrow from each word to the part of the blood where that protein is found.

2. People who suffer from hemophilia do not have normal platelets. Predict how the disease affects their bodies.

3. Define what gives a person type O blood.

Before You Read

Create a concept map in your Science Notebook using the words *Heart*, *Lungs*, and *Tissues*. Preview the lesson, and write one sentence describing how you think these three words are related.

Library books circulate. Money, rented DVDs, fresh air, and blood circulate, as well. In each case, **circulation** (sur kyuh LAY shun) means that something moves from one place to another and back again. Circulation moves blood around the body. Humans have two distinct circulations in their bodies. Blood circulates through the lungs before it circulates through the rest of the body.

Pulmonary Circulation

The **pulmonary** (PUL muh ner ee) **circulation** moves blood through the lungs. The right atrium of the heart contracts, forcing the blood into the right ventricle. The right ventricle pushes oxygen-poor blood through the pulmonary arteries and into the lungs. The pulmonary arteries are the only arteries that carry blood that is high in carbon dioxide. The pulmonary arteries divide into capillaries that surround the alveoli. As blood moves through the capillaries, it picks up oxygen and releases carbon dioxide. Then, the oxygen-rich blood moves into pulmonary veins and returns to the left atrium of the heart. The pulmonary veins are the only veins that carry oxygen-rich blood.

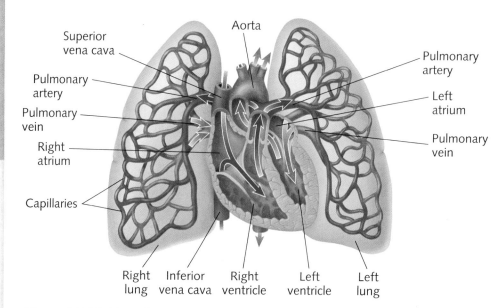

Figure 20.20 The pulmonary circulation moves blood through the lungs, where it picks up oxygen.

Systemic Circulation

The left side of the heart pumps oxygen-rich blood through the rest of the body. This circuit is called the **systemic** (sihs TE mihk) **circulation**. The left atrium contracts and forces the blood into the left ventricle. The left ventricle contracts, forcing the oxygen-rich blood out of the heart through the aorta. Arteries branch off the aorta and send blood to capillaries in all of the organs and body tissues except for the lungs. The blood exchanges nutrients and oxygen for carbon dioxide and wastes. It then moves into the veins and travels back to the right atrium of the heart through the superior and inferior venae cavae.

As You Read

On your concept map, draw arrows that show the path blood takes in the pulmonary and systemic circulations. Label each circulation on the concept map. Compare your chart with the chart of a partner, and make necessary corrections or additions.

Which of the two circulations is larger?

Figure It Out

1. What connects the arteries and the veins?
2. Predict which parts of the circuit will have the highest and the lowest blood pressure.

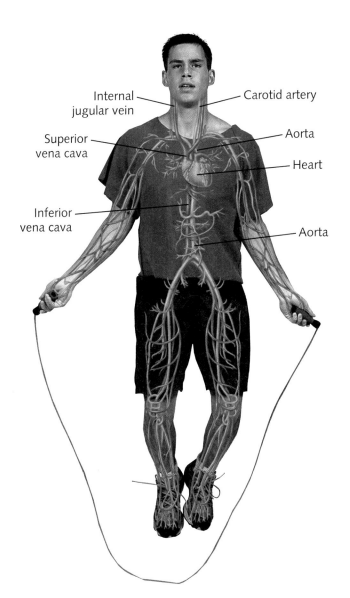

Internal jugular vein — Carotid artery
Superior vena cava — Aorta
— Heart
Inferior vena cava — Aorta

Figure 20.21 The systemic circulation carries blood to the tissues and back to the heart.

Coronary Circulation

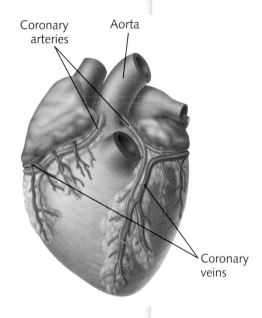

Coronary arteries · Aorta · Coronary veins

A special branch of the systemic circulation called the **coronary** (KOR uh ner ee) **circulation** makes sure that the heart gets enough oxygen. The coronary arteries branch off from the aorta immediately after it leaves the left ventricle, bringing oxygenated blood to the heart muscle. One artery brings blood to the right atrium and ventricle. Another brings blood to the left atrium and ventricle. The coronary veins drain directly into the right atrium of the heart.

Figure 20.22 The coronary circulation ensures that the heart gets oxygenated blood first.

PEOPLE IN SCIENCE William Harvey 1578–1657

Before 1628, doctors believed that there were two kinds of blood. Venous blood, they thought, was made in the liver and spread through the body in the veins. Arterial blood was made in the heart from air and a tiny amount of venous blood. William Harvey, an Englishman, changed those ideas. Harvey studied in Italy with the most famous physicians in the world. When he graduated, Harvey returned to England and soon became the king's doctor.

One of Harvey's teachers had found a series of valves inside the veins. Harvey wondered about the purpose of these valves. He studied the heart and blood vessels in animals by dissecting them and by doing experiments. Harvey's dissections indicated that all of the blood had to go through the lungs to get from one side of the heart to the other. He pushed blood through the veins and found that the valves only opened in one direction. He could push blood toward the heart, but not away from it.

All of his experiments led Harvey to conclude that blood had to circulate inside the body. He could not explain how blood got from the arteries into the veins, but he predicted that tiny vessels connected the two. Shortly after his death, capillaries were discovered.

After You Read

1. Describe the three circulations that make up the circulatory system.
2. Predict whether the blood pressure inside the pulmonary artery is larger or smaller than the pressure inside the aorta.
3. Add the coronary circulation to your concept map. Is it a specialized circuit of the pulmonary circulation or the systemic circulation?

20.7 **Cardiovascular Diseases**

Learning Goals

- Identify the symptoms of the major cardiovascular diseases.
- Evaluate the role of diet, exercise, and smoking in the development of cardiovascular disease.

New Vocabulary

hypertension
atherosclerosis
stroke

Before You Read

Draw a T-chart in your Science Notebook. On one side, write a list of the headings in this lesson. Leave several blank lines under each topic. As you read, write the symptoms for each cardiovascular disease on the other side of the chart, across from its name.

Do you know what is the number-one killer in the United States? Perhaps you think it is car accidents or sports accidents. Maybe you think it is diabetes or cancer. If so, you would be wrong. The number-one killer is cardiovascular disease. About 40 percent of all deaths in the United States each year are caused by heart disease and stroke.

Some of these deaths could have been prevented. Cardiovascular diseases develop more often in people who eat fatty foods, smoke, and avoid exercise. If a person eats a diet low in cholesterol and saturated fat, exercises regularly, and doesn't smoke, he or she can reduce the chance of developing a cardiovascular disease.

Causes of Death in the United States in 2003		
Rank	**Cause**	**Number of Deaths**
1	heart disease	685,089
2	cancer	556,902
3	stroke	157,689
4	lower respiratory disease	126,382
5	accidents	109,277
6	diabetes	74,219

Figure 20.23 Two of the top six causes of death in the United States in 2003 were cardiovascular diseases.

Figure It Out

1. How many people died of heart disease in 2003?

2. Cardiovascular diseases include both heart disease and stroke. Calculate the total number of people who died of these diseases in 2003.

Hypertension

Hypertension (hi pur TEN chun) is high blood pressure. It occurs when the heart has to pump harder than normal to push blood through the systemic circulation. Normally, systolic blood pressure is around 120 and diastolic pressure is around 80. A person with hypertension will have systolic pressure higher than 140 and diastolic pressure higher than 90.

A person with hypertension may have no symptoms. He or she may feel fine. However, consistently high blood pressure can damage other parts of the body. It is one of the major risk factors for cardiovascular disease.

Atherosclerosis

Atherosclerosis (a thuh roh skluh ROH sus) is a common cause of cardiovascular disease. It occurs when deposits made of fats and cholesterol build up on the inner walls of large arteries. The deposits are called plaques (PLAKS). Large plaques make the space inside an artery smaller, so less blood can pass through. This means that the heart has to work harder to pump blood through the arteries. Atherosclerosis can cause hypertension.

As You Read

Review the symptoms of the cardiovascular diseases you have read about so far.

What is another name for atherosclerosis?

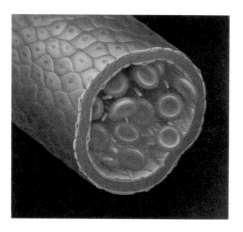 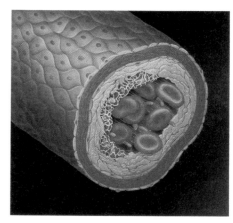

Figure 20.24 Compare the healthy artery on the left and the atherosclerotic artery on the right. Can you see how plaques have made the atherosclerotic artery narrower? Can it carry as much blood as the healthy artery?

If the plaques grow large enough, they can completely block an artery. Blood cannot flow through blocked arteries, and the tissues downstream from the blockage can starve and die. Large plaques also damage the artery walls, making them stiff. Atherosclerosis is sometimes called hardening of the arteries. When the arteries harden, they lose their ability to push the blood. This also makes the heart work harder to pump blood through the arteries.

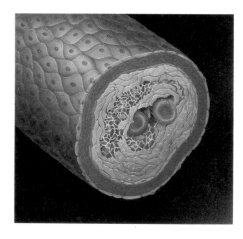

Figure 20.25 This artery is almost completely blocked by plaques. Only a little blood can squeeze through it. If the plaques get any thicker, the artery will be blocked and the tissues it supplies will die.

Platelets can stick to the plaques and form clots. Blood clots or pieces of fatty plaque sometimes break off the artery walls and float in the bloodstream. If they get stuck in blood vessels in the heart or the brain, they can cause a heart attack or a stroke.

Heart Attack

A heart attack starts when a coronary artery gets clogged by a fatty plaque, cutting off the blood supply to one part of the heart muscle. When the muscle cells do not get enough oxygen, they begin to die. The damage can make the heart beat irregularly or even stop. Even if a person survives a heart attack, the damaged heart muscle may be weaker than before. The main symptom of a heart attack is chest pain that often spreads to an arm or a shoulder. Other symptoms include shortness of breath, dizziness, and nausea. A heart attack is a medical emergency.

Stroke

A **stroke** occurs when an artery in the brain is blocked or bursts, cutting off the blood supply to part of the brain. When the nerve cells do not get enough oxygen, they die. A stroke can cause mild damage that the victim eventually recovers from, or it can cause serious disabilities or death. The effects of the stroke depend on which part of the brain was affected. Stroke victims can suffer a wide variety of symptoms, including a sudden severe headache, paralysis, blindness or blurred vision, loss of speech or balance, or amnesia. A stroke is also a medical emergency.

CONNECTION: Economics

Cardiovascular diseases are not just deadly—they are also expensive. The total cost of health care for people with these diseases and their lost time at work is approximately 403 billion dollars each year.

After You Read

1. What causes a heart attack?

2. Could a stroke be caused by an atherosclerotic blockage in the brain? Write one or two sentences to defend your answer.

3. Review the information you have recorded in the T-chart in your Science Notebook. In a well-developed paragraph, predict how people who recover from heart attacks and strokes might change their daily lives in order to avoid having a second heart attack or stroke.

Chapter 20

Summary

KEY CONCEPTS

20.1 What Is Respiration?

- Breathing moves air to the respiratory exchange surface inside the lungs.
- Oxygen and carbon dioxide move into and out of the bloodstream by diffusion.
- Cells use oxygen to carry out cellular respiration.

20.2 The Respiratory System

- The human respiratory system is a series of passages that channel air to the lungs.
- Gases are exchanged across the walls of the alveoli.

20.3 The Body's Transport System

- The cardiovascular system transports materials, regulates fluid balance and body temperature, and helps prevent infection and blood loss after an injury.

20.4 Heart and Blood Vessels

- The heart pumps blood to the lungs and the rest of the body.
- Arteries carry blood away from the heart, and veins carry blood to the heart. Capillaries connect arteries and veins.
- Blood pressure is the result of the heart pushing blood against the artery walls.

20.5 Blood

- Blood is made up of red blood cells, white blood cells, and platelets suspended in plasma.
- Plasma that enters the tissues is called lymph, and it returns to the blood vessels through the lymphatic system.
- Human blood types are the result of antigens on the surface of the red blood cells.

20.6 Circulation in the Body

- The pulmonary circulation moves blood through the lungs. The systemic circulation moves blood through other tissues.
- The coronary circulation branches off from the systemic circulation to ensure that the heart gets oxygenated blood.

20.7 Cardiovascular Diseases

- Cardiovascular diseases are the leading cause of death in the United States.
- High blood pressure and atherosclerosis can cause heart attacks and strokes.

VOCABULARY REVIEW

Write each term in a complete sentence, or write a paragraph relating several terms.

20.1
respiration, p. 351
ventilation, p. 351
lung, p. 351
diaphragm, p. 351
alveolus, p. 352
cellular respiration, p. 353

20.2
respiratory system, p. 354
pharynx, p. 354
epiglottis, p. 354
larynx, p. 354
trachea, p. 355
bronchus, p. 355

20.3
cardiovascular system, p. 357
cell metabolism, p. 358

20.4
atrium, p. 359
ventricle, p. 359
vena cava, p. 360
aorta, p. 360
artery, p. 360
capillary, p. 361
vein, p. 361
blood pressure, p. 361
pulse, p. 362

20.5
hemoglobin, p. 363
antibody, p. 363
platelet, p. 364
plasma, p. 364
lymphatic system, p. 364
lymph, p. 364
antigen, p. 365

20.6
circulation, p. 366
pulmonary circulation, p. 366
systemic circulation, p. 367
coronary circulation, p. 368

20.7
hypertension, p. 369
atherosclerosis, p. 370
stroke, p. 371

MASTERING CONCEPTS

True or False

If the statement is true, write "true." If it is false, change the underlined word or words to make the statement true.

1. Air is pushed out of your lungs when you <u>inhale</u>.

2. The <u>pulmonary circulation</u> supplies oxygenated blood to the heart.

3. <u>Red blood cells</u> carry oxygen.

4. The cardiovascular system can move <u>heat</u> around the body.

5. The <u>trachea</u> keeps food out of the lungs.

6. The cells in the heart that set the heart rate are called the <u>pacemaker</u>.

Short Answer

Answer each of the following in a sentence or brief paragraph.

7. Trace the path of an oxygen molecule as it moves from the air near the nose to the muscle cells in the thigh.

8. Describe the functions of the three main groups of blood vessels.

9. Person 1 has a blood pressure of 140 over 80 and a heart rate of 100 beats per minute. Person 2 has a blood pressure of 90 over 60 and a heart rate of 80 beats per minute. Explain which heart is working harder.

10. Describe the path of lymph as it moves from the tissues back to the bloodstream.

11. Draw a graph illustrating how the oxygen concentration in the blood changes as the blood moves around the body. Choose the axes of the graph to reflect the variables being considered.

PREPARE FOR CHAPTER TEST

To prepare for the chapter test, create a question from each Learning Goal. Use the information in your Science Notebook to answer each question. Then use these answers to write a well-developed essay about the chapter. Use the Key Concept on the first page of this chapter as your topic sentence.

Critical Thinking

Use what you have learned in this chapter to answer each of the following.

12. **Infer** "Blue babies" are born with a hole between the right and left atria. Infer what happens to the blood in people with this condition.

13. **Relate** Inhaled air is about 21 percent oxygen and 0.03 percent carbon dioxide. Exhaled air is about 17 percent oxygen and 4.5 percent carbon dioxide. Relate this change in gas composition to the process of cellular respiration.

14. **Predict** People who suffer from anemia do not make as many red blood cells as people normally do. Predict what some of their symptoms are likely to be.

Standardized Test Question

Choose the letter of the response that correctly answers the question.

15. An atherosclerotic plaque can block an artery and prevent blood from getting to the tissues the artery normally supplies. What waste products will build up in those tissues?

   A. oxygen and nitrogen

   B. oxygen and glucose

   C. carbon dioxide and glucose

   D. carbon dioxide and nitrogen

Test-Taking Tip

Use scrap paper to write notes. Sometimes making a sketch, such as a diagram or a table, can help you organize your ideas.

Chapter 21

Digestive and Excretory Systems

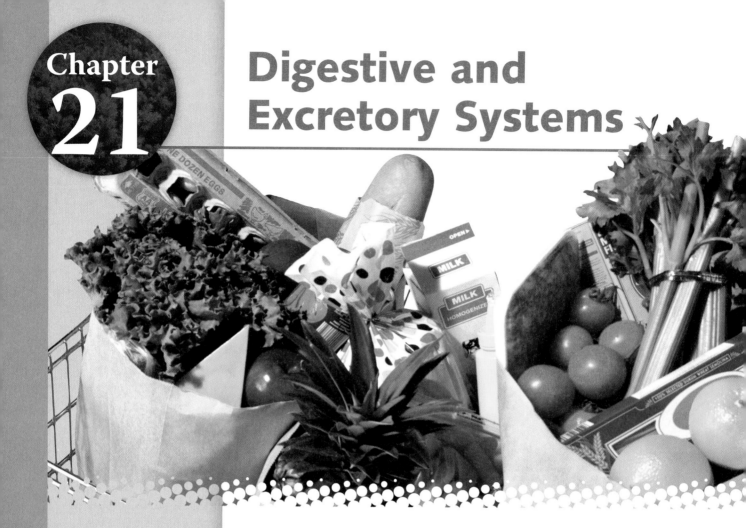

KEY CONCEPT The digestive system processes food, and the excretory system processes wastes.

Animals have lives that revolve around eating. Every animal needs to eat enough food to fuel its body, repair damaged tissues, and store some energy. When it comes to eating, humans are generalists. Humans eat grains, roots, fruits, fungi, milk, eggs, and meat. In some parts of the world, humans eat insects. No matter what people eat, everything put in a person's mouth takes the same trip. Every bit of food you eat, whether it's sushi, popcorn, or tacos, gets digested into nutrients and then is absorbed and used by your body or eventually excreted.

Think About Feeding

Like other mammals, humans chew their food. However, many animals can't chew. Lizards and alligators bite off chunks of food and swallow them whole. Snakes unhinge their jaws to swallow animals that are larger than their heads.

- In your Science Notebook, make a list of three ways a meal can be eaten without chewing.

- Think about these ways of eating. Can you think of animals that eat this way? Can you eat this way? Write what you think next to each method.

NSTA

SCI LINKS
THE WORLD'S A CLICK AWAY

www.scilinks.org
Digestive System **Code: WGB21A**
Excretory System **Code: WGB21B**

21.1 Nutrition

Learning Goals

- Identify the six nutrients the human body needs.
- Explain how to plan a balanced diet.
- Understand how to read a food label.

New Vocabulary

nutrition
nutrient
malnutrition
carbohydrate
protein
essential amino acid
fat
vitamin
mineral
Calorie

Before You Read

Read the lesson title, the headings, and the Learning Goals, and look at the photos and tables. Predict what you think you will learn in this lesson. Write your predictions as two or three sentences in your Science Notebook.

The human body is like a factory. A factory needs a steady supply of raw materials, whether it makes cars, computers, or boxed cereal. It uses some of the materials to build a product. It burns other materials for the energy to run its machines. The body uses food the same way. Food provides the energy for metabolism and the raw materials for building and repairing body parts. The process of getting the food needed to survive is called **nutrition** (new TRIH shun).

Nutrients

The raw materials the body gets from food are called **nutrients** (NEW tree unts). There are six kinds of nutrients: carbohydrates, proteins, fats, vitamins, minerals, and water. The body needs some of each of these nutrients every day in order to stay healthy. If a nutrient is missing from the diet, the body will not be able to complete some of its metabolic jobs. Over time, that missing nutrient will make a person sick. The person will suffer from malnutrition. **Malnutrition** (mal new TRIH shun) occurs when a person cannot get enough to eat or when the food eaten does not provide a balanced diet. A balanced diet contains all of the nutrients the body needs.

Carbohydrates The main sources of energy for the body are carbohydrates. **Carbohydrates** (kar boh HI drayts) are molecules made up of atoms of carbon, hydrogen, and oxygen. There are three types of carbohydrates: sugar, starch, and fiber.

Sugars are simple carbohydrates. Single sugars and short sugar chains make foods such as fruit and honey sweet. Long chains of sugars are called starch. Starches are found in cereals, bread, pasta, and vegetables such as potatoes and rice. Starch is a complex carbohydrate.

Fiber, such as cellulose, comes from the cell walls of plant cells. Fiber is a complex carbohydrate that is found in peas, beans, fruits, vegetables, and whole-grain breads and cereals. The body breaks down sugars and starches to provide energy. Fiber cannot be digested, but it is essential for the smooth running of the digestive system.

Figure 21.1 Starches are complex carbohydrates found in foods such as rice *(top)* and potatoes *(bottom)*.

Proteins A **protein** (PROH teen) is a long chain of amino acids. A protein is made up of atoms of carbon, hydrogen, oxygen, nitrogen, and sometimes sulfur. Muscles, hair, skin, fingernails, and other tissues are all built out of proteins. Each one of the thousands of proteins used in the cells contains a different combination of 20 amino acids. Nine of those amino acids are called **essential amino acids** because people cannot make them from other molecules. The essential amino acids must be eaten every day.

Foods such as meat, fish, milk, and eggs contain all 20 amino acids and are called complete proteins. Plants that are rich in protein often lack some amino acids. Because the body does not store amino acids, vegetarians—people who eat only vegetables, fruits, grains, and nuts—must carefully combine their plant foods to make sure that they get all the amino acids they need every day.

Fats A **fat**, also called a lipid, is a glycerol molecule attached to a series of fatty acid chains. There are two families of fat molecules based on their chemical structure. Saturated fats usually come from animals. They are found in foods such as butter, meat, cheese, eggs, and oily fish. Unsaturated fats usually come from plants. They are liquid at room temperature and are also called oils. Fats and other lipids are an important part of cell membranes, nerve insulation, and molecules known as hormones. They are also an important source of energy. Most of the energy the body stores is stored as fat.

Figure 21.2 Proteins are found in animal and plant foods. Animal foods such as meat, eggs, and milk contain all of the essential amino acids. Plant foods such as beans and rice are missing some essential amino acids.

As You Read

Compare the predictions in your Science Notebook with what you have learned so far about nutrition. Write a list of what you learned that is new to you. Share your list with a partner or a small group of classmates.

What nutrient is the main source of energy for the body?

Figure 21.3 There are two types of fats. Saturated fats *(left)* are solid at room temperature. Unsaturated fats *(right)* are liquid at room temperature.

CONNECTION: Social Studies

Highly spiced foods are often eaten in warm climates where meat turns bad more quickly. Scientists have shown that many of the spices have antibacterial properties and protect people from food poisoning.

Vitamins Complex organic molecules that help the body build new tissues and important molecules are called **vitamins** (VI tuh munz). Vitamins also help regulate body functions and fight disease. Only a few hundredths of a gram of each vitamin is needed every day. The body cannot make vitamins, so they must come from food.

Vitamin C and the B vitamins are soluble in water and cannot be stored in the body. Foods containing these vitamins must be eaten every day. The other vitamins are soluble in fat. They are stored in fatty tissues, and they can be toxic in large quantities. The vitamins needed to keep the body functioning properly are shown in **Figure 21.4**.

Essential Vitamins

Vitamin	Source	Function	Result of Deficiency
A	orange and dark green vegetables, eggs, dairy	used to form visual pigments	xerophthalmia (progressive blindness)
B_1	grains, meat, legumes, nuts	carbohydrate metabolism	beriberi (neurological symptoms)
B_2	milk, eggs, green leafy vegetables, grains	coenzyme for many metabolic reactions	dermatitis, sensitivity to light
B_3	grains, meat, legumes, nuts	important part of nicotinamide coenzymes	pellagra; dermatitis and neurological symptoms
B_6	meat, fruit, grains, legumes, vegetables	amino acid metabolism	skin lesions, irritability, anemia, convulsions
B_{12}	meat, eggs, dairy	used to maintain nerve cells; fatty acid and amino acid metabolism	anemia, neuropathy
pantothenic acid	liver, eggs, many other foods	used for energy metabolism	sleep disturbances, indigestion
biotin	peanuts, tomatoes, eggs, many other foods	aids fat and glycogen formation, amino acid metabolism	lack of coordination, dermatitis, fatigue
folate	vegetables, legumes	important part of DNA synthesis, red blood cell formation	anemia, digestive disorders, neural tube defects in developing embryo
C	citrus fruit, green vegetables, potatoes	used in synthesis of collagen	scurvy (bleeding gums, loss of teeth, listlessness)
D	fatty fish, eggs, fortified milk	used in skeletal formation; assists calcium absorption	rickets (unmineralized bones)
E	green vegetables, nuts, grains, oils, eggs	prevents damage to cell membranes	hemolytic anemia
K	bacteria in digestive tract, leafy green vegetables	intestinal flora; formation of clotting factors	bleeding disorders

Figure 21.4 According to the table, which vitamin is needed for blood clotting? For proper vision?

Minerals Inorganic elements that the body needs for most of its metabolic functions are called **minerals** (MIH nuh rulz). Minerals make bones stiff, enable nerves to send impulses and muscles to contract, and keep body fluids in balance. Some minerals are incorporated into proteins. The minerals needed to keep the body functioning properly are summarized in **Figure 21.5**.

Essential Minerals

Mineral	Source	Function	Result of Deficiency
calcium	milk, eggs, green vegetables, fish	mineralization of bone	rickets, osteoporosis
phosphorus	cheese, oats, liver	mineralization of bone	osteoporosis, abnormal metabolism
sodium	most foods	fluid balance; muscle contraction; nerve conduction	weakness, cramps, diarrhea, dehydration
potassium	most foods	major cellular action	muscular and neurological disorders
chloride	most foods	fluid balance; formation of hydrochloric acid in stomach	rarely occurs
iodine	seafood, iodized salt	required for synthesis of thyroid hormones	goiter (enlarged thyroid gland)
iron	red meat, liver, eggs, legumes, dried fruit	part of hemoglobin	anemia, indigestion
magnesium	green vegetables, grains, nuts, legumes	part of carbohydrate metabolism	muscular and neurological disorders, arrhythmic heartbeat
manganese	most foods	involved in many cellular processes	possible reproductive disorders
copper	meat, water	formation of hemoglobin; part of some enzymes	anemia
chromium	meat, fats, oils	regulation of blood glucose	inability to use glucose
cobalt	meat, milk, eggs	part of vitamin B_{12}	anemia
zinc	meat, grains, vegetables	part of some enzymes; involved in CO_2 transport	baldness
fluoride	fluoridated water, seafood	prevents tooth decay	dental cavities

Figure 21.5 According to the table, which mineral is found primarily in liver, red meat, and dried fruit? Which mineral is important for normal muscle and nerve functioning?

Water Humans can live for weeks without food, but they can live only a few days without water. Water is found inside and around the cells, and it is the largest component of blood. It takes part in most of the reactions of the body, and it is constantly recycled inside the cells. However, water is lost as a result of urination, sweating, and breathing. If the water is not replaced, the body becomes dehydrated.

A Balanced Diet

When you eat meals that contain all of the nutrients your body needs, you are eating a balanced diet. Most foods contain more than one nutrient. How can you be sure to eat enough of all the required nutrients?

Nutritionists are still studying what goes into a healthful diet. They have learned that people should eat more grains, fruits, and vegetables each day than meat and dairy products. Small amounts of oils and fats are healthful.

The energy stored in food is measured in **Calories** (KAL uh reez). One Calorie in heat raises the temperature of one kilogram of water by one degree Celsius. A young woman needs to eat about 1,800 to 2,000 Calories each day to maintain her weight. A young man needs to eat about 2,200 to 2,500 Calories. If a person eats more Calories than his or her metabolism burns, the body will store the extra energy as fat. Exercise burns more Calories than sedentary activities burn.

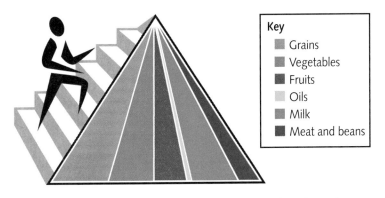

Key
- Grains
- Vegetables
- Fruits
- Oils
- Milk
- Meat and beans

Figure 21.6 The USDA's Food Guide Pyramid is a visual guide that shows how much of each kind of food is found in a healthful, balanced diet.

 **Extend It!**

Are you eating a balanced diet? In your Science Notebook, write down everything you eat for one day, including how much of it you eat. Then, collect nutrition information about the food you ate from food labels and nutrition books.

Create a table in your Science Notebook with the following columns: *Food*, *Calories*, *Carbohydrates*, *Protein*, *Fat*, *Vitamins*, and *Minerals*. Record the information for each food you ate. Calculate the amount of Calories, carbohydrates, protein, fat, vitamins, and minerals you got from your food that day.

Figure It Out

1. Which foods make up more than half of the pyramid?
2. Which foods should be consumed in the smallest amounts?

After You Read

1. Make a T-chart in your Science Notebook. On one side, list the six nutrients the body needs to function properly. On the other side, list at least two foods that contain each nutrient. Compare this chart with the predictions you recorded in your Science Notebook.

2. What is the difference between a mineral and a vitamin?

3. Why is water an important nutrient?

Learning Goals

- Identify the structures that make up the human digestive system.
- Describe how enzymes aid digestion.
- Identify where each kind of mechanical and chemical digestion occurs.

New Vocabulary

ingestion
digestion
enzyme
salivary gland
tongue
taste bud
bolus
esophagus
peristalsis
stomach
chyme
duodenum
liver
pancreas

Before You Read

Create a concept map for digestion in your Science Notebook. Start with the word *Mouth*. As you read, complete the concept map using the names of the structures that food travels through after it is ingested.

The food you eat has to be broken down into smaller molecules before your body can use it. This process occurs in a series of organs called the digestive tract. Other organs of the digestive system assist by moving food around or by producing chemicals used in digestion.

The digestive tract is a tube with a mouth at one end and an anus at the other end. Food enters the digestive tract at the mouth in a process called **ingestion** (in JES chun). Food moves through a series of organs that break it down into small particles that can pass through cell membranes, a process called **digestion** (di JES chun). The body absorbs the digested food, and any leftover indigestible remains are pushed out of the body.

The organs of the digestive tract break up food in two ways. In mechanical digestion, food is physically chopped up into smaller bits. In chemical digestion, food particles are immersed in chemicals that break the bonds of large molecules. Large molecules are broken into smaller molecules. Mechanical digestion increases the surface area of the food particles and makes chemical digestion more effective. Food can be subjected to both kinds of digestion at the same time.

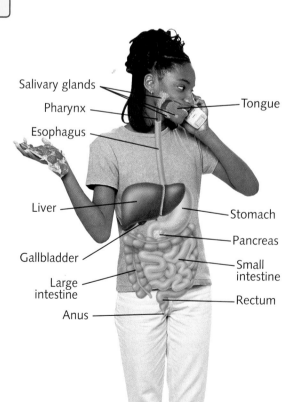

Salivary glands
Pharynx
Esophagus
Tongue
Liver
Stomach
Pancreas
Gallbladder
Small intestine
Large intestine
Rectum
Anus

Figure 21.7 The digestive system is made up of many organs that work together in the breakdown of food into nutrients.

Enzymes

Enzymes (EN zimez) are proteins that speed up chemical reactions. They make chemical digestion happen. The body could not absorb as much food as it needs without enzymes.

There are many different kinds of enzymes. Each works in only one chemical reaction. Chemical reactions do not change the enzyme. One enzyme molecule can be used over and over again.

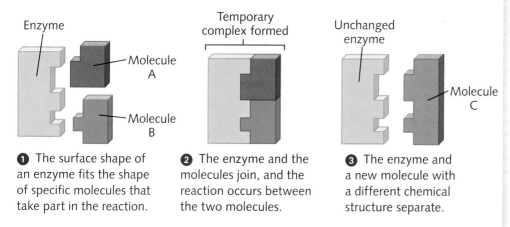

❶ The surface shape of an enzyme fits the shape of specific molecules that take part in the reaction.

❷ The enzyme and the molecules join, and the reaction occurs between the two molecules.

❸ The enzyme and a new molecule with a different chemical structure separate.

Figure 21.8 The surface of each enzyme has a shape that specific molecules fit into like keys in a lock. The reaction between these molecules occurs while they are on the enzyme. The enzyme is not changed.

The Digestive Organs

The organs of the digestive system break down food mechanically and chemically.

Mouth Mechanical and chemical digestion both start in the mouth. Teeth chop up food into smaller bits, creating new surfaces where enzymes can act. The first digestive enzymes to act are found in saliva (suh LI vuh), a fluid that is secreted into the mouth from the **salivary** (SA luh ver ee) **glands**. Saliva contains water to soften food and the enzyme amylase, which breaks starch into shorter sugars.

The **tongue** pushes food around the mouth and mixes it with saliva. The tongue is a muscular organ that can move in many different directions. The top of the tongue is covered with **taste buds**, small bumps that contain sensory receptors. The receptors tell the brain whether food is sweet, sour, salty, savory, or bitter.

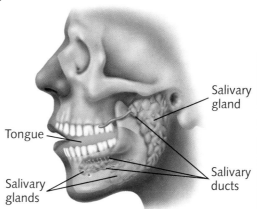

Tongue

Salivary gland

Salivary ducts

Salivary glands

Figure 21.9 Humans have three pairs of salivary glands. Each one has a duct that opens on the inside of the mouth.

Explore It!

You can test whether your mouth digests carbohydrates with crackers and a stopwatch.

Take one plain cracker, and put it in your mouth. Don't chew it. Start a stopwatch at the same time. Hold the cracker in your mouth until it starts tasting sweet. Write the time in your Science Notebook.

Take a sip of water to clear your mouth, and repeat this test with a second cracker. This time, start to chew the cracker when you start the stopwatch. When does this cracker start tasting sweet?

Figure It Out

1. What happens to the esophagus when the circular muscles contract?

2. Describe what could happen if peristalsis reversed itself in the stomach and esophagus. What might this phenomenon be called?

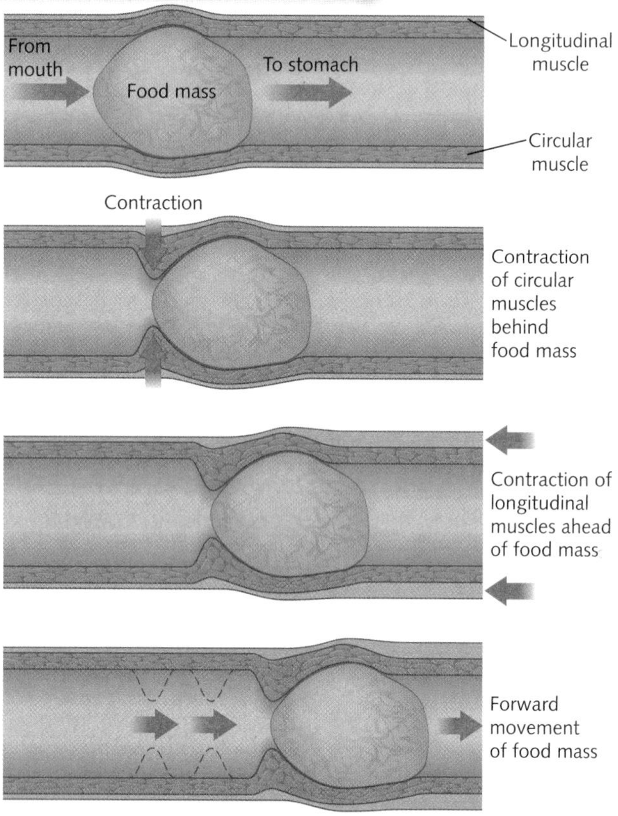

Esophagus   Chewing eventually turns food into a soft mass that is pushed into the back of the throat by the tongue. Swallowing pushes a small amount of the processed food called a **bolus** (BOH lus) through the pharynx. As the bolus moves through the pharynx, an involuntary reflex pulls a flap of cartilage called the epiglottis over the opening to the trachea. This reflex keeps food out of the airway and pushes the bolus into the esophagus.

The **esophagus** (ih SAH fuh gus) is a 25-cm-long tube that connects the pharynx with the stomach. It is lined with cells that secrete mucus to keep the bolus wet as it moves through the tube. The esophagus is also lined with smooth muscles. The muscles take turns contracting, creating a wavelike motion called **peristalsis** (per uh STAHL sus) that pushes food toward the stomach.

Stomach   As food leaves the esophagus, it enters a large, muscular bag called the **stomach**. The stomach has three jobs. It chemically digests food by secreting strong acid and the enzyme pepsin, which breaks down protein chains. It mechanically breaks the boluses of food and saliva into smaller particles and mixes them with stomach enzymes to make a semiliquid mixture called **chyme** (KIME). The stomach also stores food, slowly releasing it into the small intestine between meals.

Some cells in the stomach wall secrete mucus, which forms a protective layer that covers the stomach wall and neutralizes any acid that comes near it. If the layer is damaged, acids can burn into the stomach tissue and cause an ulcer.

From mouth — Food mass — To stomach

Longitudinal muscle

Circular muscle

Contraction

Contraction of circular muscles behind food mass

Contraction of longitudinal muscles ahead of food mass

Forward movement of food mass

Figure 21.10 Contractions of smooth muscles move food through the digestive tract.

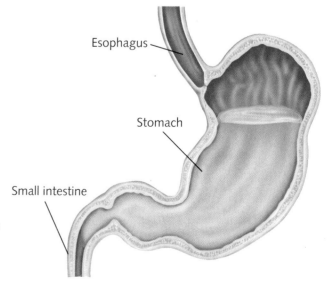

Esophagus

Stomach

Small intestine

Figure 21.11 The stomach secretes hydrochloric acid, which softens the connective tissue in meat and activates the enzyme that breaks up proteins.

Small Intestine The last stages of digestion occur in the small intestine. The small intestine is the longest part of the digestive tract. It is a muscular, 6-m-long tube that is coiled up inside the abdomen. It is called the "small" intestine because it is narrow—only about 2.5 cm wide.

The first section of the small intestine is called the **duodenum** (doo AH dun um). Most digestion happens here. Chyme enters the duodenum from the stomach. The chyme is still very acidic, so the lining of the duodenum secretes mucus for protection. Cells in the wall of the duodenum secrete enzymes that break up the carbohydrates and proteins in the chyme. The liver and pancreas add other enzymes and digestive chemicals to the chyme. Peristaltic contractions of the smooth muscle in the wall of the small intestine mix the chyme with these chemicals and push it down the digestive tract.

The Liver and Pancreas The liver and pancreas are important parts of the digestive system even though food does not move through them. They both make chemicals that aid digestion. The chemicals are sent to the small intestine through small tubes called ducts.

The **liver** is a large organ tucked underneath the diaphragm. It has many functions. Its job in digestion is to make bile, which breaks down fat. The bile made by the liver is stored in a small organ called the gallbladder. When it contracts, the gallbladder squirts bile into the small intestine.

The **pancreas** (PAN kree us) is a leaf-shaped organ that lies on top of the duodenum. It makes both digestive enzymes and hormones. The enzymes that the pancreas makes digest proteins, carbohydrates, fats, and DNA and RNA chains. The pancreas also makes large amounts of an alkaline fluid that reduces the harmful effect that the acid in the chyme would have on the small intestine.

As You Read

On the concept map in your Science Notebook, write *M* next to each organ that digests food mechanically. Write *C* next to each organ that chemically digests food. Compare your concept map with a partner's. Make adjustments or additions as needed.

What enzyme in saliva begins the digestive process? What enzyme in the stomach breaks down protein chains?

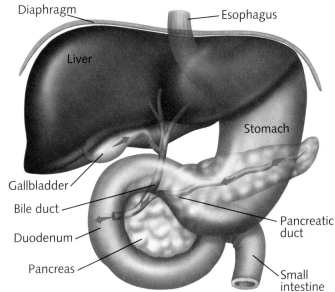

Figure 21.12 The liver and pancreas make digestive chemicals and send them into the small intestine.

After You Read

1. Think of a way to include the liver and pancreas on the concept map in your Science Notebook. Where do they release digestive chemicals?

2. What happens to amylase after it helps break starch into sugars?

3. Which organ makes bile? What does bile do?

4. How is the stomach protected from the strong acid it produces?

Learning Goals

- Describe how the area for absorption is increased in the small intestine.
- Identify what the large intestine absorbs.
- Explain how nutrients enter the bloodstream.

New Vocabulary

absorption
villus

Before You Read

Continue the concept map you started in Lesson 21.2. Add information about the last part of the digestive tract in the appropriate place.

Before the food inside the intestines can be used to provide energy and build new tissues, it has to be moved through the intestinal wall and into the bloodstream. This process is called **absorption** (ab SORP shun).

Food molecules do not just "leak" into the bloodstream. The cells that line the inside of the intestines are similar to the cells in the skin. They are bound together by tiny proteins that form a tight seal. Nothing can squeeze between those cells, so every molecule that enters the bloodstream has to move through the cells. Some molecules move by diffusion, and others are moved by active transport. Each molecule must pass through a cell membrane before it gets into the blood. This step is a filtering process that controls which molecules get in and which ones stay out.

The Small Intestine

Chyme moves through the small intestine in just a few hours. This is the only time the nutrients in food can be absorbed. The small intestine has many adaptations that give it a larger surface for absorbing nutrients.

The small intestine is long—typically three times longer than the body, as **Figure 21.13** shows. A long small intestine means that chyme moves past many absorption surfaces before it enters the large intestine.

Figure 21.13 If the small intestine were stretched out, it would be three times longer than the body. Its enormous length creates a large absorptive area for food to pass through.

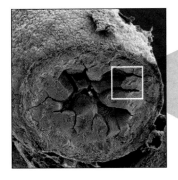
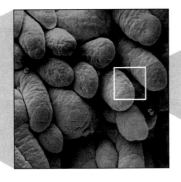

Figure 21.14 The inner surface of the small intestine is very large because it is folded on three levels. Tissue folds into the inside of the small intestine *(left)*. Villi cover the surface of the folds *(center)*. Each cell on the villi has microvilli on its surface *(right)*.

The inner surface of the small intestine is not smooth and flat. It has folds and millions of fingerlike projections called **villi** (VIH li, singular: villus) on each of those folds. Each villus contains a small artery and vein connected by capillaries and a small lymph vessel. Each villus is also covered with a layer of tightly bound cells. Each of these cells has a brush of tiny folds called microvilli on its surface. The folding inner surface and the presence of villi and microvilli, shown in **Figure 21.14**, greatly increase the surface area of the small intestine.

How Absorption Works

Nutrients are absorbed through the cells that line the surface of the villi. Sugars, short proteins, some vitamins and minerals, and simple fatty acids move through the cell membranes of the villi. Some nutrients move passively by diffusion, and others move by active transport. All of the nutrients move through the intestinal cells and are absorbed by capillaries on the other side.

Other nutrients take a different path into the bloodstream. Fats and fat-soluble vitamins are both absorbed by lymphatic vessels in the villi. These nutrients move through the lymphatic system and enter the blood near the heart. After the nutrients enter the blood, they are carried to all the other cells in the body.

As You Read

Review the concept map in your Science Notebook. Predict what would happen to a person's weight if half of the villi in his or her small intestine disappeared. Write a sentence describing your prediction.

Figure It Out

1. Which molecules are moving into the blood and lymph vessels in the villi?

2. Reflect on what could happen if a tumor blocked off the arteries that bring blood to the intestines.

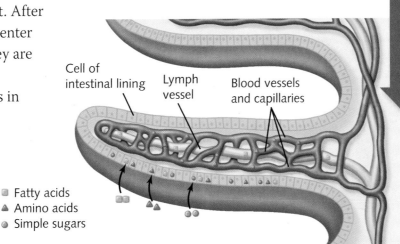

Cell of intestinal lining

Lymph vessel

Blood vessels and capillaries

▪ Fatty acids
▲ Amino acids
● Simple sugars

Figure 21.15 The cells of the villi transport nutrients from the inside of the intestine to blood and lymph vessels. From there, they enter the bloodstream.

Nutrients in the Body

Cells break down some of the food molecules to release the chemical energy stored in their bonds. Molecules that are used for energy are first converted to glucose. Glucose, a simple sugar, is the main energy source for the body. Cells break down glucose bit by bit during a process called cellular respiration. The first part of the process occurs in the cytoplasm, but it does not release much energy. The rest of the process, which releases large quantities of energy, occurs inside organelles called mitochondria. The energy released by cellular respiration is stored in molecules of a chemical called adenosine triphosphate, or ATP.

Cells use some nutrients as the building blocks for larger and more complex molecules. Amino acids are strung into proteins, building functional molecules such as enzymes as well as those that make up tendons, muscles, and skin. Nucleic acids are used to make new strands of DNA and RNA. Unused energy is stored in fats.

Figure 21.16 Nutrients in food are used by the body for energy and tissue repair.

CONNECTION: Zoology

It may surprise you to learn that there are many organisms that live inside the intestines and absorb the food the body has digested. To these organisms, the body is a habitat!

Some organisms provide something in return for the habitat the body supplies. For example, the bacteria that live inside the intestine are symbiotes. The body provides food and shelter, and the bacteria eat the material that the body cannot digest and secrete vitamins that the body can use.

Other organisms give nothing in return. Parasites such as roundworms and flatworms can live inside the digestive system and absorb food, sometimes causing sickness.

Flatworms from the genus *Taenia* are the source of most human tapeworm infections. They also infect pigs and cows, so eating undercooked meat can be a means of transmitting the tapeworm. When a larval tapeworm gets inside a person, it attaches to the intestinal wall with its spiky head and starts to grow. Adult worms can grow to 7 m long. They have no digestive or circulatory systems, so they simply absorb nutrients directly from the chyme.

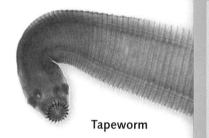

Tapeworm

The Large Intestine

The large intestine is wider than the small intestine and much shorter—it is only 1.5 m long. It surrounds the small intestine, starting at the lower right portion of the abdomen and running up, across, and down the left side of the abdominal cavity.

By the time the chyme reaches the large intestine, most of the nutrients have been removed from it. It now contains indigestible materials, such as cellulose, and a lot of water. The large intestine reabsorbs most of the water—between 400 and 1,000 mL. This process allows the body to reuse the water and keeps it from getting dehydrated.

Some of the bacteria that live inside the large intestine can digest cellulose. The bacteria give off gas as a waste product of metabolism. They also secrete vitamin K and two B vitamins, thiamine (B_1) and niacin (B_3), that are absorbed by the large intestine.

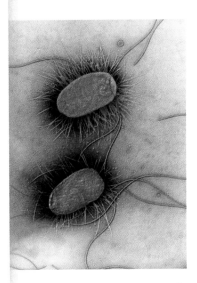

Figure 21.17 Bacteria such as these *E. coli* cells live symbiotically inside the large intestine. Humans give them food and shelter. They give humans vitamins K, B_1, and B_3.

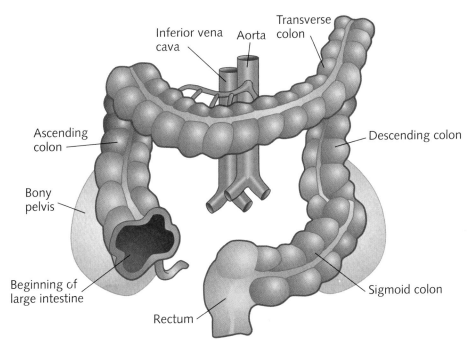

Inferior vena cava · Aorta · Transverse colon · Ascending colon · Descending colon · Bony pelvis · Beginning of large intestine · Rectum · Sigmoid colon

Figure 21.18 The large intestine absorbs water, salts, and some vitamins.

After You Read

1. On the concept map in your Science Notebook, draw a box around the organs that digest food. In a different color, draw another box around the organs that absorb food. Where do the boxes overlap? Write a well-developed paragraph describing the structures and functions of the digestive system.

2. Name two things the large intestine absorbs.

3. Why do the intestinal cells act as filters when they absorb nutrients?

Before You Read

Reword the headings of this lesson so that they form questions. Write the questions in your Science Notebook. Write answers to the questions as you read.

Every factory makes wastes. The human body is no exception. Some wastes can be recycled and reused in other processes. Many wastes, however, are useless to the body. These molecules become poisonous if they build up in the tissues.

Fortunately, the body has systems that remove wastes from the tissues and dump them outside of the body. The process of waste removal is called **excretion** (ihk SKREE shun). Waste is excreted as solids, liquids, and gases.

Solid Wastes

Not everything a person eats can be digested. The solid materials left over after the digestion and absorption of food are called **feces** (FEE seez). Human feces contain water, dead bacteria, bile, salts, and indigestible material such as cellulose.

Most plant tissue is made of cellulose fibers, shown in **Figure 21.19**. Although people can chop up cellulose with their teeth, they do not make enzymes that can digest it. Cellulose passes through the human digestive system unchanged. When the material that holds cellulose fibers together gets digested, the fibers separate into individual strands. These strands absorb water, making feces softer and easier to excrete. This is the reason fiber is an important part of a balanced diet.

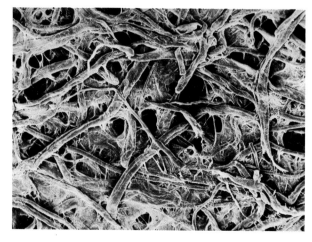

Figure 21.19 Humans cannot digest the cellulose in plant tissues. What is cellulose?

Liquid Wastes

Liquid wastes are excreted in **urine** (YOOR un) and in **sweat**. Both of these wastes are more than 90 percent water. Waste nitrogen and salts are suspended in the water. The exact amount of water in the urine depends on how hydrated the body is. A person excretes more water when he or she is well hydrated than when he or she is dehydrated.

Amino acids are absorbed from food, but the body has no way to store them from day to day. If they are not used immediately in building a new protein, they are broken down for energy. This process releases waste nitrogen as ammonia. Ammonia is toxic, so the body converts it to a less harmful chemical called **urea** (you REE uh). The kidneys filter urea out of the blood for excretion.

Urine also contains salts that the body no longer needs and pigments that make the urine look yellow. Some of the pigments come from the breakdown of old red blood cells. The kidneys filter these molecules out of the blood. Urine is sterile, or free of microorganisms, in the upper part of the urinary tract. As urine flows through the tubes that take it out of the body, the urine cleanses the tubes of microorganisms that could produce infection.

Sweat is excreted from glands in the skin. It cools the body as it evaporates. These glands make sweat by filtering the fluid surrounding the cells. Like urine, sweat contains small molecules such as salts, but it does not contain larger molecules such as proteins. The salts and a small amount of urea are left behind on the skin as the water in the sweat evaporates.

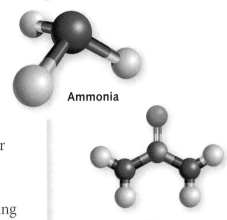

Ammonia

Urea

Figure 21.20 Amino acid breakdown releases ammonia, which the body converts into less toxic forms such as urea.

As You Read

Use the answers to the questions you wrote in your Science Notebook to write a few sentences that explain the following statement: Solid wastes contain materials that cannot be chemically processed. Liquid wastes contain the end result of chemical processing.

Figure 21.21 Liquid wastes are also excreted when a person sweats.

Gaseous Wastes

Cellular respiration takes glucose and turns it into energy, water, and carbon dioxide. The body can reuse the water in many of its cellular processes. The carbon dioxide, however, is a waste gas. It is carried to the lungs by the blood and is excreted during exhalation.

Figure 21.22 Carbon dioxide gas is excreted when a person exhales.

Figure It Out

1. This person is releasing carbon dioxide during exhalation. What goes into the body during inhalation?

2. Hypothesize why a person breathes faster after exercising than when resting.

Explain It!

Think about the way a mirror or a windowpane fogs up when you breathe on it. In your Science Notebook, write a sentence explaining why this happens.

Carbon dioxide is carried by the blood in three ways. A little bit of the gas dissolves in the plasma. Some of the gas binds to hemoglobin and gets carried in the red blood cells. But most of the carbon dioxide reacts with water to form bicarbonate, which gets carried by the cytoplasm of red blood cells and plasma. Because the blood can carry carbon dioxide in three different ways, it can move more of the gas out of the body.

After You Read

1. Name the two liquid wastes the human body makes.

2. Define excretion. Explain how it differs from ingestion.

3. Review the information you have recorded in your Science Notebook. Explain whether or not gaseous waste is the result of chemical processes.

21.5 Organs of Excretion

Learning Goals

- Name the organs of excretion.
- Identify where urea is made.
- Understand how the kidney produces urine.

New Vocabulary

rectum
anus
nephron
glomerulus
reabsorption
ureter
bladder
urethra
sweat gland

Before You Read

Preview the lesson by reading the headings. In your Science Notebook, create a three-column chart. Label the first column *Organ*, the second column *Waste Produced*, and the third column *Origin of Waste*. Fill in the chart as you read the lesson.

Many organs in the human body are involved in excretion. Solid wastes are processed and expelled by the large intestine. Carbon dioxide diffuses into the lungs as gas to be exhaled. The liver is an important site for filtering the blood and converting poisons into less toxic materials. The kidneys filter the blood to remove liquid wastes from the body.

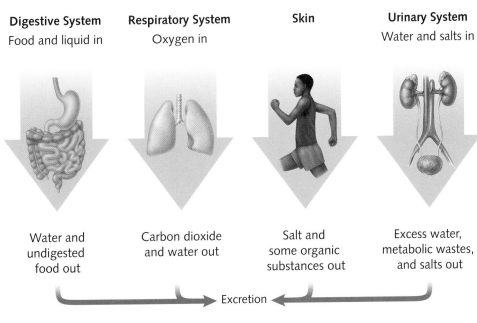

Digestive System
Food and liquid in

Respiratory System
Oxygen in

Skin

Urinary System
Water and salts in

Water and undigested food out

Carbon dioxide and water out

Salt and some organic substances out

Excess water, metabolic wastes, and salts out

Excretion

Figure 21.23 The excretory system is made up of several different body systems.

Large Intestine

The large intestine reabsorbs water and salts from chyme. It is also the organ where indigestible material from food is compacted into feces. Chyme is pushed through the large intestine by peristaltic contractions of the smooth muscle in its wall. As the chyme moves up, across, and down the body, water is pulled out of it and indigestible material such as cellulose is left behind.

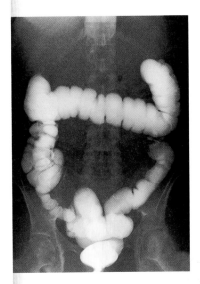

Figure 21.24 This X-ray image of the large intestine shows the path of the organ around the abdomen and down to the rectum.

Indigestible material collects near the end of the large intestine and gets compressed into solid feces. Solid feces are next pushed into the **rectum** (REK tum), a chamber at the end of the large intestine that stores feces until they can be excreted. When the rectum fills, a person gets the urge to defecate, or push the solid wastes out of the body. Feces leave the body through the opening at the far end of the large intestine, called the **anus** (AY nus).

Lungs

The lungs excrete gaseous wastes. Carbon dioxide is a waste product of cellular respiration. It diffuses into the blood and is pumped into the alveoli of the lungs by the heart. The air inside the alveoli contains far less carbon dioxide than does the deoxygenated blood coming from the heart. Because molecules diffuse from areas where they are concentrated to areas where they are less concentrated, carbon dioxide leaves the blood and diffuses across the alveoli and into the air. Air that is rich in carbon dioxide is exhaled, releasing the gaseous waste outside the body.

Liver

The liver is the largest internal organ. It is a huge gland with four lobes, each of which is built out of many small, roughly hexagonal units called lobules. Lobules contain liver cells surrounded by blood vessels, lymphatic vessels, and bile ducts.

Liver cells have many functions. They make bile that dissolves fats in the digestive system. They break down hemoglobin from old red blood cells. They store glucose, iron, and some vitamins. They make cholesterol and many proteins that help blood clot. They also play an important part in removing toxins from the blood.

Liver cells break down toxins such as alcohol, turning them into less poisonous molecules that the kidneys can remove from the blood for excretion. The most important molecule they change is ammonia, which they combine with carbon dioxide to make urea. Urea is much less toxic than ammonia.

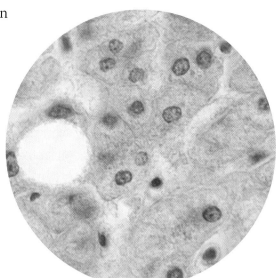

Figure 21.25 Liver cells are the chemical factories of your body.

As You Read

Under each heading in the chart in your Science Notebook, write what you have learned so far about the organs of the excretory system. Share your chart with a partner. Add to or edit the information in your chart as needed.

How does the liver aid in excretion? Describe the structure of a nephron.

Kidneys

The kidneys filter the blood, remove urea and other metabolic wastes, and make urine for excretion. The human body has two kidneys, one on each side of the spine. They sit against the back in the upper part of the abdomen, partly covered by the ribs. Blood enters each kidney through a large artery that branches off the aorta. Inside the kidney, the artery divides into many networks of capillaries that surround the functional units of the kidney, the **nephrons** (NE frahnz).

Each kidney has about one million nephrons. Each nephron looks like a long coiled tube with a cup at one end. The cups of the nephrons are found in the outer rim of the kidney. Each tiny tube, or tubule, extends into the center of the kidney.

The cup of the nephron surrounds a ball of capillaries called the **glomerulus** (gluh MER uh lus). When pressurized blood enters the glomerulus, fluid is forced through the walls of the capillaries and into the nephron. Small molecules such as urea, salt ions, amino acids, and glucose are pushed into the nephron with the fluid. Large proteins and blood cells, however, are too big to fit through the capillary walls. They remain inside the capillaries.

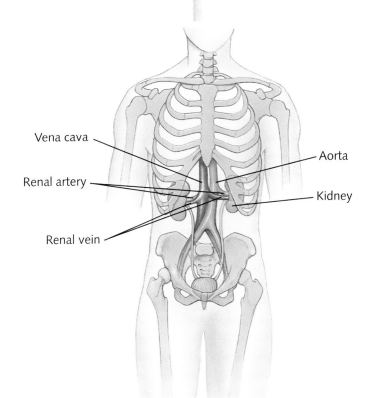

Figure 21.26 The kidneys are located just above the waist, behind the stomach.

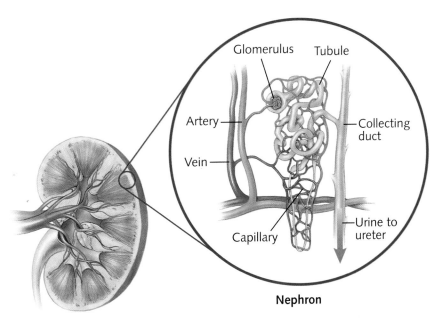

Figure 21.27 This kidney has been split in half so its internal structure can be seen *(left)*. Each nephron *(right)* has a cup and a long tubule.

Figure It Out

1. What is the single largest source of water loss in the body?
2. Predict which system will lose more water if a person exercises outdoors on a hot day.

Major Routes for Water Loss	
Source	Amount (mL)
urine	1,500
skin	500
lungs	300
feces	150
Total	**2,450**

Figure 21.28 This table shows where the body loses water each day.

The filtered fluid in the nephron now contains wastes, but it also contains molecules that the body needs. The kidney must put these molecules back into the circulatory system. The process in which this occurs is called **reabsorption**. It occurs in the nephron tubule. During reabsorption, amino acids and glucose are returned to the blood.

The nephrons also reabsorb water. Because all of the blood in the body is filtered through the kidneys every five minutes, the kidneys make about 170 L of filtered fluid each day. Most of the water is reabsorbed by the nephrons to form urine. The human body excretes about one liter of urine each day.

The end of each nephron is connected to a collecting duct that brings urine to the **ureters** (YOO ruh turz). Each kidney has a ureter, a long, thin tube that directs urine to the bladder. The **bladder** is a muscular bag that can stretch like a balloon as it fills with urine. It can hold about half a liter of urine. When the bladder fills, it stretches and sends signals to the brain that stimulate the urge to urinate. When a person urinates, urine leaves the body through a short tube called the **urethra** (yoo REE thruh).

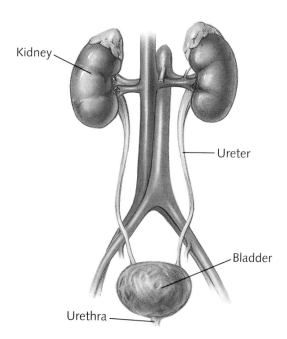

Kidney

Ureter

Bladder

Urethra

Figure 21.29 The ureters connect the kidneys to the bladder. The urethra connects the bladder to the outside of the body.

CONNECTION: Medicine

The first human organ transplant was performed in 1954. The organ transplanted was a kidney, and the operation was performed at Peter Bent Brigham Hospital in Boston, Massachusetts. Today, organ transplants take place often and are performed in most countries of the world.

Skin

The skin is the complex organ that covers your body. In Chapter 19, you learned that the skin is the largest organ of the body. It covers and protects the body's organs and functions as a sense organ and as the body's insulator. It keeps moisture in and disease-causing pathogens out. Besides all of this, the skin is an important organ of excretion.

The skin contains two kinds of **sweat glands**. Both kinds are coiled tubes inside the dermis that open onto the surface of the skin. They make different kinds of sweat. Most sweat glands make watery sweat containing water, salts, and a little urea. This sweat cools the body when it is too hot. Sweat glands found at the base of hairs in the armpits and groin release sweat that also contains some fatty acids and proteins. Bacteria break down the proteins and fatty acids, creating body odors.

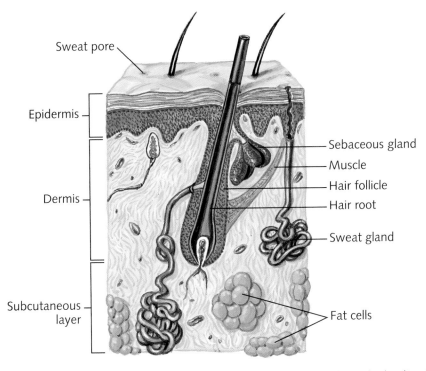

Figure 21.30 A typical sweat gland looks like a knotted ball with a tube leading to the surface of the skin.

After You Read

1. Where is ammonia converted to urea?
2. Name the two main functions of a nephron.
3. Explain why a blocked ureter is a serious medical problem.
4. Refer to the chart of the excretory organs you made in your Science Notebook. Draw a sketch of the human body showing the location of each organ. Label each organ, and identify the waste it is associated with.

Summary

KEY CONCEPTS

21.1 Nutrition

- The basic nutrients are carbohydrates, proteins, fats, vitamins, minerals, and water.
- A balanced diet contains the correct amount of every nutrient.
- People need to eat more of some nutrients than others.

21.2 The Process of Digestion

- The human digestive tract is made up of a series of organs that starts at the mouth and ends at the anus.
- The digestive system breaks up food mechanically and chemically.
- Enzymes are specialized proteins that speed up chemical reactions.

21.3 Absorption of Food

- Most nutrients are absorbed through the wall of the small intestine.
- The large intestine absorbs water, salts, and some vitamins.

21.4 What Is Excretion?

- Excretion is the process of removing wastes from the body.
- Wastes can be excreted as solids, liquids, or gases.

21.5 Organs of Excretion

- The large intestine excretes solid residues of digestion.
- The lungs excrete carbon dioxide gas.
- The liver transforms nitrogenous wastes into urea.
- Most urea is excreted by the kidneys, but some is released in sweat.

VOCABULARY REVIEW

Write each term in a complete sentence, or write a paragraph relating several terms.

21.1
nutrition, p. 375
nutrient, p. 375
malnutrition, p. 375
carbohydrate, p. 375
protein, p. 376
essential amino acid, p. 376
fat, p. 376
vitamin, p. 377
mineral, p. 378
Calorie, p. 379

21.2
ingestion, p. 380
digestion, p. 380
enzyme, p. 381
salivary gland, p. 381
tongue, p. 381
taste bud, p. 381
bolus, p. 382
esophagus, p. 382
peristalsis, p. 382
stomach, p. 382
chyme, p. 382
duodenum, p. 383
liver, p. 383
pancreas, p. 383

21.3
absorption, p. 384
villus, p. 385

21.4
excretion, p. 388
feces, p. 388
urine, p. 389
sweat, p. 389
urea, p. 389

21.5
rectum, p. 392
anus, p. 392
nephron, p. 393
glomerulus, p. 393
reabsorption, p. 394
ureter, p. 394
bladder, p. 394
urethra, p. 394
sweat gland, p. 395

True or False

If the statement is true, write "true." If it is false, change the underlined word or words to make the statement true.

1. <u>Saturated</u> fats are solid at room temperature.

2. Carbohydrate digestion starts in the <u>stomach</u>.

3. Humans <u>can</u> digest cellulose.

4. The process in which the nephrons return molecules to the bloodstream is called <u>filtration</u>.

5. A human can live for a few <u>weeks</u> without water.

6. The inner surface of the small intestine has many <u>folds</u>.

Short Answer

Answer each of the following in a sentence or brief paragraph.

7. A glucose molecule is ingested as part of a dinner roll. Trace its path through the digestive system to a muscle cell.

8. Explain how mechanical digestion aids chemical digestion.

9. Name the functional units of the lungs, the liver, and the kidneys.

10. Discuss why people need to eat cellulose.

11. If an hour-long walk burns 480 Calories, how long will it take to burn off one serving of a snack that contains 576 Calories?

PREPARE FOR CHAPTER TEST

To prepare for the chapter test, create a question from each Learning Goal. Use the information in your Science Notebook to answer each question. Then use these answers to write a well-developed essay about the chapter. Use the Key Concept on the first page of this chapter as your topic sentence.

Critical Thinking

Use what you have learned in this chapter to answer each of the following.

12. **Infer** During the eighteenth century, sailing ships would set out for long voyages with plenty of meat and bread for everyone to eat. But, after several months, sailors would begin to get malnourished. They typically had no fresh vegetables or fruit on board. Infer the cause of their malnutrition.

13. **Relate** You are studying a new species of mammal from a tropical rain forest. When you dissect its kidneys, you find that it has very short nephrons. Relate its kidney anatomy to its habitat.

14. **Predict** Some adults do not make the enzyme lactase, which digests milk sugar, or lactose. The bacteria in the large intestine can digest lactose. Predict what might happen if a lactase-deficient person ate a large bowl of ice cream.

Standardized Test Question

Choose the letter of the response that correctly answers the question.

15. Which of the following nutrients is the most important source of energy for the body?

    A. proteins

    B. carbohydrates

    C. vitamins

    D. none of the above

Test-Taking Tip

If "none of the above" is one of the choices in a multiple-choice question, be sure that none of the choices is true.

Chapter 22

Nervous and Endocrine Systems

KEY CONCEPT The nervous and endocrine systems—the body's two communication systems—work together to control and coordinate all the body's other systems.

Think of all the different ways you communicate with people every day. You chat with friends. You write answers to homework questions. You call, email, text message, or send letters to friends and family members. Even your body language is a form of communication. Communication is an essential part of your everyday life.

Just as communication with other people is important, communication within the body is essential to staying alive and healthy. The nervous and endocrine systems are the body's two communication systems. Together, they help control and coordinate all the body's systems.

NSTA

*SCI*LINKS.
THE WORLD'S A CLICK AWAY

www.scilinks.org
Nervous System **Code: WGB22A**
Endocrine System **Code: WGB22B**

Think About Communication Systems

Think about all the activities the body does every day.

- Make a list in your Science Notebook of activities that require control and coordination. Can you think of any that do not? Write these as well.

- Choose two of the activities, and name all the body parts that must work together to complete them. Explain how these parts must be coordinated.

22.1 The Nervous System

Learning Goals

- Describe the functions of the nervous system.
- Identify the structure and main classes of neurons.
- Explain how nerve impulses are sent through the body.
- Compare the functions of the central and peripheral nervous systems.

New Vocabulary

stimulus
neuron
dendrite
axon
synapse
central nervous system
peripheral nervous system

Recall Vocabulary

tissue (p. 332)
cell (p. 331)

Before You Read

In your Science Notebook, create a concept map for the nervous system. Write *The Nervous System* in a large central circle, and draw several smaller circles around it. Then, look at the Learning Goals, headings, and pictures in this lesson. What do you predict you will write in the smaller circles of your concept map?

As you read this book, countless things are happening inside your body. Blood is being circulated through arteries and capillaries to deliver oxygen and food to cells. The intestines and liver are working to break down food and process waste. The skin is taking in information about the temperature of the environment. Other information about the environment is being collected by the ears, which are also helping the body keep its balance. Stomach, back, and neck muscles are working to hold the body upright. How can the body do so many things at the same time? What controls and coordinates all of the different functions of the body systems?

Functions of the Nervous System

The nervous system is the body's command system. Like a computer's CPU, the body's nervous system processes and stores information. It communicates information throughout the body and controls the body's response to stimuli (STIHM yuh lye, singular: stimulus). A **stimulus** is any change that happens inside the body or in the external environment.

During a soccer game, for example, the nervous system enables the body to take in and process information about the environment and the players on the field. It coordinates the action of the body's skeletal muscles, which allow the body to move around obstacles and toward the goal. It enables the body to communicate with other players. It even increases the heart and breathing rates so that the body has enough blood flow and oxygen to participate in the game. The nervous system regulates all the tasks that the body performs.

The nervous system does much more than this, however. Unlike a computer's CPU, the body's nervous system allows people to have feelings. It controls a person's mood and shapes a person's personality. It is responsible for emotions and dreams. It helps make us who we are. It helps make us uniquely human.

Working along with the nervous system is the endocrine system. The endocrine system is discussed in Lesson 22.4.

Figure 22.1 The nervous system coordinates the internal changes that enable a soccer player to perform individually and with other players.

Organization of the Nervous System

The main organs of the nervous system are the brain, spinal cord, and sensory organs. The eyes, ears, nose, tongue, and skin are the sensory organs. They provide people with the ability to see, hear, smell, taste, and touch, respectively.

Nervous tissue, one of the four main types of tissues that make up organs, is the tissue of the nervous system. It is made up of cells called **neurons** (NOOR ahnz). As shown in **Figure 22.2**, a neuron has a cell body that contains a nucleus. Unlike other cells, however, a neuron has structures that branch out from the cell body. These structures are the **dendrites** (DEN drites) and one long **axon** (AK sahn). A dendrite takes in information from the environment or from other cells in the body. The axon mainly transmits information to other cells.

Neurons are arranged in the body so that the dendrites of one cell connect with the axon of one or more other neurons. Most of the neurons do not actually touch one another. Instead, they are separated by a small fluid-filled space called a **synapse** (SIH naps).

There are three main classes of neurons. Each class has a different function in the body. Sensory neurons carry information from the body or the environment to the spinal cord and brain. Motor neurons carry information from the brain and spinal cord to the muscles or glands. Interneurons are relay neurons. They transmit information to and from sensory and motor neurons.

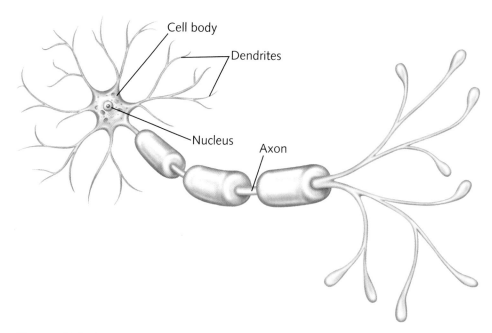

Figure 22.2 Neurons can vary in size, shape, number of dendrites, and location in the body. However, all neurons have the same basic structures, and they all function in essentially the same way.

As You Read

In the outer circles of your concept map, fill in words that describe the nervous system. Add details by connecting and filling in more circles around the outer circles. Work with a partner to be certain that you include all the important information.

What are some functions of the nervous system?

Sending Signals

Neurons are responsible for every sight you see, every odor you smell, and every flavor you taste. They are responsible for your hearing and your thoughts. Each of these experiences is complex and very different from the others. However, all information is communicated and processed by neurons in a similar way.

Neurons communicate by changing information from a stimulus into electrical and chemical impulses. Suppose something touches the skin. This stimulus triggers an electrical impulse in the dendrites of a sensory neuron. The electrical impulse travels through the cell body and down the cell's axon. When it reaches the end of the axon, it causes the axon to release chemical messengers called neurotransmitters. The chemicals move away from the axon and across the synapse. Then, they bind to the dendrites of a neighboring neuron. This triggers an electrical impulse in that neuron. The whole process begins again. Neurons are adapted in such a way that all nerve impulses travel in only one direction.

Information travels through one or more neurons in this way until it reaches the brain. The brain processes the information. Then, it sends electrical and chemical impulses back through the body. These impulses let the body know how to respond to the stimulus. All this happens in less than one second!

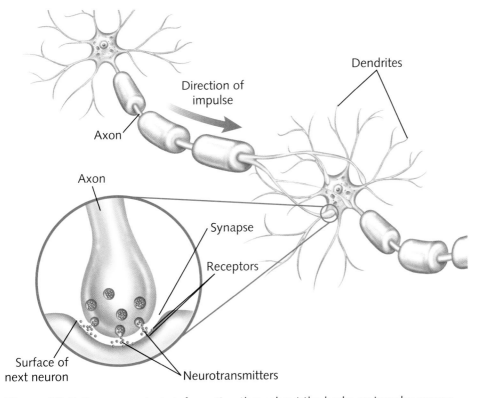

Figure 22.3 To communicate information throughout the body, an impulse moves from the axon of one cell to the dendrites of a neighboring cell.

Two Systems in One

The nervous system has two divisions: the central nervous system and the peripheral nervous system. The **central nervous system** (CNS) is made up of the brain and spinal cord. It forms the largest part of the nervous system. It is responsible for such things as communication, memory, and thought.

The **peripheral** (puh RIH frul) **nervous system** (PNS) connects the brain and spinal cord with the rest of the body. It is made up of 12 pairs of cranial nerves and 31 pairs of spinal nerves. Cranial nerves are nerves in the brain. Spinal nerves are nerves that branch off the spinal cord.

The PNS can be further divided into the somatic and autonomic nervous systems. The somatic nervous system controls voluntary actions, which are actions the body consciously controls. Jogging, lifting a book, and opening a door are examples of voluntary actions. The autonomic nervous system controls involuntary actions, which are actions the body does not consciously control. Digestion, breathing, and heart rate are examples of involuntary actions.

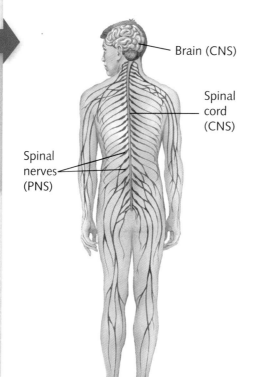

Brain (CNS)

Spinal cord (CNS)

Spinal nerves (PNS)

Figure 22.4 The brain and spinal cord make up the central nervous system. The spinal nerves are part of the peripheral nervous system. The CNS and PNS work together to control and coordinate all the actions of the body.

After You Read

1. How does the nervous system help keep the body healthy and active?

2. How are neurons similar to other cells in the body? How do they differ?

3. Sequence and describe the process through which neurons send information to the brain.

4. Review the concept map in your Science Notebook. Work with a partner to answer the following questions. What are two divisions of the nervous system? How do these divisions differ? Record your answers in your Science Notebook.

22.2 The Brain and Spinal Cord

Learning Goals

- Describe the structures and functions of the brain and spinal cord.

- Explain what a reflex is and how reflexes are transmitted through the body.

New Vocabulary

cerebrum
cerebellum
brain stem
vertebra
reflex
reflex arc

Before You Read

Create a K-W-L-S-H chart in your Science Notebook. Think about the title of this lesson. Then look at the pictures and headings in the lesson. In the column labeled *K*, write what you already know about the brain and spinal cord. In the column labeled *W*, write what you would like to learn.

You're in the video store trying to decide what movie you'd like to watch tonight. Are you in the mood for an action movie? No, you've already seen the popular ones at least twice, and none of the others look interesting. Perhaps you'll try a comedy or romance instead. You end up picking a drama and leaving the store.

What happened in this seemingly simple scenario? You thought about something you would like to do in the future. You read and recognized the titles of films. You thought about your mood and how you were feeling. You remembered things you had seen in the past. You reasoned and made choices. In only a short time, you did a number of tasks that no other living or nonliving thing can do. For all this, you can thank the human brain.

The Brain

Humans are among the most complex beings on Earth. This is due in large part to the structure and function of the human brain. The human brain is responsible for reason and intelligence. It enables people to analyze situations, communicate, create, feel, and learn.

The mass of the human brain is only about 1.4 kg. That is about 2 percent of an average adult's body mass. The brain uses between 20 and 25 percent of the body's energy, however. It contains more than 100 billion neurons. Each of these neurons connects with up to 10,000 other neurons to transmit information throughout the body.

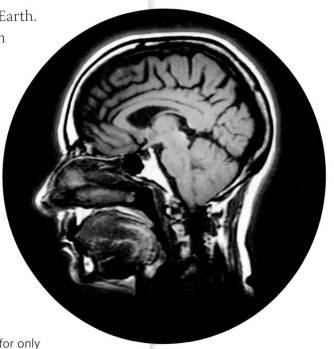

Figure 22.5 The human brain, as seen in this MRI, accounts for only about 2 percent of an average adult's mass. Yet its billions of neurons function to make humans among the most complex organisms on Earth.

Structure and Function The brain is divided into three major parts. These are the cerebrum (suh REE brum), the cerebellum (ser uh BEL um), and the brain stem. The **cerebrum** is the largest part of the brain. It is where memories are stored. It is also where thinking occurs. The cerebrum processes information from the sensory organs. It allows for communication and learning.

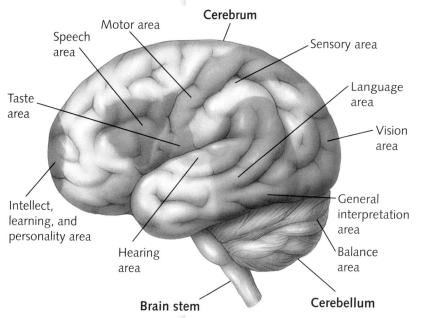

Cerebrum

Motor area

Speech area

Sensory area

Taste area

Language area

Vision area

Intellect, learning, and personality area

General interpretation area

Hearing area

Balance area

Brain stem

Cerebellum

Figure 22.6 The three major parts of the human brain are the cerebrum, cerebellum, and brain stem. Different areas of the cerebrum are thought to be responsible for different activities. What part of the brain is responsible for balance?

The **cerebellum** is located below and toward the back of the cerebrum. It is responsible for coordinating voluntary actions. It regulates skeletal muscles and maintains muscle tone. It also helps the body maintain balance.

The **brain stem** connects the brain to the spinal cord. It is a thick, cylindrical structure that extends from the cerebrum. The brain stem is made up of three smaller structures—the midbrain, pons, and medulla. The midbrain and pons help connect different parts of the brain. The medulla helps control involuntary actions.

The skull, a thin layer of bone, surrounds the brain. The skull helps protect the brain from injury. A layer of protective fluid fills the space between the brain and the skull. This fluid acts as a shock absorber and helps deliver nutrients to the brain.

CONNECTION: Computer Science

Artificial intelligence is a branch of science that aims to build intelligent machines—machines that can think and reason as humans do. Such machines would, in a sense, mimic the processes of the human brain.

Scientists began searching for ways to build intelligent machines in the 1940s. At that time, computers were becoming more sophisticated; they were able to do more complicated tasks with ease. They were also getting smaller. Scientists were learning more about the brain and nervous system. Mathematicians were learning to solve mathematical problems that are related to human reasoning. With continuing advances in these three fields, the science of artificial intelligence experienced great progress.

Scientists have made many advances toward artificial intelligence. Today, computers can mimic many thought processes. They can play a game of chess, for example, and can solve algebraic equations. Yet, they have not advanced to the level of human intelligence. For now, those kinds of computers are still a fictional invention.

The Spinal Cord

The spinal cord helps connect the brain with the rest of the body. As shown in **Figure 22.7**, the spinal cord is made up of bundles of neurons held together by connective tissue. Sensory neurons and interneurons connect inside these bundles to carry information to the brain. Motor neurons and interneurons connect to carry information away from the brain. Blood vessels within the connective tissue bring food and oxygen to the cells.

The spinal cord is surrounded by strong, flexible bones that together form the spine, or backbone. The spine is divided into 33 irregularly shaped bones called **vertebrae** (singular: vertebra). The vertebrae protect the spinal cord from injury. A layer of protective fluid surrounds the spinal cord within the spine. The extra protection for the spinal cord is crucial.

If the spinal cord is severed or damaged, paralysis can occur. Paralysis is a condition in which a person cannot move parts of his or her body. Paralysis can "freeze" the whole body from the neck down, or it might affect just the legs and pelvis. Paralysis is caused when damage to the spinal cord stops the brain from being able to communicate with parts of the body. The brain can form a signal, but the signal cannot reach its destination.

As You Read

In the column labeled *L* in your K-W-L-S-H chart, write what you have learned about the brain.

Why is the spinal cord an important organ?

Did You Know?

Once the spinal cord has been severed, it is unlikely that the connection between the brain and other parts of the body will be reestablished. Scientists hope that technology will one day enable them to re-form connections between the brain and spinal cord.

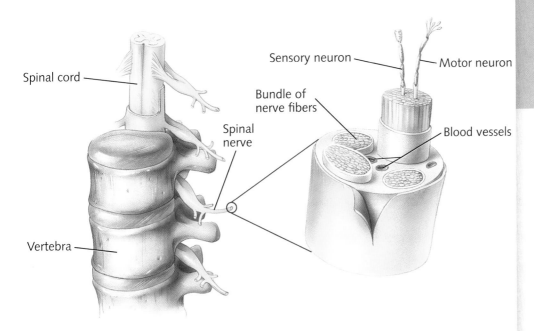

Figure 22.7 A column of vertebrae protects the spinal cord. The spinal cord, which provides the connection between the brain and the rest of the body, is about the width of an adult thumb and about 43 cm long. It is very flexible.

Spinal cord

Vertebra

Spinal nerve

Bundle of nerve fibers

Sensory neuron

Motor neuron

Blood vessels

Figure It Out

1. Why are reflexes automatic responses rather than conscious responses?

2. How does the withdrawal reflex help keep the body healthy?

Reflexes

Have you ever touched something hot and pulled your hand away quickly without thinking about it? Such an action is an example of a reflex. A **reflex** is an automatic, involuntary response to a stimulus.

During a reflex, an impulse is sent through a simple network called a **reflex arc**. Most reflex arcs are made up of just three neurons: a sensory neuron, an interneuron, and a motor neuron. **Figure 22.8** shows how an impulse flows through a reflex arc. First, a stimulus triggers an impulse in a sensory neuron. The impulse is sent from the sensory neuron to an interneuron in the spinal cord. The interneuron sends the impulse directly to a motor neuron. Then, the motor neuron causes the muscles to move.

Reflexes are controlled in the spinal cord rather than in the brain. That is why people are not aware of them until after they have happened. Reflexes allow the body to quickly respond to stimuli that may be harmful. After the response has happened, the brain processes the information so the body knows what to do next.

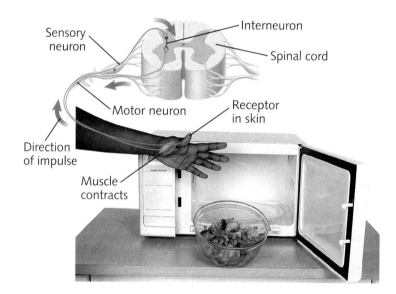

Figure 22.8 Reflex responses are controlled by the spinal cord. The reflex response shown here is known as the withdrawal reflex.

After You Read

1. What are three parts of the brain? What role does each one play in the body?

2. Explain how reflexes occur. How do they differ from other impulses?

3. Review the K-W-L-S-H chart in your Science Notebook. Then, summarize what you learned about the brain and spinal cord. What new questions do you have? How could you find the answers to these questions? Record your questions and ideas in the S and H columns of your chart.

22.3 The Senses

Before You Read

Create a lesson outline in your Science Notebook. Use the title and headings of this lesson to organize the outline. What do you think you will learn about in this lesson? Use the outline and the pictures to help answer this question.

Smell, sight, hearing, taste, and touch are the body's senses. The senses connect people with the outside world. They enrich people's everyday lives and experiences. They help make people aware of the things around them. Without the senses, people would not be able to communicate with one another or to recognize the beauty that surrounds them.

Making Sense of Things

Each of the five senses allows people to experience a different aspect of their surroundings. The senses give the brain information. The brain turns this information into a perception.

The body's sensory organs are the eyes, ears, nose, tongue, and skin. These organs have special cells called **sensory receptors**. These cells are sensitive to specific types of energy, summarized in **Figure 22.10**. The sensory receptors change the energy given off by a stimulus into electrical impulses that can be sent to the brain. The brain then processes these impulses into a sight, a sound, a taste, a smell, or a touch.

The Senses and Energy		
Sense	**Sensory Organ**	**Energy Detected**
sight	eyes	light energy
hearing	ears	mechanical energy
smell	nose	chemical energy
taste	tongue	chemical energy
touch	skin	mechanical energy, chemical energy, and thermal energy

Figure 22.10 The sensory receptors within a sensory organ allow the organ to respond to specific forms of energy. This table shows the five senses and the forms of energy responsible for each sense.

Learning Goals

- Identify the five sense organs.
- Explain what a sensory receptor is and which stimuli humans are sensitive to.
- Describe how the sensory organs allow the body to see, hear, smell, taste, and touch.

New Vocabulary

sensory receptor
cone
rod
retina
cochlea
cilia

Recall Vocabulary

taste bud (p. 381)

Figure 22.9 How are these teenagers using their senses? What sights, sounds, and smells do you think they are noticing? Describe each example using as many descriptive words as you can.

Sight

Sight allows people to see the things around them. It allows people to detect light energy from the environment and turn it into images. Sight is among the most studied senses and is probably the best understood. The eyes are the sensory organs of sight.

The eye contains two types of sensory receptors that are sensitive to light energy. These receptors are often called photoreceptors, the prefix *photo–* meaning "light." The receptor cells are called **cones** and **rods** for their distinctive shapes. Cones are mainly responsible for day vision, or seeing in bright light. Cones are sensitive to red, blue, and green light. Together, these three types of cones let people see all the colors around them. Rods are responsible for night vision, or seeing in dim light. Rods are not sensitive to different colors of light. This is why images appear in black and white in dim light.

Cone cells and rod cells are located on the **retina**, a thin layer of tissue on the back of the eyeball. Both types of cells change light energy that enters the eye into electrical impulses. These electrical impulses can be sent through the nervous system to the brain. **Figure 22.11** shows how light travels through the eye.

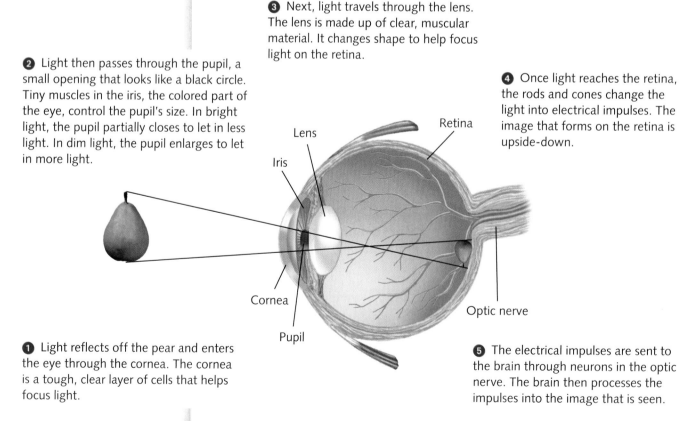

❸ Next, light travels through the lens. The lens is made up of clear, muscular material. It changes shape to help focus light on the retina.

❷ Light then passes through the pupil, a small opening that looks like a black circle. Tiny muscles in the iris, the colored part of the eye, control the pupil's size. In bright light, the pupil partially closes to let in less light. In dim light, the pupil enlarges to let in more light.

❹ Once light reaches the retina, the rods and cones change the light into electrical impulses. The image that forms on the retina is upside-down.

❶ Light reflects off the pear and enters the eye through the cornea. The cornea is a tough, clear layer of cells that helps focus light.

❺ The electrical impulses are sent to the brain through neurons in the optic nerve. The brain then processes the impulses into the image that is seen.

Labels: Lens, Iris, Retina, Cornea, Pupil, Optic nerve

Figure 22.11 The parts of the eye work together with the brain to enable people to see. How are electrical impulses sent from the eye to the brain?

Hearing

Hearing involves the detection of sound waves. A sound wave is a vibration. Vibrations are mechanical energy produced when an object moves back and forth quickly. The ears are the sensory organs of hearing.

The ear contains sensory receptors that are sensitive to mechanical energy. These receptor cells are found in the **cochlea**. They contain tiny hairs called **cilia** that change vibrations entering the ear into electrical impulses. These electrical impulses are sent to the brain, where they are processed into sound. **Figure 22.13** shows how mechanical energy travels through the ear.

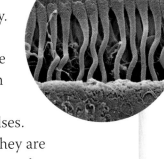

Figure 22.12 Tiny cilia are responsible for detecting sound. They are shown here magnified by about 400x.

Balance

The ears also sense the body's balance. Structures in the inner ear called the semicircular canals contain tiny hair cells and jellylike fluid. Inside this fluid are tiny grains called ear stones. As the body moves, the ear stones roll back and forth, bending the tiny hair cells. The hair cells respond by sending nerve impulses to the brain. If the brain interprets the impulses to mean the body is losing its balance, it will send signals to the muscles to either contract or relax to restore balance.

❶ Vibrations enter the outer ear and travel toward the eardrum.

❷ The vibrations cause the eardrum to vibrate, which in turn causes three tiny bones—the hammer, anvil, and stirrup—to vibrate.

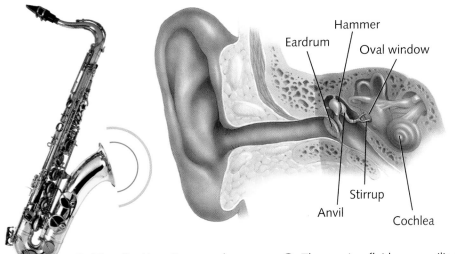

Hammer
Eardrum
Oval window
Stirrup
Anvil
Cochlea

❸ The vibrating stirrup pushes and pulls on the oval window. This causes fluid inside the cochlea to vibrate and move.

❹ The moving fluid causes cilia to sway back and forth and create an electrical impulse that is sent to the brain. The brain processes the impulse into a sound.

Figure 22.13 The parts of the ear work together with the brain to enable people to hear and to sense balance.

Smell

Humans can detect thousands of smells, including the sweet scent of perfume, the unpleasant odor of a skunk, and the delicious aroma of a freshly baked pie. The nose is the sensory organ of smell. Chemicals enter the nose when people inhale. They bind to sensory receptors and trigger an electrical impulse. When the impulse reaches the brain, it is processed and interpreted as a smell.

Taste

Like the sense of smell, the sense of taste is a chemical sense. The tongue is the sensory organ of taste. The sense of taste is probably the least understood of the senses. What scientists do know is that people experience taste when chemicals enter the mouth and bind to sensory receptors. Most sensory receptors for taste are found on the tongue, bundled together in structures called taste buds. Humans are sensitive to five tastes—sweet, salty, bitter, sour, and umami, which is a taste sensation described as meaty or savory. Scientists believe that each receptor cell is sensitive to only one of the five tastes.

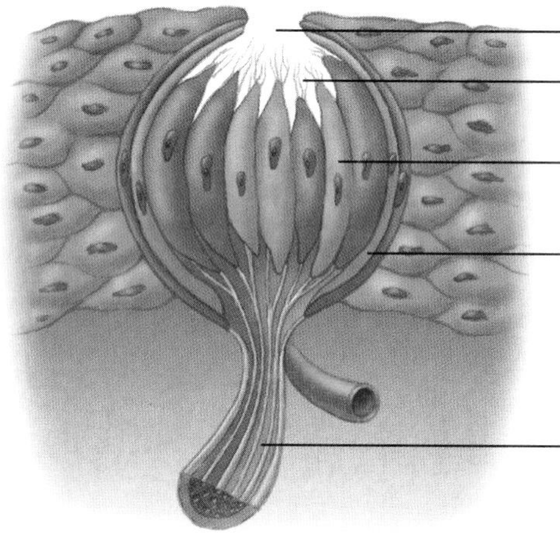

Tongue

Taste pore
Taste hairs
Sensory cells
Supporting cells
Nerve fibers

Figure 22.14 Each taste bud contains many receptor cells. The cells have taste hairs projecting from them. As food dissolves in the saliva in the mouth, the mixture stimulates the taste hairs. An impulse is sent to the brain and interpreted as a specific taste.

Touch

The sense of touch enables people to feel things such as pressure, temperature, and pain. The skin is the body's sensory organ of touch. Some sensory receptors in the skin are sensitive to chemicals. Other sensory receptors are sensitive to heat. Still other sensory receptors are sensitive to mechanical energy.

After You Read

1. What are the body's five sensory organs?
2. Describe the process by which people see. Why are the eyes, not the nose or tongue, the organs of sight?
3. Compare the sense of smell with the sense of taste. How are these senses alike? How do they differ?
4. Use the lesson outline in your Science Notebook to explain why you think the body has different senses.

22.4 The Endocrine System

Learning Goals

- Describe the functions of the endocrine system, and compare these with the functions of the nervous system.
- Identify the body's endocrine glands, and explain what each one does in the body.
- Explain the role of hormones and their control by feedback systems.

New Vocabulary

hormone
gland
negative feedback system

Before You Read

Create a Venn diagram in your Science Notebook. Write *The Endocrine System* above one of the circles and *The Nervous System* above the other. Then, read the headings and subheadings in this lesson and look at the pictures. What are some ways in which the two systems might be alike? What are some ways in which they might differ?

Suppose it were time for your school to have a fire drill. The goal of the fire drill is to make sure that every student and teacher gets out of the building quickly and safely. How should the principal let each class know that a drill is taking place? She could walk from classroom to classroom to tell the teachers and students individually, but that could take a long time. A faster and easier way might be for her to use the public address and alarm systems. These systems would inform all the students and teachers at the same time that a drill was occurring. Everyone would get out of the building quickly.

The endocrine system is a little like the public address and alarm systems in your school. It allows the body to send a mass message to many cells at once.

A Messenger System

Like the nervous system, the endocrine system is a messenger system. It works with the nervous system to help control many of the body's internal functions. The two systems are very different, however. Whereas the nervous system makes specific connections in the body, much like the principal traveling from classroom to classroom, the endocrine system sends mass messages throughout the body. It does this by releasing chemicals called **hormones** (HOR mohnz) directly into the bloodstream. Because hormones are released into the bloodstream, they can travel throughout the entire body. Chemicals released by neurons of the nervous system, you will recall, are directed to one cell or a small group of cells. They do not travel throughout the body.

The speed at which hormones travel in the bloodstream is slow compared with the speed of nerve impulses. Thus, the endocrine system tends to regulate body processes that happen slowly over a long period of time. These processes include cell growth and development, metabolism, sexual function, and reproduction. The endocrine system is also involved in regulating the body's response to danger or stress.

Structure and Function

The main organs of the endocrine system are glands. A **gland** is an organ that produces and releases a substance. In the case of the endocrine system, this substance is a hormone. Some glands release substances directly to the organ that uses them through tubelike structures called ducts. Endocrine glands release hormones into the bloodstream. Because endocrine glands do not have ducts, they are often called ductless glands.

Figure 22.15 shows the major glands of the endocrine system. Endocrine glands are scattered throughout the body. In general, they are not directly connected with one another.

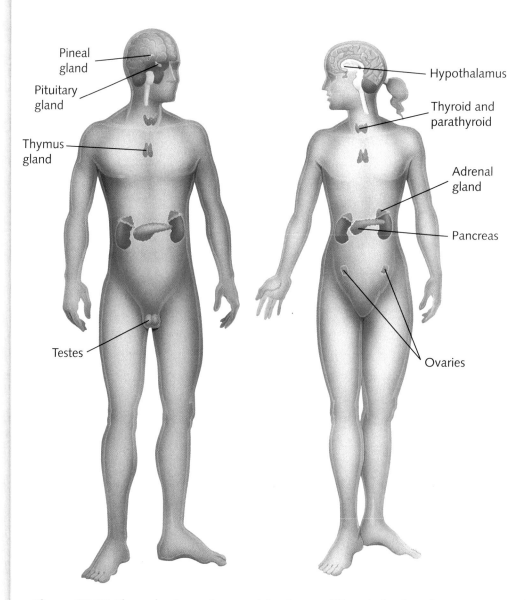

Figure 22.15 The endocrine system consists of many different glands and organs that are involved in regulating and coordinating various body functions. What is the name of the chemicals produced by endocrine glands?

Hypothalamus The hypothalamus, an area in the brain, connects the nervous system with the endocrine system. Its major function is to control the pituitary gland.

Pituitary Gland The pituitary gland is directly connected to the hypothalamus. It secretes nine different hormones. Some of these hormones directly regulate body processes. Others indirectly regulate body processes by controlling other glands. A few of the pituitary's many hormones and functions are shown in **Figure 22.17**.

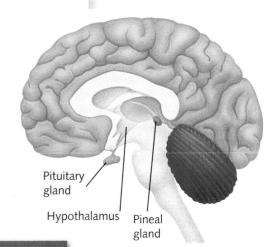

Figure 22.16 The hypothalamus, pituitary gland, and pineal gland are located in the brain. What is the primary function of each of these structures?

Figure It Out

1. Which hormones regulate body processes indirectly?

2. How would you describe the functions of the pituitary gland?

Pituitary Hormones

Hormone	Function in the Body
growth hormone	triggers the body to produce proteins; regulates growth in cells
follicle-stimulating hormone	triggers the body to produce sperm and mature egg cells
prolactin	triggers the production of milk in females after giving birth
thyroid-stimulating hormone	triggers the thyroid gland to produce and release a hormone called thyroxine

Figure 22.17 For a gland the size of a bean, the pituitary has many functions.

Pineal Gland The pineal gland is about the size of a bean and is buried deep within the brain. It produces and releases a hormone called melatonin. Melatonin is involved in regulating the body's rhythms, such as the sleep-wake cycle.

Thyroid Gland The two-lobed thyroid gland, located at the base of the neck, releases a hormone called thyroxine. Thyroxine regulates the process through which cells break down chemicals and get energy.

Parathyroid Glands Located on the back surface of the thyroid gland are four parathyroid glands. These tiny glands help regulate the amount of calcium that is in the body. Calcium is important for proper functioning of neurons. It is also important for bone development and muscle control.

As You Read

Use the Venn diagram in your Science Notebook to record important information about the endocrine system.

What are the main organs of the endocrine system? What chemicals do they release?

Figure 22.18 Tiny clusters of endocrine tissue called the islets of Langerhans are scattered throughout the pancreas. The millions of cells that make up the islets produce the hormones insulin and glucagon. What is the function of these hormones?

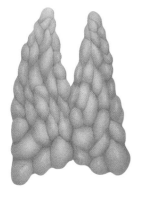

Figure 22.19 Located just behind the sternum in the upper chest, the thymus gland produces infection-fighting cells that are part of the body's immune response.

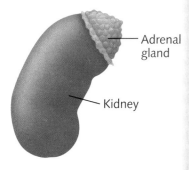

Adrenal gland

Kidney

Figure 22.20 One adrenal gland sits atop each kidney. The two adrenal glands are complex glands that produce a variety of hormones, some of which help stabilize the levels of sugar in the blood.

Pancreas   The pancreas, an organ of the digestive system, is also an endocrine gland. The pancreas releases hormones called insulin and glucagon. These hormones work together to control the amount of sugar that is in the blood.

Thymus Gland   The thymus is located in the upper chest. It produces hormones that aid in the body's ability to fight disease. In particular, it helps the body form T cells, which are found in the blood and fight infection.

Adrenal Glands   The adrenal glands are located above the kidneys. The adrenal glands release a hormone that is commonly called adrenaline. Adrenaline is responsible for the rush you feel when you are particularly nervous or excited. It helps the body cope with stressful situations. The adrenal glands also release more than 24 other hormones that help regulate other processes in the body.

Reproductive Glands   The testes in males and the ovaries in females are the body's reproductive glands. The reproductive glands produce hormones that control secondary sex characteristics, such as pubic hair, a male's deep voice, and a female's breasts. The hormone testosterone plays a role in the production of sperm cells in males. The hormones estrogen and progesterone aid in the development of ova, or egg cells, in females. These hormones also help regulate a female's menstrual cycle.

Controlling the Endocrine System

The endocrine system sends messages throughout the body. It does this by releasing hormones that travel through the bloodstream to target cells. A target cell is a cell that has receptors for a particular hormone. Once the hormone reaches a target cell, it triggers the cell to carry out a specific task. How are the actions of hormones controlled in the body? What causes a gland to start or stop releasing a particular hormone?

The body uses feedback systems to help control the production and release of hormones. Regulation of the endocrine system is most often controlled through a type of internal feedback system called a **negative feedback system**. In such a system, the hormones or the products of their effects are fed back to stop the original signal. Thus, if homeostasis is disrupted, the endocrine system can act to restore it.

Figure 22.21 shows an example of how negative feedback controls a function of the endocrine system. First, low blood sugar signals the hypothalamus to stimulate the pituitary gland. The pituitary gland then releases human growth hormone. The growth hormone (hGH) travels through the bloodstream to target cells in the liver. It triggers the liver to release glucose into the bloodstream. Increased blood sugar signals back to the hypothalamus to stop stimulating the pituitary gland. The pituitary gland stops releasing growth hormone, and the liver stops releasing glucose. When the blood sugar level drops again, the process will repeat itself.

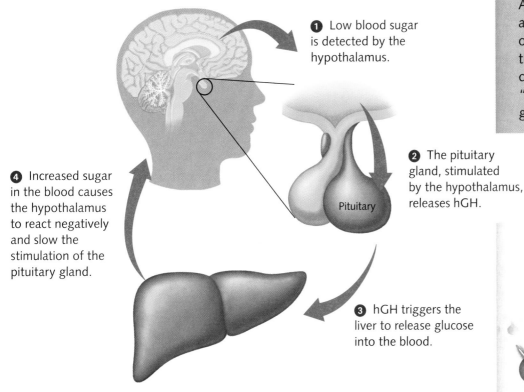

❶ Low blood sugar is detected by the hypothalamus.

❷ The pituitary gland, stimulated by the hypothalamus, releases hGH.

❹ Increased sugar in the blood causes the hypothalamus to react negatively and slow the stimulation of the pituitary gland.

Pituitary

❸ hGH triggers the liver to release glucose into the blood.

Figure 22.21 The endocrine system is a control system, but it too must be controlled. The body does this through a negative feedback system.

After You Read

1. What are the main functions of the endocrine system?
2. List three glands of the endocrine system, and describe what each one does in the body.
3. Explain how the endocrine system is controlled.
4. Complete the Venn diagram in your Science Notebook by filling in ways in which the endocrine and nervous systems are alike and different. Use your Venn diagram to write a well-developed paragraph that compares and contrasts the two body systems.

Explain It!

The operation of a thermostat in your home or school is an example of a negative feedback system. Explain how a thermostat works. Use a diagram similar to the one in Figure 22.21 to illustrate your explanation. Record your work in your Science Notebook.

KEY CONCEPTS

22.1 The Nervous System

- The nervous system helps control all the other body systems.

- Neurons are the cells of the nervous system. They have three main segments: a cell body, dendrites, and an axon.

- Information is sent through the nervous system in the form of electrical and chemical impulses.

- The nervous system can be divided into two smaller systems. The central nervous system (CNS) is made up of the brain and spinal cord. The peripheral nervous system (PNS) is made up of all the other nerves in the body.

22.2 The Brain and Spinal Cord

- The brain is divided into three parts: the cerebrum, cerebellum, and brain stem. The brain controls all of the body's voluntary actions, as well as many of the body's involuntary actions. It is the site of thought, reason, and emotion.

- The spinal cord connects the brain with the body's other organs. It is made up of bundles of nerves held together by connective tissue, and it is surrounded by bone and fluid.

- A reflex is an automatic, involuntary response to a stimulus. Reflexes involve short networks called reflex arcs. They are controlled in the spinal cord rather than in the brain.

22.3 The Senses

- The eyes, ears, nose, tongue, and skin are the organs responsible for the senses.

- Sensory receptors are able to change energy from a stimulus into an electrical impulse.

- When stimulated, the eyes, ears, nose, tongue, and skin produce electrical impulses that can be sent to the brain.

- The ears also sense the body's balance.

22.4 The Endocrine System

- The endocrine system helps control body processes such as growth, metabolism, and reproduction.

- The endocrine system is made up of glands and tissues that release hormones into the bloodstream.

- Hormones travel through the bloodstream to target cells. Their production and release is controlled by a negative feedback system.

VOCABULARY REVIEW

Write each term in a complete sentence, or write a paragraph relating several terms.

22.1
stimulus, p. 399
neuron, p. 400
dendrite, p. 400
axon, p. 400
synapse, p. 400
central nervous system, p. 402
peripheral nervous system, p. 402

22.2
cerebrum, p. 404
cerebellum, p. 404
brain stem, p. 404
vertebra, p. 405
reflex, p. 406
reflex arc, p. 406

22.3
sensory receptor, p. 407
cone, p. 408
rod, p. 408
retina, p. 408
cochlea, p. 409
cilia, p. 409

22.4
hormone, p. 411
gland, p. 412
negative feedback system, p. 414

PREPARE FOR CHAPTER TEST

To prepare for the chapter test, create a question from each Learning Goal. Use the information in your Science Notebook to answer each question. Then use these answers to write a well-developed essay about the chapter. Use the Key Concept on the first page of this chapter as your topic sentence.

True or False

If the statement is true, write "true." If it is false, change the underlined word or words to make the statement true.

1. The <u>endocrine system</u> helps control growth.

2. The <u>ears</u> are sensitive to light.

3. Receptors inside the nose are sensitive to <u>mechanical energy</u>.

4. The <u>nervous system</u> is responsible for people's ability to think and reason.

5. The <u>spinal cord</u> connects the brain with the rest of the body.

6. The endocrine system is made up of organs called <u>glands</u>.

Short Answer

Answer each of the following in a sentence or brief paragraph.

7. What is a reflex?

8. What kinds of cells allow the eyes to sense light?

9. What are the main functions of the cerebellum?

10. What do hormones do? How do they work in the body?

11. What are two divisions of the nervous system?

Critical Thinking

Use what you have learned in this chapter to answer each of the following.

12. **Compare and Contrast** How are reflexes similar to voluntary actions controlled by the nervous system? How do they differ?

13. **Explain** Why is it important to care for your ears and other sensory organs?

14. **Classify** What are three types of neurons? How do they differ?

15. **Infer** Why might the body have two different communication systems?

Standardized Test Question

Choose the letter of the response that correctly answers the question.

16. All of the following structures are essential to the process of hearing *except* the _____.

    A. semicircular canal

    B. cochlea

    C. eardrum

    D. cilia

Test-Taking Tip

Be careful of words such as *except*. This word means that all the answers are correct *except* for one.

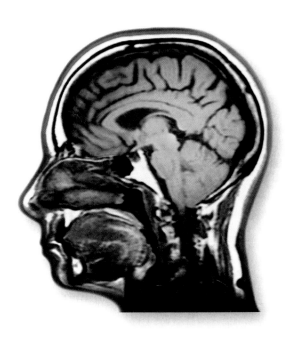

Chapter 23

Reproduction and Development

KEY CONCEPT The reproductive system is responsible for making new individuals.

One of the defining characteristics of living things is their ability to make new living things. Most animals—including humans—reproduce sexually. In sexual reproduction, a male and a female from the same species mix their genes to make a new, genetically unique individual.

Mixing genes has its advantages. By having offspring that are not exact copies of each other, the parents reduce the chance that all of their babies can be wiped out by a passing bacterium or parasite. To mix their genomes, however, animals need to make special cells that can fuse together. Each sex has a specialized system to produce those cells.

Think About Life Cycles

Every animal has a life cycle—a series of stages that it passes through from the beginning of its life until it has offspring of its own.

- In your Science Notebook, list all the human life stages you can think of.

- How many stages have you passed through already? Draw an arrow marking where you are now. Add arrows with labels for your parent(s) or guardian(s) and for any aunts, uncles, and grandparents.

NSTA
SCiLINKS
THE WORLD'S A CLICK AWAY

www.scilinks.org
Reproductive System **Code: WGB23**

23.1 The Reproductive System

Before You Read

Draw a T-chart in your Science Notebook. Label one side *Male* and the other side *Female*. As you read this lesson, write the characteristics of males and females in the correct columns.

Reproductive systems are essential to the survival of the species but not to individual survival. Men and women are different, and their reproductive systems contain different organs. The physical differences between men and women reflect their different roles in reproduction.

Gametes Make Zygotes

A man and a woman each contribute one **gamete** (GA meet), or sex cell, to make a baby. Male gametes are called **sperm**. Each of these small cells has a head and a long tail. A portion of the tail just behind the head is packed with mitochondria that produce energy a sperm uses to swim toward an egg. The head contains the nucleus of the cell and enzymes that help the sperm enter the egg. Female gametes are called **ova** (OH vuh, singular: ovum), or eggs. Ova are large cells that are filled with nutrients and other materials to help new life grow. **Figure 23.1** shows an ovum and several sperm.

Gametes are haploid cells. They contain only one set of chromosomes. When a haploid sperm fuses with a haploid egg, a cell with the diploid number of chromosomes results. The diploid number is the number of chromosomes an adult has. Thus, the chromosome number remains the same from one generation to the next.

Fertilization (fur tuh luh ZAY shun) occurs when a sperm cell fuses with an ovum to form a new cell called a **zygote** (ZI goht). The zygote combines the chromosomes from the sperm and the egg in its nucleus, making a new diploid cell. The zygote will develop into a new human being.

A human zygote contains 46 chromosomes. Half of them come from the sperm, and the other half are from the ovum. One pair of the chromosomes controls whether the zygote becomes a boy or a girl. If the zygote contains two large sex chromosomes, which are also called X chromosomes, the zygote becomes a girl. If the zygote contains one X chromosome and a small chromosome called the Y chromosome, the zygote becomes a boy. The Y chromosome carries fewer and different genes than the X chromosome does. This means that males have only one copy of the genes on the X chromosome. Recessive genetic traits on the X chromosome are always expressed in males. These traits, such as hemophilia and color blindness, are called sex-linked traits.

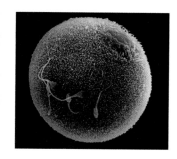

Figure 23.1 Males and females make different kinds of gametes. Five tiny sperm can be seen on the surface of this much larger ovum.

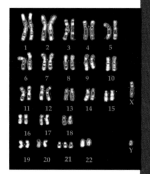

PEOPLE IN SCIENCE Patricia Jacobs 1934 –

Sometimes, chromosomes do not separate correctly during meiosis. Sometimes, they form gametes that contain either too many or too few chromosomes. Patricia Jacobs has spent her career studying disorders that develop when these gametes become part of a zygote.

Jacobs was born in London in 1934. During the 1950s, she began to study human chromosomes by observing cells from different people. She organized the chromosomes into a karyotype (KER ee uh tipe), which shows the chromosomes arranged in order from largest to smallest. In a karyotype, the chromosomes are given numbers based on their size and are paired by matching their patterns of light and dark bands.

By comparing karyotypes from different people, Jacobs found that some people had an abnormal number of chromosomes. In 1959, she discovered that people with Down syndrome had an extra copy of chromosome 21.

Explore It!

Get a set of paper human chromosomes from your teacher. Use their sizes, colors, and patterns of light and dark bands to match up the pairs of like chromosomes and arrange them in rows and from largest to smallest. You just made a karyotype.

Male Reproductive System

Males make sperm in a pair of organs called the **testes** (TES teez, singular: testis). The testes also make the male sex hormone **testosterone** (tes TAHS tuh rohn).

Males constantly make new sperm cells. A grown man can make about 300 million sperm every day. The new sperm cells are collected in the **epididymis** (eh puh DIH duh mus). The epididymis stores sperm cells while they mature and develop the ability to swim. Each testi and its epididymis is suspended in a pouch of skin, called the **scrotum** (SKROH tum), that hangs between the thighs. The testes are found outside the main body cavity because sperm can only mature at temperatures that are cooler than the normal human body temperature.

A **vas deferens** (VAS • DEF uh runz) moves sperm out of each testis and into the urethra. As sperm move toward the urethra, a series of glands add fluids that help the sperm move toward an ovum. The sperm combine with these fluids to form **semen** (SEE mun), which leaves the man's body.

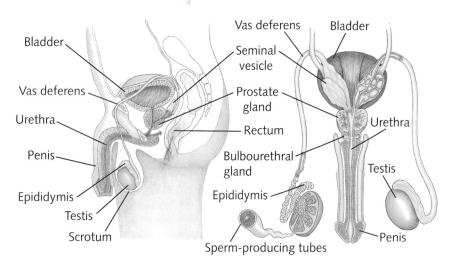

Side view   Rear view

Figure 23.2 Men have both external and internal reproductive organs.

Semen leaves a man's body through the urethra after it travels through a specialized reproductive organ called the penis. The penis places sperm inside the female reproductive tract.

Female Reproductive System

Women make ova in a pair of organs called the **ovaries** (OH va reez, singular: ovary). Unlike men, women do not continually make new ova. A newborn girl's ovaries already contain all of the cells that will mature into eggs over her lifetime.

A tube called the **oviduct** (OH vuh duct), or fallopian tube, is wrapped around one end of the ovary. When an ovum is released from the ovary, cilia inside the oviduct sweep the egg into the tube and push it toward the uterus. Fertilization usually occurs inside the oviduct.

The **uterus** (YEW tuh rus) is a pear-shaped organ located between the ovaries and oviducts. It has a thick lining in which fertilized eggs implant and develop. Its muscular walls are capable of stretching as the embryo grows and producing strong contractions to push the baby out at birth.

The lower end of the uterus has an opening called the **cervix** (SUR vihks). The cervix opens into a stretchy muscular tube, called the **vagina** (vuh JI nuh), that opens to the outside of the body. The vagina is also called the birth canal.

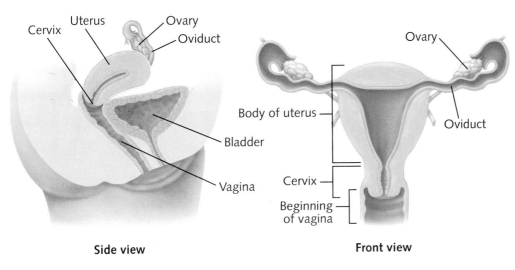

Figure 23.3 Unlike some male reproductive organs, all female reproductive organs are internal.

The Menstrual Cycle

Every month, sexually mature women shed their uterine lining and release a new egg from an ovary. These events repeat over and over again, month after month. The cycle lasts about 28 days on average. The cycle is only broken when an egg gets fertilized and starts to develop into an embryo.

As You Read

Your T-chart shows the differences between male and female reproductive systems. Write a few sentences in your Science Notebook describing similarities between the two systems.

 **CONNECTION: History**

The person who discovered the oviduct, or fallopian tube, was Gabriele Falloppio (1523–1562), an Italian anatomist. Although his family was noble, Falloppio had to struggle to obtain an education. He received his medical degree in 1548. Falloppio studied the male and female reproductive organs, and he described the tube that connects the ovary to the uterus in females. That tube—the fallopian tube—was named after him.

1. According to this chart, on which day does a woman ovulate?

2. If a woman begins to ovulate at age 13 and continues to release one egg each month until she is 50, how many eggs will she release during her reproductive lifetime?

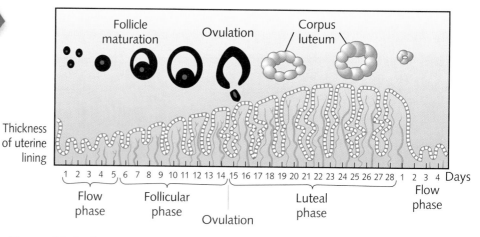

Figure 23.4 The menstrual cycle is caused by a regular pattern of hormones that prepare the uterus for a pregnancy each month.

The female reproductive cycle is called the menstrual cycle, and the process of shedding the uterine lining is called **menstruation** (men stroo AY shun).

For one week after menstruation, a woman's pituitary gland releases hormones that make an egg develop in one of her ovaries. The two ovaries usually take turns releasing eggs. An egg matures inside a case called a follicle (FAH lih kul). The follicle secretes the female sex hormone **estrogen** (ES truh jun), which makes the uterus grow a new lining. On day 14 of the cycle, pituitary hormones make the follicle rupture, releasing the mature egg into the oviduct. The release of the mature egg is called **ovulation** (ahv yuh LAY shun).

After ovulation, cilia in the oviduct push the ovum toward the uterus. The empty follicle becomes a structure called the corpus luteum (KOR pus • LEW tee um). It continues to release estrogen and also begins to release a second hormone called **progesterone** (proh JES tuh rohn). These hormones make the uterine lining grow even thicker so it can support a fertilized egg. If the egg is not fertilized, it dies and disintegrates. The woman will shed the uterine lining at the start of the next menstrual cycle.

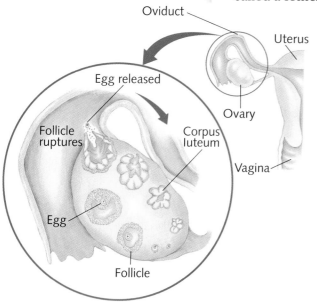

Figure 23.5 Each month, an egg matures inside a follicle. When the follicle ruptures, the egg is released into the oviduct.

After You Read

1. Under the T-chart in your Science Notebook, compare and contrast the testes and ovaries using a Venn diagram. Then, write a paragraph that summarizes the similarities and differences in the two organs.

2. What is formed when an ovum and a sperm cell fuse?

3. Predict what will happen to the corpus luteum if an egg is not fertilized.

23.2 Development Before Birth

Learning Goals

- Describe the stages of human development.
- List the events that occur during childbirth.
- Understand what causes identical and fraternal twins.

New Vocabulary

blastocyst
pregnant
placenta
umbilical cord
embryo
development
amniotic sac
fetus
labor
fraternal twins
identical twins

Before You Read

Create a concept map in your Science Notebook for the content of this lesson. Start by writing the word *Zygote* and circling it. Draw four circles vertically beneath this circle. Draw an arrow from each circle to the one below it. Write the word *Newborn* in the bottom circle. As you read, fill in the remaining circles with the names of the stages that the offspring moves through during pregnancy.

Two hundred million sperm will try to penetrate an ovum. Only one will succeed. None of the other sperm will be allowed to enter. The

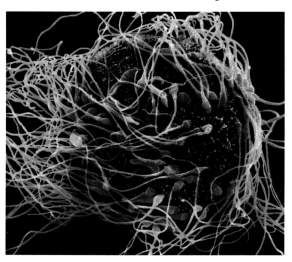

zygote that forms when a sperm and an egg fuse will divide, change, and grow for nine months. In the end, the zygote becomes a baby that is ready to be born.

Figure 23.6 Many sperm reach the egg, but only one will fertilize it.

Fertilization

When a sperm cell enters an ovum, it triggers a series of events that prevent any other sperm from entering the egg. The haploid nuclei of the egg and sperm fuse together to form a new diploid zygote.

Fertilization usually occurs in the oviduct. The zygote starts to divide as it is pushed toward the uterus. One cell becomes two, two cells become four, and so forth, until a hollow ball of cells called a **blastocyst** (BLAS toh sist) is formed. When the blastocyst reaches the uterus, it burrows into the uterine lining. The woman is now **pregnant**, or carrying a developing offspring.

Some of the cells in the blastocyst will continue to develop into a new individual. The rest of the cells form an exchange organ called the **placenta** (plu SEN tuh). The placenta exchanges nutrients, oxygen, and wastes between the mother and her offspring. The main part of the placenta grows into the uterine lining and makes contact with the mother's circulatory system. The embryo is attached to the placenta by a stalk of tissue that eventually becomes the **umbilical** (um BIH lih kul) **cord**.

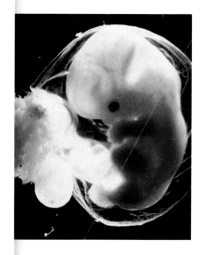

Figure 23.7 After several weeks of pregnancy, a developing human floats inside the amniotic sac. Part of the placenta is visible on the left side of the photo.

As You Read

On the concept map in your Science Notebook, draw a box around the stages of development that take place inside the oviduct. Label the box *Oviduct*. Draw another box around the stages of development that take place inside the uterus. Label this box *Uterus*.

What is an embryo called eight weeks after fertilization?

Figure It Out

1. Does the three-week-old embryo have arms or legs?

2. Describe how the appearance of the head region changes from week 5 to week 9.

Pregnancy

After it implants itself in the uterus, the offspring is called an **embryo** (EM bree oh). For the next few weeks, the embryo grows and changes. It gets larger, and the cells inside its body move and change in a process called **development**. **Figure 23.8** shows how a human embryo changes early in development.

When it first enters the uterus, the embryo is a ball of cells that all look the same. After one week, it is 2 mm long with a distinctive head region and an internal tube that will become the digestive system. After two weeks, it has doubled in size and has grown a simple tube-shaped heart. After a month, it is about 10 mm long and has begun to develop four limb buds that will become its arms and legs. The embryo is now surrounded by tissues that make a fluid-filled bag called the **amniotic** (am nee AH tihk) **sac**. The amniotic sac cushions the embryo as it grows.

Eight weeks after fertilization, the embryo becomes a **fetus** (FEE tus). Although it has developed all of its organ systems, it is still only 4 cm long. It cannot live outside of its mother. Over the next seven months, the fetus grows larger. Its brain and other organs mature. It starts making its own blood cells. Fat develops underneath its skin, making the fetus look less wrinkled. It grows hair, fingernails, and toenails.

The fetus also starts to move. It practices sucking and swallowing, breathing, and moving its arms and legs. After about the fourth month of pregnancy, the mother can feel the fetus move. By the ninth month, the fetus has changed position so that it is head-down against the cervix. It has enough fat to help it stay warm. Its lungs are ready to expand and breathe air. It is about 50 cm long and has a mass of about 3 kg. It is ready to be born.

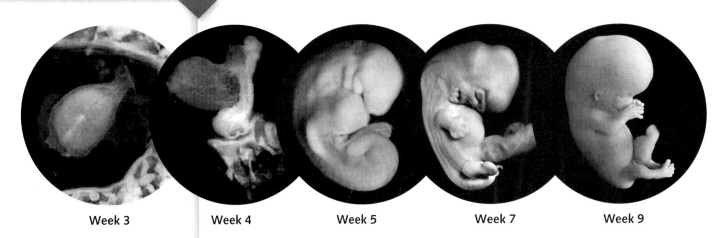

| Week 3 | Week 4 | Week 5 | Week 7 | Week 9 |

Figure 23.8 Human embryos change dramatically during early development.

Birth

Birth is the process that pushes a fetus out of the uterus and into the outside world. A protein from the fetus's lungs crosses the placenta and enters the mother. When the mother's body detects the protein, her brain releases a hormone that makes her body push out the fetus. The fetus only makes the protein when it is ready to breathe on its own.

The process of giving birth is called **labor**. Labor begins after the woman releases a hormone called oxytocin (ahk sih TOH sun). The hormone makes the muscles in the uterus contract regularly. The contractions pull open the cervix. The amniotic sac usually ruptures during labor, and the amniotic fluid flows out through the vagina, or birth canal.

When the cervix is fully open, the uterine contractions become very forceful. They push the baby through the birth canal and the cervix and the vagina until it is out of the mother's body. After the baby is born, the placenta separates from the uterine wall and is pushed out through the birth canal. The uterus keeps contracting to close off blood vessels and to prevent bleeding.

After birth, the baby is still attached to the placenta by its umbilical cord. Doctors clamp and cut the umbilical cord near the baby's belly. After a few days, the stump of cord dries up and falls off, leaving a scar called the navel (NAY vul)—commonly known as the belly button.

Figure 23.9 Labor has three stages. **1)** During the first stage, uterine contractions open the cervix. **2)** During the second stage, the baby is pushed out through the birth canal. **3)** During the third stage, the placenta is pushed out through the birth canal.

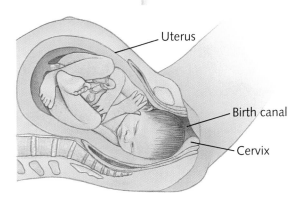

1 Dilation Labor contractions open the cervix.

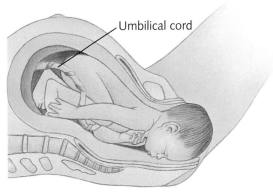

2 Expulsion The baby rotates as it moves through the birth canal, making expulsion easier.

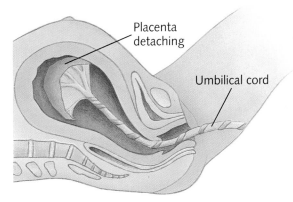

3 Placental stage During the placental stage, the placenta and umbilical cord are expelled.

⟨⟩ CONNECTION: Physics

When a newborn baby takes its first breath, its lungs inflate for the first time. The pressure change in the lungs immediately reroutes blood through the heart. In the fetus, most blood bypasses the lungs. In the newborn baby, most blood moves through the lungs.

Multiple Births

Only one baby usually develops during a pregnancy, because a woman usually releases one egg at a time. But sometimes, a woman can have more than one baby from a single pregnancy.

If a woman releases two eggs at the same time, they can both get fertilized and implant in the uterus. They will develop into **fraternal** (fruh TUR nul) **twins**. Fraternal twins come from different eggs and sperm cells. They can, but do not always, look alike—as with any other pair of siblings. They can be two girls, two boys, or a boy and a girl.

Sometimes, a single zygote will split in two early in its development. The two zygotes grow into **identical twins**. Identical twins have the exact same genes. They look exactly the same. They are either two girls or two boys.

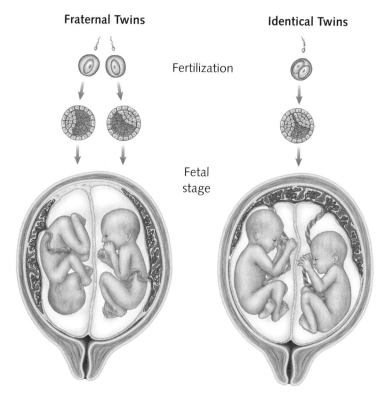

Figure 23.10 Fraternal twins and identical twins develop differently.

After You Read

1. Review the concept map you created in your Science Notebook. You should have one arrow in your concept map that is not in a box. What physical process does that arrow represent? Write it in your Science Notebook.

2. What is a blastocyst? What two things develop from it?

3. Explain the difference between identical twins and fraternal twins.

23.3 Human Life Stages

Before You Read

Look at the five main headings in this lesson. Make a chart in your Science Notebook with five columns. Write one heading in each column. Then, write a sentence summarizing what you know about each one.

People continue to grow and change after birth. Over time, a helpless newborn baby learns to walk and talk, becoming an active toddler. The toddler grows into a child, the child grows into a teenager, and the teenager eventually becomes an adult.

Infancy

A newborn baby cannot do very much. It cannot hold up its own head, roll over, or smile. It can only see about as far as its mother's face from her breast. It spends most of its time asleep. But for the next 18 months, it will go through some dramatic changes. This period of great physical and mental change is called **infancy**.

Infants grow quickly. Most newborn babies are about 50 cm long and have a mass of between 3 and 4 kg. Within the first six months of their lives, they will double their weight. By the end of their first year, they will have tripled their birth weight and will have grown about 50 percent in height. Imagine what you would look like if you grew that much in the next year!

Infant brains also develop and change quickly. Babies learn to coordinate their nervous system with their muscles, rapidly learning to roll over, sit up, grab toys, and crawl. Most infants learn the complex balancing act of walking by the time they are 18 months old. Their ability to think and reason also grows during infancy. Most children speak their first words by the end of this period.

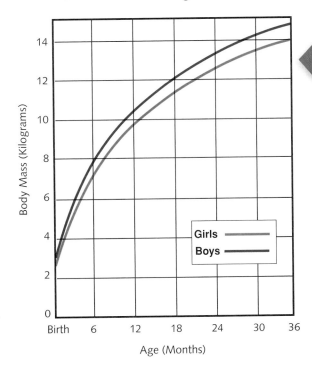

Growth Rate of Average Human Infants

Figure 23.11 This graph illustrates the rapid growth of infants during their first year of life.

Learning Goals

- Name the human life stages.
- Describe the physical changes that occur at puberty.
- Identify some of the physical changes of old age.

New Vocabulary

infancy
puberty
fertile
secondary sex characteristic
menopause

Figure It Out

1. What is the body mass of an average three-month-old girl?

2. How do the masses of the average three-month-old boy and the average three-month-old girl compare?

CHAPTER 23 **427**

As You Read

Record important information about the physical changes in each life stage in your chart. In your Science Notebook, write your hypothesis about the following question.

Why do the most dramatic changes occur during infancy?

**CONNECTION:
Social Studies**

Many cultures mark the onset of puberty with special events that show the child is becoming an adult. There are many different coming-of-age rites. They can be religious, such as Roman Catholic confirmation and Jewish bar or bat mitzvah, or cultural, such as the Latino *quinceañera*.

Childhood

The period between 18 months of age and the onset of sexual maturity at around age 12 to 13 is called childhood. Growth is not as rapid as in infancy, but a child still grows quickly and steadily.

Children continue to develop physically and mentally. Their control over their bodies improves, and they learn to speak in complex sentences. Formal education begins later in childhood.

Adolescence

Adolescence begins with **puberty** (PYEW bur tee) and lasts through the teen years. During puberty, a child's body changes into an adult body. These physical changes start when the brain begins to release hormones that make the testes or ovaries release sex hormones. These hormones mature the organs of the reproductive system and make the person **fertile**, or capable of having children. They also induce the development of adult physical characteristics, or **secondary sex characteristics**. Boys begin to look like men, and girls start to look like women.

In boys, puberty begins when the testes start releasing testosterone. The penis and testes grow larger, and the testes begin to make sperm. The larynx grows, deepening the voice, and the boy grows facial and body hair. The muscles in the shoulders and chest get bigger than those in women. In girls, puberty begins when the ovaries start releasing estrogen. The menstrual cycle begins as the eggs in the ovaries start to mature regularly. The hips grow wider, and breasts develop. Boys and girls both grow underarm and pubic hair and go through a rapid growth spurt to reach their adult height. Girls finish growing around age 16, and boys finish around age 18. Puberty in both boys and girls can be a time of intense emotional shifts as the sex hormones are released into their bodies. Many of the changes associated with puberty occur rapidly and simultaneously, making this a difficult time for adolescents.

Figure 23.12 A child's body grows faster than its head.

Adulthood

Adulthood starts at the end of adolescence and continues through middle and old age. Adults stop growing taller and remain approximately the same size for the remainder of their lives. Most people work and raise families during this life stage.

Aging begins to change the adult body after age 45. The body's metabolism slows down and physical strength declines, sometimes leading to weight gain. The bones begin to become more brittle. The skin becomes less elastic and begins to get wrinkled. Hair loses its pigment and becomes gray. In women, the ovaries lose the ability to make estrogen and the menstrual cycle stops. This change in the reproductive system is called **menopause** (MEN uh pawz). Menopausal women stop ovulating and lose the ability to have children. The lack of estrogen also signals other changes, such as greater risk of heart attack and increased loss of bone density. Menopause means women must adjust their lifestyles to accommodate these changes and stay healthy.

Older Adulthood

After adults pass the age of 60, their cells do not divide and the tissues do not repair themselves as well as they did in youth. This makes it harder for the body to respond to external stresses caused by injury or disease. Skin wrinkles deepen, eyesight and hearing diminish, and reflexes slow down. Nerves and muscles work more slowly, leading to slower response times for physical tasks such as driving. The lenses in the eyes get harder, making it more difficult to focus on things that are nearby. However, older adults can still continue to learn new things, and many remain physically active.

Figure 23.13 Humans spend most of their lives as adults.

Figure 23.14 Aging slowly changes most of the tissues in the body.

 **Extend It!**

With a partner, spend an hour with a person in each of the five human life stages. Using the chart in your Science Notebook, write each person's age and what you observe about him or her during that hour.

After You Read

1. Name the human life stages.
2. Using the information you have recorded in your Science Notebook, create a Venn diagram to compare and contrast the physical changes boys and girls go through during puberty.
3. What are some physical characteristics of adulthood and older adulthood?
4. What is menopause?

Summary

KEY CONCEPTS

23.1 The Reproductive System

- Men and women have different reproductive anatomies.
- Sperm are the male gametes, and ova, or eggs, are the female gametes.
- A new individual results when a sperm and an egg cell fuse. The resulting zygote has the diploid number of chromosomes.
- Males have a pair of sperm-producing organs called the testes.
- Females have a pair of egg-producing organs called the ovaries.
- Female reproduction is characterized by a cycle of hormonal changes.

23.2 Development Before Birth

- A zygote becomes an embryo and then a fetus before being born.
- A woman is pregnant once the developing embryo has become attached to the wall inside her uterus.
- During a woman's pregnancy, the developing embryo becomes a fetus. Nurients and wastes are exchanged between mother and baby within the placenta.
- Childbirth has three stages: cervical dilation, birth of the child, and expulsion of the placenta.
- Twins can result from two zygotes or a single zygote that splits in two.

23.3 Human Life Stages

- Infancy and childhood are periods of rapid growth and mental development.
- People become sexually mature during puberty.
- Aging is linked to the degeneration of cells.

VOCABULARY REVIEW

Write each term in a complete sentence, or write a paragraph relating several terms.

23.1
gamete, p. 419
sperm, p. 419
ovum, p. 419
fertilization, p. 419
zygote, p. 419
testis, p. 420
testosterone, p. 420
epididymis, p. 420
scrotum, p. 420
vas deferens, p. 420
semen, p. 420
ovary, p. 421
oviduct, p. 421
uterus, p. 421
cervix, p. 421
vagina, p. 421
menstruation, p. 422
estrogen, p. 422
ovulation, p. 422
progesterone, p. 422

23.2
blastocyst, p. 423
pregnant, p. 423
placenta, p. 423
umbilical cord, p. 423
embryo, p. 424
development, p. 424
amniotic sac, p. 424
fetus, p. 424
labor, p. 425
fraternal twins, p. 426
identical twins, p. 426

23.3
infancy, p. 427
puberty, p. 428
fertile, p. 428
secondary sex characteristic, p. 428
menopause, p. 429

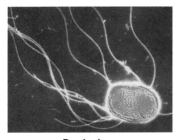
Bacterium
Salmonella

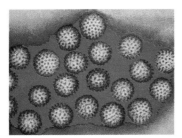
Virus
Rotavirus

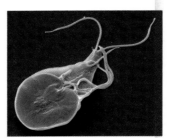
Protozoan
Giardia lamblia

Fungus
Epidermophyton

Figure 24.3 Many types of pathogens can cause human diseases. *Salmonella* is a food-borne bacterium that causes diarrhea. Viral infections caused by rotavirus kill 600,000 children in developing countries each year. *Giardia* is a protozoan transmitted by drinking unclean water. Fungal infection by *Epidermophyton* causes the itchy, peeling skin of athlete's foot.

Pathogens and Disease

Bacteria and viruses cause many serious diseases. Fungi also cause some mild diseases. Protozoans and parasitic worms are common pathogens in other parts of the world. Each type of pathogen invades the body in a different way and causes specific diseases. Most infectious diseases in the United States are caused by bacteria and viruses. **Figure 24.4** lists some of the diseases caused by bacteria, viruses, protozoans, and fungi.

Bacteria Bacteria are single-celled organisms found in every imaginable place on Earth. Only a small fraction of the thousands of species of bacteria cause human diseases. However, these pathogenic bacteria lead to about five million deaths each year. The main danger from bacteria comes from the toxins, or poisons, they produce. Bacterial toxins can cause fever, diarrhea, and cell death.

Viruses Viruses are not living organisms. They consist of genetic material enclosed in a protein shell. Viruses can reproduce only with the help of a host cell, which they destroy in the process. The destruction of host cells is what causes the symptoms of viral diseases. The single deadliest infectious disease, AIDS, is caused by a virus called HIV.

Pathogens that Cause Infectious Diseases	
Type of Pathogen	**Diseases Caused**
bacterium	food poisoning, strep throat, sinus infection, tooth decay, tuberculosis, plague, anthrax
virus	cold, influenza, viral hepatitis, viral pneumonia, AIDS, SARS, Ebola, chicken pox, measles, rabies, smallpox, diarrhea
protozoan (protist)	malaria, traveler's diarrhea, amebic dysentery
fungus	athlete's foot, ringworm
animal (worm)	tapeworm, trichinosis, hookworm

Figure 24.4 Each infectious disease is caused by a different pathogen.

Figure It Out

1. Which type of pathogen causes ringworm?
2. Which kingdoms of life do not cause infectious diseases in humans?

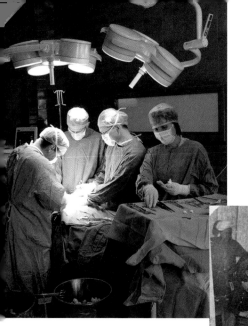

Figure 24.5 These photos show how surgery was performed before *(right)* and after *(left)* the acceptance of the germ theory. How many differences can you find between the two scenes?

Medicine and Disease

Pathogens were discovered in the mid-nineteenth century. Before that time, physicians and common people alike had no idea what caused infectious diseases. Most people believed that sickness came from wrongful actions, clouds of diseased air, or even glances from diseased people. Diseases were treated by bleeding victims or feeding them poisonous mixtures of chemicals. Many patients were more hurt than helped by doctors' treatments.

Louis Pasteur and other scientists discovered that pathogens, commonly called germs, cause many diseases. This idea is called the germ theory. Pasteur had a difficult time convincing doctors that washing hands and using clean surgical instruments could save lives. Medicine became much more effective after the germ theory was accepted. Doctors were able to slow the spread of infectious diseases.

Antibiotics   Chemicals that kill bacteria are called **antibiotics** (an ti bi AH tihks). Sir Alexander Fleming first discovered antibiotics in 1928 when spores from the airborne mold *Penicillium notatum* accidentally fell on a sample of bacteria he was growing in his lab. All around the mold, the bacteria died. Fleming was able to purify the substance the mold secreted—penicillin—and use it to treat patients with bacterial diseases. The use of antibiotics has saved millions of lives.

Antibiotics have their limits in fighting disease, however. First, antibiotics are only useful against bacteria. Viruses and other pathogens are not affected. Second, bacteria can develop resistance to commonly used antibiotics. Bacteria evolve quickly through mutation and natural selection. Scientists are concerned that a "superbug" bacterium will evolve that is resistant to all known antibiotics. This is why it is important to use antibiotics only when needed and to finish all of the medicine prescribed by a doctor.

After You Read

1. Describe how medicine changed after the acceptance of the germ theory.

2. Do bacteria and viruses make you sick in the same way? Explain your answer.

3. According to your lesson outline, what are three ways in which pathogens can be transmitted? Answer this question in a well-developed paragraph in your Science Notebook.

24.2 The Nonspecific Defense System

Learning Goals

• Identify the role and actions of the nonspecific defense system.

• Explain the purpose and effects of inflammation.

New Vocabulary

nonspecific defense system
inflammation
phagocyte
natural killer cell

Before You Read

Some processes take place in a specific order. Think about a process that is routine to you. In your Science Notebook, describe the order of events involved in the process. Preview the lesson, and note the order of events as a pathogen invades the body.

All of the surfaces of the body are coated with bacteria. Bacteria also live and thrive inside the body. For the most part, the bacteria on the skin and in the gut are harmless. They compete with pathogenic bacteria and prevent infections. The symbiotic bacteria in the intestines help digest food and manufacture vitamins. As long as these microorganisms do not enter body tissues, they cause no harm. Problems begin when symbiotic bacteria or more dangerous bacteria enter body tissues.

The Role of the Nonspecific Defense System

Microorganisms can become pathogens once inside body tissues. Pathogens disrupt normal body functions and release toxins. It is the job of the **nonspecific defense system** to keep foreign cells out of body tissues and destroy them if they do get inside. The skin and mucous membranes keep most pathogens away from body tissues. White blood cells fight pathogens that succeed in invading the body.

Babies are born with a functioning nonspecific defense system. That is why it is often called the innate, natural, or inborn defense system. This system attacks all cells and large molecules that it recognizes as foreign to the body. Pathogens, pollen, splinters, and transplanted tissues are all treated the same way.

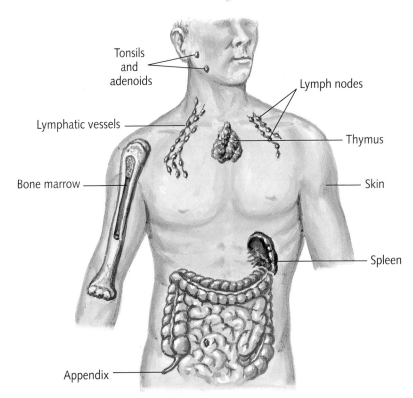

Figure 24.6 Many different parts of the body function in the nonspecific defense system.

There is a saying that "the best offense is a good defense." That is certainly true when it comes to fighting disease. The energy required to keep pathogens out of the body is much less than the energy it takes to defeat an active infection. The nonspecific defense system of the body consists of the skin, mucous membranes, inflammation, and white blood cells.

Skin The skin is designed to keep bodily fluids in and foreign material out. The outer layers of skin are made of dead cells that pathogens cannot invade. The skin is a dry, acidic environment that is unsuitable for most microorganisms. Sweat and oils produced by glands in the skin contain proteins that disrupt bacterial cell walls. The symbiotic bacteria that normally live on the skin are adapted to these hostile conditions. The symbiotic bacteria compete with dangerous microorganisms and produce chemicals that kill them.

Mucous Membranes Skin is not the only part of the body that comes in contact with pathogens. Pathogens can also enter the body through the eyes, respiratory tract, digestive tract, and any other part that comes in contact with the air. Body cavities that are exposed to the outside world are lined with mucous membranes. The mucus, or sticky liquid, that these membranes produce traps germs and other foreign particles before they can invade cells. In the nose and throat, tiny hairs called cilia sweep mucus to the throat, where it can be swallowed. Once in the stomach, most pathogens are destroyed by stomach acid. Other ways in which the body expels pathogens include tears, sneezing, coughing, urination, and defecation.

The body would quickly be overwhelmed by invading pathogens without the skin and mucous membranes. Any gap in these defenses allows microorganisms into body tissues. At that point, the next level of bodily defenses is activated.

Figure 24.7 Sneezing is one way that the body expels pathogens. Sneezing can spread pathogens from one person to another.

Inflammation: The Body's First Response

Cells respond immediately to an injury or invasion by pathogens. Cells in the damaged area release chemicals that cause inflammation. **Inflammation** (ihn fluh MAY shun) is redness, heat, swelling, and pain at the site of an injury or infection. Chemicals released by cells near the injury or infection cause blood vessels to expand. The blood vessels become leaky, allowing fluids and cells to pass from them into the injured tissues.

Figure 24.8 Inflammation is the body's first response to injury or infection. An insect sting can cause the skin to swell, become red, and feel warm to the touch. Inflammation may be painful, but it speeds the body's repairs.

Inflammation starts a series of body defenses. Blood platelets flow through gaps in the blood vessels. The platelets seal out microorganisms from an injury or infection. A fever may develop, speeding body repairs and slowing pathogen reproduction. Various types of defensive cells are also attracted to the inflammation. All of these defensive cells are types of white blood cells.

Too much inflammation can cause damage to the body. The chemicals involved in inflammation have powerful effects on cells and their DNA. Constantly inflamed tissues cannot function properly. Their cells may change and become cancerous. Scientists suspect that inflammation is also important in the development of heart disease, arthritis, and other diseases.

Figure It Out

1. What temperature range is ideal for the body's defense response?

2. Is it a good idea to take a fever reducer as soon as the body temperature rises above 37°C?

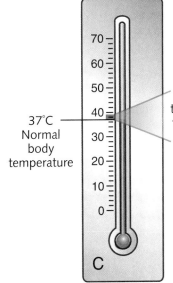

37°C Normal body temperature

Best temperature for immune response

Figure 24.9 A fever can be a helpful part of the body's defense against infection. A low fever slows pathogen reproduction. The cells of the body's defense system also function best under fever conditions.

Fighting Invaders Inside Body Tissues

Defensive cells respond quickly to the alarm signal of inflammation. Some of the defenders are **phagocytes** (FA guh sites), white blood cells that recognize and eat pathogens. The phagocytes can identify pathogens by their pattern of antigens, or proteins on their surfaces. Most phagocytes die after consuming pathogens. The dead cells build up as a pocket of pus at the infection site. The body absorbs this pus within a few days.

Just as pathogens have an identifiable pattern of antigens on their surface, so too does an infected cell. The infected cell shows the antigens of an invader on its surface. Another type of defender called a **natural killer cell** recognizes these antigens and attacks the infected cell. The natural killer cell releases chemicals that destroy the infected cell's outer membrane.

Viruses activate a different kind of defense. A cell infected by a virus releases chemicals called interferons (ihn tuhr FIH rahns). The interferons stimulate nearby cells to resist viral attacks. Natural killer cells recognize viral antigens on the infected cells and destroy the cells. A third type of defensive cell attacks larger parasites, such as hookworms. These cells attach to the parasite's body and release chemicals that damage its tissues.

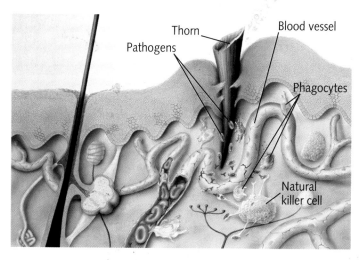

Figure 24.10 The body's nonspecific defenses respond to an injury within seconds.

After You Read

1. Explain what features make mucous membranes difficult for pathogens to invade.

2. According to the sequence chart in your Science Notebook, what is the body's first response to an infection?

3. Describe how defensive cells distinguish between healthy body cells and infected cells.

24.3 The Immune System

Learning Goals

- Describe the four major features of the immune system.
- Explain immunity and the ways the body can acquire it.
- Describe the different types of immune disorders.

New Vocabulary

immune system
incubation period
memory cell
immunity
active immunity
passive immunity
vaccine
artificial immunity
allergic reaction
autoimmune disorder
cancer

Before You Read

Preview the lesson, and look for examples of immune functions and disorders. In your Science Notebook, record your examples by using the lesson headings and subheadings as organizing titles. Share your examples with a partner. Add any necessary information to your list.

The nonspecific defense system works all the time to prevent pathogens from invading body tissues. Despite these efforts, however, pathogens are sometimes successful at creating an infection. At that point, the body uses its system of defense against specific pathogens: the **immune system**.

Features of the Immune System

The immune system has several unique features. First, the immune system responds to specific pathogens. The antigens displayed on the outsides of pathogens and other foreign substances alert the immune system to the presence of a threat to the body. White blood cells called B cells and T cells that target the specific invader are produced. B cells produce proteins, called antibodies, that attach to the antigens on the pathogen. These antibodies interfere with the functions of the pathogen and mark it for destruction. Killer T cells seek out infected cells and pathogens marked by antibodies and destroy them.

A second feature of the immune system is its ability to respond to millions of different threats. The B cells that respond to each invader are present in the bone marrow of a newborn baby. The immune cells are inactive until each one is awakened by the presence of a specific antigen in the body.

Third, the immune system is capable of distinguishing between body cells and cells from other organisms. Immune cells respond to pathogens, cancerous cells, transplanted tissues, and insect venoms. Finally, the immune system is able to remember pathogens and develop immunity.

Features of the Immune Response	
specific	responds only to the pathogen involved
adaptive	responds to millions of different pathogens
self-recognizing	can distinguish between body cells and foreign or cancerous cells
able to remember	creates memory cells that respond quickly to a second invasion by a pathogen

Figure 24.11 The immune response is very different from the nonspecific body defenses.

Explain It!

People who suffer from certain hereditary diseases cannot produce B cells or T cells. In your Science Notebook, explain what effects this condition would have on a person's immunity and general health. Include the following terms in your explanation: *antigen, nonspecific defense system, antibodies, memory cells, active immunity, infection, pathogens,* and *incubation period.*

Did You Know?

Venomous snakebites can be deadly. In order to save snakebite victims, doctors give them antivenin. The antivenin consists of antibodies produced by a horse or other large animal that has been exposed to snake venom. The antivenin gives the victim passive immunity to snakebites that lasts for a few weeks.

Figure 24.12 After a viral infection, memory cells for the virus remain in the body. The immune system responds much faster to a second exposure to the same virus.

Immunity

Pathogens in the body must race to reproduce themselves before the immune system is ready to attack them. Each disease has an **incubation period** of several days to years before the first symptoms appear. For example, chicken pox has an incubation period of 14 to 16 days. The first time the body is exposed to chicken pox, it takes the immune system a week or two to start producing B cells and T cells. During that time, the population of the chicken pox virus multiplies rapidly. A patient may get very sick before the immune response can start to defeat the disease.

Some of the immune cells developed against the chicken pox virus become memory cells. The **memory cells** remain in the body after infection and respond immediately if the pathogen invades a second time. Memory cells allow the body to fight off a second chicken pox infection before the incubation period ends. **Immunity** means that a person does not suffer symptoms of disease when he or she is exposed to a pathogen. The type of immunity that memory cells provide is called **active immunity**. In active immunity, the body makes its own antibodies in response to an antigen.

Passive Immunity Babies are born without any memory cells in their immune systems. However, they do have antibodies given to them by their mother's body during pregnancy and nursing. The transfer of antibodies from another's body is called **passive immunity**. Passive immunity sometimes involves injecting antibodies into a person's body. This is often done to combat snake venom and the rabies virus. Passive immunity lasts only for a few weeks or months, until the antibodies break down in the body.

Figure It Out

1. After how many days is the first immune response strongest?
2. Why doesn't a person feel any symptoms of a viral disease after being exposed to it for the second time?

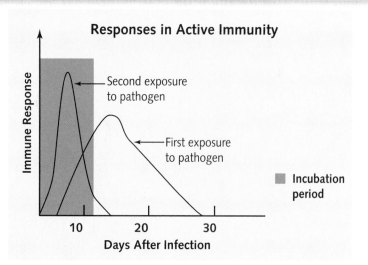

Vaccines and Artificial Immunity   The development of vaccines against pathogens was a great step forward in medicine. **Vaccines** (vak SEENZ) contain weakened, dead, or incomplete portions of pathogens or antigens. When injected into the body, the pathogens or antigens cause an immune response without causing serious disease. Vaccines produce immunity because they cause the body to react as if it were naturally infected. Thus, the immune system produces a response to the pathogen's antigens and creates memory cells against it. Immunity produced by a vaccine is called **artificial immunity**. Vaccines against polio, smallpox, and measles have saved millions of lives around the world. Newly developed vaccines against chicken pox and the virus that causes cervical cancer hold great promise for the coming decades.

As You Read

Compare the examples of immune functions and disorders you recorded at the beginning of the lesson with what you know now. Make corrections or additions to your entries.

What is one example of a way in which a person can gain passive immunity to a disease?

Causes of Disease Before and After Vaccine Availability in the U.S.

Disease	Average Number of Cases per Year Before Vaccine Available	Cases in 1998 After Vaccine Available
measles	503,282	89
diphtheria	175,885	1
tetanus	1,314	34
mumps	1,152,209	606
rubella	47,745	345
pertussis (whooping cough)	147,271	6,279

Figure 24.13 There were many more serious childhood illnesses in the United States before vaccines became available. Which disease was almost completely eradicated by 1998 as a result of a vaccine?

⚙ CONNECTION: Medicine

Polio, measles, mumps, and tetanus are dangerous childhood diseases that have something in common: they are now rare in the United States. Widespread use of vaccines has reduced the number of deaths caused by infectious diseases to low levels. Children who live in developing countries are not so lucky. One-quarter of the children born each year in these countries are not vaccinated. Three million of them will die of diseases that could be prevented using $30 worth of vaccines per child. Unfortunately, many developing nations spend as little as $3 per citizen on health care each year.

Many international organizations are striving to raise vaccination rates around the world. The GAVI Alliance (formerly known as the Global Alliance for Vaccines and Immunisation) is a worldwide group that helps developing nations set up and run vaccination programs. Americans have pledged billions of dollars to GAVI in an effort to someday make measles, rotavirus, and hepatitis as rare in Africa as they are now in the United States.

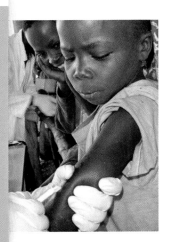

Immune System Disorders

The immune system has a very difficult job to do. It must recognize and respond to millions of potential pathogens. At the same time, the immune system must not attack the healthy cells of its own body. At times, this delicate balance falls apart.

Allergic Reaction   Sometimes, the immune system mistakes harmless molecules for pathogens. An **allergic reaction** is an immune response to a food, medication, or other chemical. Minor symptoms of an allergy include itching, sneezing, runny nose, and coughing. More serious allergies can lead to rashes, asthma, and low blood pressure. About 700 people in the United States die each year from allergic reactions to antibiotics, foods, and insect stings.

The percentage of Americans with allergies and asthma is increasing. There is evidence that the rise in allergies is due to our cleaner homes. The immune system is designed to seek out and destroy pathogens. A century ago, the immune system was kept busy fighting parasites and infectious diseases. Now, with our much cleaner homes, the immune system may target and attack previously harmless molecules.

Autoimmune Disorders   When the immune system attacks the cells of its own body, **autoimmune disorders** result. The disease multiple sclerosis (MS) is caused by an immune attack on the tissues of the nervous system. Type 1, or juvenile onset, diabetes results when the insulin-producing cells of the pancreas are mistaken for pathogens. In some forms of arthritis, the cartilage in the joints is the target. Autoimmune diseases are not curable, although medications can slow their progress and ease symptoms.

Extend It!

Research a disease and the efforts that are being made to combat it. Record your findings in your Science Notebook. Then, prepare a class presentation that answers the following questions:

- What causes the disease? Is the disease preventable?

- How does the body normally defend against this disease?

- What are the symptoms and signs of the disease?

- What medical treatments are available? What new treatments might be available in the near future?

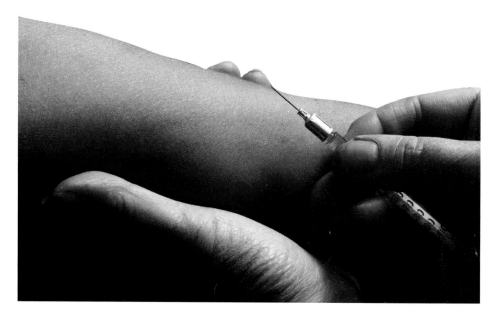

Figure 24.14 One in five people in the United States suffers from allergies. A skin-prick test is one way to find out which allergens are causing allergic symptoms.

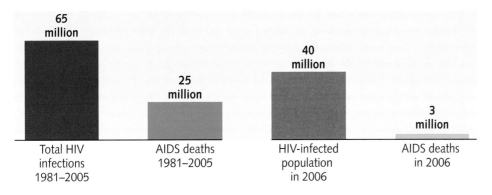

65 million	25 million	40 million	3 million
Total HIV infections 1981–2005	AIDS deaths 1981–2005	HIV-infected population in 2006	AIDS deaths in 2006

Figure 24.15 The AIDS epidemic is already one of the worst in human history. Although there is currently no cure for the disease, antiretroviral treatment can prolong life.

AIDS The human immunodeficiency virus (HIV) causes disease by invading the cells of the immune system. HIV disables T cells and prevents the body from defending itself. Untreated HIV infection leads to acquired immunodeficiency syndrome (AIDS). Pathogens invade AIDS patients and cancers develop. AIDS patients generally die within a few years without treatment. Good health care can prolong the lives of patients infected with HIV for many years.

HIV can only be passed from one person to another directly through bodily fluids. The most common means of transmission are unprotected sexual relations and the sharing of needles by drug users. Thousands of scientists and health professionals around the world are working to develop a vaccine against HIV, improve treatments, and prevent HIV infections.

Cancer Cells that grow and divide uncontrollably lead to **cancer**. Cancer results from a series of mutations in the cells' DNA. One of the jobs of the immune system is to detect cancerous cells and destroy them. Cancerous cells produce tumors if they are allowed to multiply. These tumors can grow and spread to any part of the body. Untreated cancer disrupts body functions. Cancer is the second leading cause of death in the United States today.

Fortunately, cancer treatment is improving rapidly. Tumors can be surgically removed or treated with poisonous chemicals or radiation. As scientists learn more about how cancerous cells develop, they can help people avoid cancer. Some of the best ways to prevent certain kinds of cancer are to stay away from tobacco products and to use sunscreen when exposed to sunlight.

After You Read

1. How do killer T cells recognize their targets?

2. Give an example from your Science Notebook of a disease that can be prevented by vaccination.

3. Explain why so much time and money is being spent on research to combat AIDS.

KEY CONCEPTS

24.1 Disease

- Infectious diseases are caused by pathogens that invade the body.
- Pathogens can be transmitted from human to human; through air, water, and food; and by vectors.
- Each pathogen has different effects on the body and causes a different disease.
- The germ theory and antibiotics have greatly improved medicine and human health.

24.2 The Nonspecific Defense System

- The body's nonspecific defense system works to destroy pathogens and keep them out of the body.
- The body's nonspecific defense system includes the skin, mucous membranes, inflammation, and several types of white blood cells.
- Inflammation is the body's first defense against pathogens that get inside the body.
- Phagocytes are white blood cells that recognize and attack pathogens inside body tissues.
- Phagocytes can identify pathogens by the pattern of antigens on the pathogens' surfaces.
- Natural killer cells recognize and destroy infected body cells that show the antigens of their invaders on their surfaces. The killer cells also recognize viral antigens on infected cells and destroy these cells.

24.3 The Immune System

- The immune system recognizes and responds to each specific pathogen.
- Memory cells provide immunity to a second attack by a pathogen.
- Active immunity develops as a result of having a disease and building up antibodies to it. Passive immunity develops when antibodies are transferred from another person's body or from an animal that is already immune to the disease.
- Vaccines, which contain weakened, dead, or incomplete portions of pathogens or antigens, provide artificial immunity to infectious diseases.
- Immune disorders can result in allergies, autoimmune diseases, AIDS, and cancer.

VOCABULARY REVIEW

Write each term in a complete sentence, or write a paragraph relating several terms.

24.1
disease, p. 433
noninfectious disease, p. 433
pathogen, p. 433
infectious disease, p. 433
symptom, p. 433
vector, p. 434
epidemic, p. 434
antibiotic, p. 436

24.2
nonspecific defense system, p. 437
inflammation, p. 439
phagocyte, p. 440
natural killer cell, p. 440

24.3
immune system, p. 441
incubation period, p. 442
memory cell, p. 442
immunity, p. 442
active immunity, p. 442
passive immunity, p. 442
vaccine, p. 443
artificial immunity, p. 443
allergic reaction, p. 444
autoimmune disorder, p. 444
cancer, p. 445

PREPARE FOR CHAPTER TEST

To prepare for the chapter test, create a question from each Learning Goal. Use the information in your Science Notebook to answer each question. Then use these answers to write a well-developed essay about the chapter. Use the Key Concept on the first page of this chapter as your topic sentence.

MASTERING CONCEPTS

True or False
If the statement is true, write "true." If it is false, change the underlined word or words to make the statement true.

1. An immune response to a food or other chemical is a(n) <u>autoimmune disorder</u>.

2. The <u>immune system</u> in a newborn baby is ready to fight any foreign molecules it detects in the body.

3. A(n) <u>epidemic</u> is the spread of a disease to many victims.

4. <u>Artificial immunity</u> lasts only a few weeks or months.

5. The common cold is caused by a(n) <u>bacterium</u>.

6. Immune cells and phagocytes recognize pathogens by the pattern of <u>antigens</u> they display on their surface.

Short Answer
Answer each of the following in a sentence or brief paragraph.

7. What clues do doctors use to determine which pathogen is making a patient sick?

8. Compare and contrast natural killer cells, phagocytes, and memory cells.

9. Explain why cancer and AIDS cannot be treated with antibiotics.

10. Describe how a restaurant employee could start an epidemic.

11. Explain where mucous membranes are found in the body, and describe their role in preventing infection.

12. Explain the relationship among an incubation period, immunity, and memory cells.

13. How can bacteria help the body?

Critical Thinking
Use what you have learned in this chapter to answer each of the following.

14. **Predict** Which of these people would benefit from a measles vaccination: a healthy child, a person with measles, a person with AIDS, or a person who had measles a year ago? Explain your answer.

15. **Infer** What disease-fighting advantages might a breast-fed baby have over a formula-fed baby?

16. **Generalize** What are the major reasons why people in poor countries suffer from more infectious diseases than people in wealthy countries?

Standardized Test Question
Choose the letter of the response that correctly answers the question.

Percentage of Deaths Due to Major Disease				
Disease	Year			
	1950	1980	1990	2000
heart disease	37.1	38.3	33.5	29.6
cancer	14.6	20.9	23.5	23.0
stroke	10.8	8.6	6.7	7.0
diabetes	1.7	1.8	2.2	2.9

17. According to the table, which disease showed a steady increase in percentage of deaths from 1950 to 2000?

   A. heart disease

   B. cancer

   C. stroke

   D. diabetes

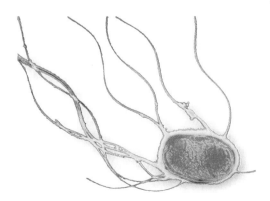

Test-Taking Tip

After you read a multiple-choice question, answer it in your head before reading the choices provided. This way the choices will not confuse you or trick you.

Chapter 25

Health and Wellness

KEY CONCEPT Body systems can be cared for by practicing healthful habits and avoiding harmful substances.

Although you may not realize it, you make decisions almost every minute of each day that affect your body and overall health. You decide whether or not to eat breakfast. You decide whether or not to shower and brush your teeth. You decide when to cross the street, how loud to play the music you listen to, who to socialize with, and what to do after school. Each of these decisions affects your health in some way. Making decisions that are healthful starts with being aware of the choices you make and how those choices affect your body.

Think About Health and Wellness

Think about the things you do that affect your health and how you feel physically and emotionally.

- Make a list in your Science Notebook of such types of actions.

- Classify the actions on your list as having a positive effect or a negative effect on your health. Choose one or two actions that have a negative effect. How could you change these actions so that they have a positive effect? What could you do differently? Record your thoughts in your Science Notebook.

NSTA

SCiLINKS.
THE WORLD'S A CLICK AWAY

www.scilinks.org
Nutrition **Code:** WGB25

Healthful Living

- Describe how healthful habits are important to one's health.

- Identify five healthful habits a person can practice.

- Explain ways in which healthful habits can be incorporated into daily life.

New Vocabulary

aerobic exercise
stress
hygiene
first aid

Recall Vocabulary

nutrition (p. 375)
nutrient (p. 375)

Before You Read

Create a concept map in your Science Notebook for the term *Healthful Habits*. Then, look at the pictures, headings, and Learning Goals in this lesson. What do you predict you will include in your concept map?

The organs and systems of the body constantly work together to keep you alive and well. For example, the skin protects the internal organs and helps prevent disease. The circulatory and respiratory systems supply oxygen to the body's cells. The digestive system breaks down food into energy the body can use. Muscles get you out of bed in the morning and to school. Your brain takes in and remembers information and helps you do well on the next science exam. Your immune system fights off disease and infections. Each body organ and system carries out functions you depend on to stay healthy.

Healthful Habits

Most people demand a lot from their bodies. Like all relationships, however, your relationship with your body is one that involves giving and taking. You must practice healthful habits so that your body can do all the things you need and expect from it. If you don't take care of your body, it can't take care of you.

Healthful habits include eating a well-balanced diet, getting exercise and rest, managing stress, practicing good hygiene, and keeping safe. Healthful habits should be a part of your daily life.

Figure 25.1 Practicing healthful habits, such as getting plenty of exercise, can help you look and feel your best.

Nutrition

Eating healthful foods ensures that you are getting good nutrition. Nutrition is the process of getting the food needed to survive.

The human body needs more than 40 nutrients for good health. As shown in **Figure 25.2**, these nutrients can be grouped into six main categories: carbohydrates, proteins, fats, vitamins, minerals, and water.

Essential Nutrients

Nutrient	Function	Food Sources
carbohydrates	provide the body with its main source of energy	bread, rice, pasta, cereals, grains
proteins	help the body build, repair, and maintain tissue	red meat, chicken, fish, cheese, beans
fats	provide the body with energy; help the body use certain vitamins	oils, nuts, meats, dairy products
vitamins	aid the body in the production of certain chemicals and cells; help the body process carbohydrates, proteins, and fats	vegetables, fruits, eggs, cereals, meats
minerals	help the body grow and develop; control chemical reactions in the body; keep bones strong	meats, grains, dairy products, fruits, vegetables
water	makes up more than 50 percent of the body's cells; helps the body digest food and eliminate waste; helps regulate the body's temperature; cushions joints	water, fruits, vegetables

Figure 25.2 There are six groups of nutrients the body needs for good health.

Explore It!

Food labels indicate the types and amounts of nutrients that are found in foods. Collect food labels from a variety of foods. Try to include foods you think are healthful as well as foods you consider to be unhealthful. Compare the nutritional information for each food. Using *MyPyramid* and the information on your food labels, work with the class to develop a well-balanced menu for one day.

Eating a balanced diet ensures the body receives all the nutrients it needs. *MyPyramid* from the United States Department of Agriculture (USDA) can help you make the right food choices.

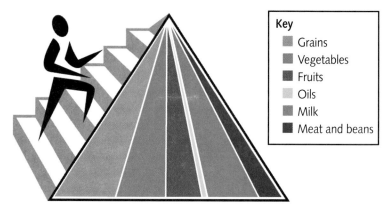

Key
- Grains
- Vegetables
- Fruits
- Oils
- Milk
- Meat and beans

Figure 25.3 The USDA released *MyPyramid* on April 19, 2005. It is a modification of the original U.S. Food Guide Pyramid. The new pyramid stresses activity along with a proper mix of food groups in one's diet.

Exercise and Rest

Keeping the body healthy requires more than just good nutrition. It requires plenty of exercise and plenty of rest.

Exercise Exercise strengthens bones and muscles. It reduces the risk of disease and keeps the heart and lungs working efficiently. It keeps joints from becoming stiff and improves flexibility. It can also improve mental health and make a person feel better.

Exercising four to six times per week for about 30 to 60 minutes each time is a good goal for most healthy young adults. A healthful exercise routine can include a combination of aerobic exercise and weight training. **Aerobic** (er ROH bihk) **exercise** is any exercise that increases the heart and breathing rates. Running, swimming, speed walking, and dancing are some examples of aerobic exercise. Weight training helps strengthen bones and muscles. Lifting weights is a weight-training exercise.

As You Read
Add to your concept map words describing healthful habits you can incorporate into your life.

Why are nutrition and exercise important to your health?

Figure 25.4 There are many ways to incorporate exercise into one's daily life. Walking or biking instead of driving and taking the stairs instead of the elevator are just a few things people can do to stay active and fit.

Rest Rest is as important to the body as exercise is. Resting gives the body a chance to recover from the day and to get ready for a new day. Without sleep, people can feel tired, cranky, and irritable. Some people can have trouble thinking clearly and following directions. The immune system and growth can be affected, too. The body is more likely to get sick when it does not get enough rest. Most teens need about nine or more hours of sleep each night. It is important to schedule time for rest so that the body can look, feel, and function at its best.

Managing Stress

Most people are familiar with the term *stress*. In science, **stress** is defined as the body's response to the demands of everyday life. Distress is negative stress. People experience distress when they perceive a situation as difficult, dangerous, or painful and they do not have the resources to cope. Eustress is positive stress. A moderate amount of stress can help people improve their performances and achieve their goals. Research shows that some stress is actually necessary for life.

When stress is not managed, it can lead to anxiety, withdrawal, sleeplessness, fatigue, aggression, depression, and physical illness. Here are some things you can do to manage stress.

- Identify your own sources of stress, and make choices that help control the amount of stress you experience.

- Exercise, eat healthful foods, and get plenty of rest.

- Avoid caffeine. It can increase feelings of anxiety.

- Talk with friends, parents, or other respected adults about how you are feeling.

- Try a hobby that you enjoy.

- Manage your time wisely.

- Break large tasks into smaller, more manageable chunks.

Hygiene and Posture

Most healthy people look healthy. That healthy look has a lot to do with good hygiene and good posture. **Hygiene** (HI jeen) describes the things people do to keep clean and to prevent disease. Here are some things you can do to have good hygiene.

Figure 25.5 Keeping neat and clean is part of good hygiene. Why is good hygiene important to good health?

- Bathe or shower daily and after exercise.

- Wash your hair with shampoo at least two times per week.

- Brush your teeth after meals and before you go to bed each night. Floss to remove plaque and food that can get stuck between teeth.

- Keep fingernails and toenails trimmed and clean.

- Have regular physical and dental checkups.

Extend It!

Many soaps and detergents are currently marketed as "antibacterial." They are described as being an effective way to protect people from common illnesses. Some researchers, however, argue that the antibacterial agents contained in these soaps and detergents may actually kill off normal bacteria, creating an environment for mutated bacteria that are resistant to antibiotics.

Conduct research to learn more about antibacterial soaps. Are they better than regular soaps? Could they do more harm than good in the long term? Record your findings in your Science Notebook.

Posture is also important to health. Standing, sitting, and walking erect allow your backbone to protect your spinal cord and enable the organs of the respiratory system to work more effectively.

Figure 25.6 Practice good posture by consciously standing straight with your shoulders down and your stomach held in.

Safety and First Aid

More than 100,000 people in the United States lose their lives each year in accidents. Accidents can happen anywhere. When accidents do happen, first aid can help minimize the damage. **First aid** is the immediate care given to a person who is injured or sick. First aid skills can be learned by taking a course offered at a hospital or other health care organization, such as the American Red Cross.

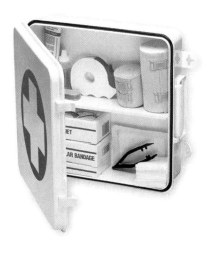

Figure 25.7 A first aid kit includes things such as adhesive bandages, medical tape, aspirin, a thermometer, a breathing mask, tweezers, and disposable gloves.

PEOPLE IN SCIENCE Clara Barton 1821–1912

Clara Barton, known as the "angel of the battlefield," was the founder of the American Red Cross. Clara began what would become her life's work and passion during the American Civil War. Despite dangerous conditions, she aided the sick and injured on the battlefield. Off the battlefield, she gathered supplies and comforted returning soldiers.

After the war, Clara worked to find and identify missing soldiers. She also worked tirelessly to educate the public about the Red Cross, which had been established in Switzerland in 1869. She gave speeches and handed out brochures. In 1881, she began the American Red Cross and acted as its first president for 23 years. The American Red Cross helps people prevent, prepare for, and respond to emergencies.

After You Read

1. Review your concept map. In a well-developed paragraph, describe some healthful habits people can practice to achieve and maintain good health.

2. Explain why healthful habits are important to have. What could happen to a person's body if he or she does not practice healthful habits?

3. What are some ways in which you can make healthful habits a part of your daily life?

Before You Read

Create a K-W-L-S-H chart in your Science Notebook. Think about the title of this lesson. Then, look at the pictures, headings, and Learning Goals in the lesson. In the column labeled *K*, write what you already know about drugs. In the column labeled *W*, write what you would like to learn.

Aspirin, cough syrup, and asthma medication probably don't come to mind when you hear the term *drugs*. However, they are all examples of drugs used as medicines.

What Is a Drug?

A **drug** is any substance other than food that, when taken into the body, alters one or more chemical processes in the body. Drugs change the way the body or mind works. All medicines are drugs, from the medicines prescribed by a doctor to the medicines people buy over the counter to treat a cold or flu. Medicines are drugs that are used to fight disease or to treat, prevent, or relieve symptoms of disease.

Caffeine, alcohol, and nicotine are drugs that can be harmful to one's health. Caffeine is a chemical found in coffee, tea, chocolate, cola, and other foods and beverages. It causes the body's heart rate to increase. Alcohol is a chemical found in beer, wine, and liquors such as vodka. Nicotine is found in cigarettes, cigars, and chewing tobacco.

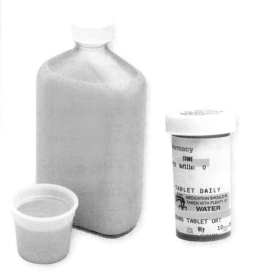

Figure 25.8 An over-the-counter (OTC) drug *(left)* is a drug people can buy without a doctor's prescription. OTC drugs include aspirin, ibuprofen, some types of cold remedies, and some types of cough syrup. A prescription drug *(right)* is a drug that requires a doctor's permission to take.

Drug Abuse

Medicines can keep you healthy, but they can also harm your health when they are abused. **Drug abuse** is the intentional misuse of a legally purchased drug or the use of an illegal drug. Drug abuse can be life threatening. It can harm internal organs and can cause your body systems to fail.

Drug abuse can lead to drug dependence. Drug dependence, or drug addiction, is the continued need for the effects of a drug even when the effects are harmful. Drug dependence can be caused by tolerance, physical dependence, and/or psychological dependence. **Tolerance** is what happens when the body gets used to a drug and needs higher and higher doses of the drug to produce the original effects. **Physical dependence** is a condition in which the user relies on a drug and requires the effects of the drug in order to feel well. **Psychological** (si kuh LAH jih kul) **dependence** is a condition in which the user has an overwhelming, emotional desire to continue using a drug.

Withdrawal symptoms are a sign of physical dependence. These symptoms occur when a drug is no longer in the body.

As You Read

In the column labeled *L* in your K-W-L-S-H chart, write what you have learned about drugs.

Are medicines drugs? Explain your answer.

Types of Drugs and Their Effects

Drug Type	Effects on the Body	Examples
stimulants	• increased activity of the central nervous system • increased heart rate, breathing rate, and blood pressure	caffeine, cocaine, amphetamines
depressants	• decreased activity of the central nervous system • relaxed muscles and sleepiness	alcohol, barbiturates, tranquilizers
narcotics	• decreased pain (when used as painkillers) • increased activity of the central nervous system	morphine, codeine, heroin
hallucinogens	• interference with the senses and perception • hallucinations in which people see or hear things that are not real	LSD, PCP, ecstasy
marijuana	• mood changes • decreased short-term memory	

Figure 25.9 The drugs in this table, classified into groups based on their effects on the body, are commonly abused.

Figure It Out

1. What types of drugs speed up the activity of the central nervous system?

2. Explain why a dependence on hallucinogens is harmful to one's health.

After You Read

1. What is a drug? What is drug abuse?

2. What are some drugs that are commonly abused?

3. Use your completed K-W-L columns to summarize this lesson in a well-developed paragraph in your Science Notebook. Then, write in the *S* column what you would still like to know and in the *H* column how you can find this information.

Explain It!

Write a public service announcement in your Science Notebook warning people about the dangers of alcohol or tobacco use. Include pictures from magazines, or draw your own pictures. Remember that an effective announcement makes an impact on its readers.

Did You Know?

Alcohol contains many Calories, but it has little nutritional value. Its Calories are considered "empty," or wasted.

Before You Read

Make a Venn diagram to compare and contrast alcohol and tobacco. Predict what you will write in the Venn diagram.

Count out eight seconds. During those eight seconds, somewhere in the world, someone died from tobacco use. Tobacco is responsible for more than 500,000 deaths in the United States every year. That is about one out of every five American deaths. More than 100,000 other American deaths are caused by alcohol use. Nearly 40 percent of all traffic fatalities are alcohol related. Alcohol and tobacco are the most abused substances in the world. They are also the deadliest.

Alcohol

Alcohol is the oldest-known drug. **Alcohol** is a depressant that slows down the central nervous system and can cause both physical dependence and psychological dependence. While present in the body, it can alter people's emotions, perceptions, judgment, and coordination. It can affect people's vision, hearing, and speech. It can even cause people to lose consciousness.

Alcohol enters the bloodstream when a person drinks alcoholic beverages such as beer, wine, or liquor. It travels through the body and begins blocking messages from reaching the brain. In time, it travels to the liver. There, it is broken down into the waste products carbon dioxide and water. Before it is broken down, however, it builds up in the bloodstream. The higher the concentration of alcohol in a person's blood, the more the person's nervous system is affected. In higher concentrations, alcohol can kill brain cells.

Blood alcohol concentration (BAC) describes the amount of alcohol in a person's blood. In most states, a BAC between 0.08 and 0.10 percent is high enough for a person to be considered intoxicated. When a person is intoxicated, he or she is physically or mentally impaired. How much and how quickly a person drinks, how much a person weighs, and how much a person has eaten are some factors that affect a person's BAC. In general, the more a person drinks, the higher his or her BAC becomes and the more intoxicated he or she becomes.

Alcohol can have long-term effects on the body, such as immune system suppression, liver damage, high blood pressure, heart failure, and infertility. These effects are usually caused when a person abuses alcohol over a prolonged period of time.

Tobacco

Tobacco is a broad-leafed plant whose leaves can be dried and then smoked in cigarettes or cigars. Its leaves can also be chewed. Smokeless tobacco is tobacco that is chewed.

Tobacco contains a drug called nicotine. **Nicotine** (NIH kuh teen) is a highly addictive stimulant. When tobacco is smoked, chewed, or sniffed, nicotine enters the bloodstream. Within seconds, it enters the brain and alters the way the brain works. It speeds up the activity of the central nervous system and increases blood pressure, heart rate, and breathing rate. It causes glucose to be released into the bloodstream, making the body feel more alert. It causes chemicals that are responsible for feelings of pleasure to be released in the brain. Within 40 minutes of entering the body, nicotine's effects are reduced by half. As the effects wear off, the body has an overwhelming desire for more nicotine. After continued use, the body builds up a tolerance for nicotine. Nicotine use can quickly lead to both physical and psychological dependence.

When tobacco is burned, a thick, sticky fluid called **tar** is produced. Tar can stick in the bronchioles of the lungs. As it builds up in the lungs, it decreases the lungs' ability to pass oxygen into the blood. Tar causes lung cancer, emphysema, and bronchial diseases.

Tobacco smoke also contains at least 4,000 other chemicals. One of these chemicals is carbon monoxide, a substance that can cause heart disease. Finally, tobacco smoke harms the cilia lining the organs of the respiratory system.

As You Read

Use your Venn diagram to record important information about alcohol and tobacco.

What is alcohol? What are its effects on the body?

Figure It Out

1. How are a smoker's lungs different from a nonsmoker's lungs?
2. Infer the effects of smoking on a person's respiratory system.

Figure 25.10 Compare the lung of a nonsmoker *(left)* with the lung of a smoker *(right)*. Smoking can also cause mouth cancer.

After You Read

1. What is alcohol?
2. What substance in tobacco leaves is a drug? What does this drug do to the body?
3. Why are alcohol consumption and tobacco use harmful to one's health?
4. Use your completed Venn diagram to write in your Science Notebook a well-developed paragraph that compares and contrasts alcohol and tobacco.

KEY CONCEPTS

25.1 Healthful Living

- Practicing healthful habits is an important part of maintaining one's body systems and overall health.

- Good nutrition, plenty of exercise and rest, stress management, good hygiene, and safety are the goals of practicing healthful habits.

25.2 Drugs and Drug Abuse

- A drug is a substance other than food that affects how the body and mind work.

- Drug abuse is the intentional misuse of a legal drug or the use of an illegal drug.

- Drug abuse can lead to drug dependence. Drug dependence can be caused by tolerance, physical dependence, and/or psychological dependence.

- Some commonly abused drugs include stimulants, depressants, narcotics, hallucinogens, and marijuana.

25.3 Alcohol and Tobacco

- Alcohol and tobacco are harmful substances that are commonly abused.

- Alcohol is a depressant—it slows down the activities of the nervous system. It has both short-term and long-term effects on one's health.

- Blood alcohol concentration (BAC) describes the amount of alcohol present in a person's blood. A BAC between 0.08 and 0.10 percent is high enough in most states for a person to be considered intoxicated.

- Nicotine is a drug found in tobacco. It is a highly addictive stimulant. Tobacco smoke also contains tar, carbon monoxide, and many other chemicals. Tar can cause cancer and lung disease; carbon monoxide can cause heart disease.

VOCABULARY REVIEW

Write each term in a complete sentence, or write a paragraph relating several terms.

25.1
aerobic exercise, p. 451
stress, p. 452
hygiene, p. 452
first aid, p. 453

25.2
drug, p. 454
drug abuse, p. 454
tolerance, p. 455
physical dependence, p. 455
psychological dependence, p. 455

25.3
alcohol, p. 456
blood alcohol concentration, p. 456
tobacco, p. 457
nicotine, p. 457
tar, p. 457

PREPARE FOR CHAPTER TEST

To prepare for the chapter test, create a question from each Learning Goal. Use the information in your Science Notebook to answer each question. Then use these answers to write a well-developed essay about the chapter. Use the Key Concept on the first page of this chapter as your topic sentence.

True or False

If the statement is true, write "true." If it is false, change the underlined word or words to make the statement true.

1. <u>All</u> medicines are drugs.

2. Running is a <u>weight-training</u> exercise.

3. Eating a well-balanced diet helps you be sure that you get the right <u>first aid</u>.

4. Alcohol is the <u>oldest</u> drug known to humans.

5. Nicotine is a <u>depressant</u>.

6. Caffeine is a <u>drug</u> found in coffee and tea.

Short Answer

Answer each of the following in a sentence or brief paragraph.

7. What are some healthful ways that a person can manage stress?

8. What is good posture?

9. What happens as a person's blood alcohol concentration increases?

10. At what BAC is a person considered intoxicated?

11. What is an over-the-counter drug? What are some examples?

Critical Thinking

Use what you have learned in this chapter to answer each of the following.

12. **Compare and Contrast** How are stimulants and depressants alike? How do they differ?

13. **Explain** Why is rest important to the body?

14. **Infer** How might drug abuse affect one's relationships?

15. **Evaluate** Should people avoid taking all types of drugs? Explain your answer.

Standardized Test Question

Choose the letter of the response that correctly answers the question.

16. Nicotine is a drug found in which of the following?

    A. caffeine

    B. tobacco

    C. alcohol

    D. tar

Test-Taking Tip

If you don't know an answer and you won't be penalized for guessing, make an educated guess. Use context clues if you don't understand the question.

Technology Leads the Way

MEDICAL TECHNOLOGY promises to help many people live healthier lives. Here are just a few new technologies being developed.

Laboratory-Grown Organs

After 16 years of research and development, scientists are now growing artificial bladders in the laboratory. Bladders are simpler to make than many other organs because they have no blood vessels. People who need new bladders donate cells from their damaged organs. Scientists put these cells into "scaffolds," organ-shaped foundations that the cells can grow on. The scaffolds are then transplanted into patients. Bladder cells grow into new tissue over time, and the scaffolds dissolve. Eventually, completely new bladders form.

Researchers can now grow artificial bladders. They hope to make hearts and kidneys one day, too.

Vegetarian Ice Cream

Most people love to eat ice cream. People with dairy allergies cannot eat ice cream. A solution to these two problems just might be an all-vegetable ice cream made from turnips and flower seeds. A special protein made in the laboratory is added to give the product a creamy texture. It even melts like real ice cream and has lots of vitamins but no cholesterol.

New Prosthetics

About half of the two million Americans who have lost arms or legs use prosthetics, or artificial limbs. New technology has made prosthetics more natural and useful. Researchers are working on artificial limbs that will be attached directly into existing bone with rods. Artificial nerves made of glass capsules containing electrodes will be attached to real nerve endings. Nerve impulses will be able to travel between the artificial limb and microchips embedded in the brain. All it will take to move an artificial arm or leg will be thinking about it!

Transplanted Rods and Cones

Today, most loss of sight is permanent. However, that may not be the case in the future if experiments on mice have application in humans. Recently, scientists transplanted rod and cone cells from the retinas of sighted mice into the eyes of blind mice. The cells lined themselves up correctly, connected to other nerve cells, and transmitted signals to the brain, restoring some eyesight. It will be many years before this procedure can be done in humans, but its success in mice could mean a brighter future for many people.

CAREER CONNECTION LAB TECHNICIAN

YOU DON'T SEE lab technicians when you visit the doctor, but doctors could not do their jobs without them. The doctor might take blood or urine samples. He or she might swab an infection or take a tissue sample. These samples all go to lab technicians.

Lab technicians prepare microscope slides, examine blood and other body fluids, and test for certain drugs. They analyze tissue samples to detect bacteria, fungi, or other disease-causing organisms. They also examine tissue samples to look for abnormal cells, such as those that are cancerous. Doctors depend on the information they get from lab technicians to diagnose and treat disease.

Most lab technicians work in hospitals or private labs. They need good problem-solving skills, computer expertise, and lots of patience. Many lab technicians have a two-year associate's degree from a community college or a certificate from a hospital program or a vocational school. Some technicians learn their skills on the job.

New Ways to **Fight** Disease

DISEASE-CAUSING AGENTS BEWARE! You are also the targets of new medical technologies.

Portable Blood Purifier

Once blood is contaminated with a deadly virus, it's not easy to get rid of the infection. But a portable, pen-sized gadget may soon do the job. This device is a blood purifier, and it hooks into two arteries in a person's arm. Contaminated blood flows from one artery into a tube and filter in the device. The holes in the filter are so tiny that red blood cells cannot pass through, but tiny viruses can. Antibodies that coat the filter stick to the viruses and keep them from passing back into the blood. Clean, filtered, virus-free blood returns to the person's body.

Disease-Fighting Viruses

Listeria is a deadly bacterium that can lurk in processed foods, such as sandwich meat. One of every three people who eats food contaminated with this bacterium will die, so eliminating the threat is important. Today, scientists are enlisting an unusual helper: viruses that get rid of the bacteria. Some viruses destroy bacteria. Scientists have identified harmless, bacteria-destroying viruses. The viruses can be sprayed on the meat at meatpacking plants, and the meat will stay safe for months.

Nanoshells

Chemotherapy is a cancer treatment that involves the use of strong toxins to kill cancer cells. Many people get sick during treatment because the poisons also kill healthy cells. Now, there is a new way to kill cancer cells while leaving healthy cells untouched. It involves the nanoshell—a carbon polymer sphere that is 1,000 times smaller than the period at the end of this sentence. Nanoshells filled with cancer drugs are injected into the body. Like guided missiles, they seek out and destroy cancer cells by attaching to proteins on the cells of malignant tumors. They burrow into the tumors and release their drugs, destroying them from the inside but not making a patient ill. It will take several more years to do further tests, but scientists are hopeful that these tiny spheres can become a giant weapon against cancer.

Time Line of Progress

Before 1900

1590	The microscope is invented.
1796	The first vaccine (smallpox) is developed.
1816	The first stethoscope is developed.
1818	The first successful blood transfusion takes place.
1849	The first American woman earns a medical degree.
1870s	Germs are identified as the cause of disease.
1895	X rays are discovered.
1896	Aspirin is developed.
1897	Penicillin is discovered.

Since 1900

1937	The first blood bank is created.
1952	The first heart pacemaker is developed.
1953	Scientists first describe the structure of DNA.
1954	The first kidney transplant takes place.
1967	The first human heart transplant takes place.
1978	The first "test tube baby" is born.
1980	Smallpox is wiped out worldwide.
1982	The first artificial heart transplant takes place.
1983	The virus that causes AIDS is identified.
1996	The first mammal (sheep) is cloned.

Research and Report
What type of medical breakthrough would you like to see happen in the future? Is there a disease you hope will be cured? Are there new things you hope doctors will be able to do? Identify this future advancement, and write a brief news story that announces it. Base part of your story on information you have learned in this unit. Share what you write with the class.

LAB SAFETY

Safety Symbols

These safety symbols are used in laboratory and field investigations in this book to indicate possible hazards. Learn the meaning of each symbol and refer to this page often. *Remember to wash your hands thoroughly after completing laboratory procedures.*

Safety Symbols	Hazard	Examples	Precaution	Remedy
Disposal	Special disposal procedures need to be followed.	certain chemicals, living organisms	Do not dispose of these materials in the sink or trash can.	Dispose of wastes as directed by your teacher.
Biological	Organisms or other biological materials that might be harmful to humans	bacteria, fungi, blood, unpreserved tissues, plant materials	Avoid skin contact with these materials. Wear mask or gloves.	Notify your teacher if you suspect contact with material. Wash hands thoroughly.
Extreme Temperature	Objects that can burn skin by being too cold or too hot	boiling liquids, hot plates, dry ice, liquid nitrogen	Use proper protection when handling.	Go to your teacher for first aid.
Sharp Object	Use of tools or glassware that can easily puncture or slice skin	razor blades, pins, scalpels, pointed tools, dissecting probes, broken glass	Practice common-sense behavior and follow guidelines for use of the tool.	Go to your teacher for first aid.
Fume	Possible danger to respiratory tract from fumes	ammonia, acetone, nail polish remover, heated sulfur, moth balls	Make sure there is good ventilation. Never smell fumes directly. Wear a mask.	Leave foul area and notify your teacher immediately.
Electrical	Possible danger from electrical shock or burn	improper grounding, liquid spills, short circuits, exposed wires	Double-check setup with teacher. Check condition of wires and apparatus.	Do not attempt to fix electrical problems. Notify your teacher immediately.
Irritant	Substances that can irritate the skin or mucous membranes of the respiratory tract	pollen, moth balls, steel wool, fiberglass, potassium permanganate	Wear dust mask and gloves. Practice extra care when handling these materials.	Go to your teacher for first aid.
Chemical	Chemicals that can react with and destroy tissue and other materials	bleaches such as hydrogen peroxide; acids such as sulfuric acid, hydrochloric acid; bases such as ammonia, sodium hydroxide	Wear goggles, gloves, and an apron.	Immediately flush the affected area with water and notify your teacher.
Toxic	Substance may be poisonous if touched, inhaled, or swallowed.	mercury, many metal compounds, iodine, poinsettia plant parts	Follow your teacher's instructions.	Always wash hands thoroughly after use. Go to your teacher for first aid.
Open Flame	Open flame may ignite flammable chemicals, loose clothing, or hair.	alcohol, kerosene, potassium permanganate, hair, clothing	Tie back hair. Avoid wearing loose clothing. Avoid open flames when using flammable chemicals. Be aware of locations of fire safety equipment.	Notify your teacher immediately. Use fire safety equipment if applicable.

Eye Safety	**Clothing Protection**	**Animal Safety**	**Radioactivity**
Proper eye care should be worn at all times by anyone performing or observing science activities.	This symbol appears when substances could stain or burn clothing.	This symbol appears when safety of animals and students must be ensured.	This symbol appears when radioactive materials are used.

Safe Laboratory Practices

Your personal safety in the laboratory is your responsibility. Following standard safety procedures will ensure your safety and that of your classmates. Before performing any experiment, read the entire procedure so you are familiar with the steps you will be following. Make note of any CAUTION statements and safety symbols displayed. Familiarize yourself with the safety guidelines and rules given here. Doing all these things will ensure a safe and successful laboratory experience.

Preventing Accidents

- Always wear chemical splash safety goggles (not glasses) in the laboratory. Goggles should fit snugly against the face to prevent any liquid from entering the eyes. Put on your goggles before beginning the lab and wear them throughout the entire activity, cleanup, and hand washing. Only remove goggles with your teacher's permission.
- Wear protective aprons and the proper type of gloves as instructed by your teacher.
- Keep your hands away from your face and mouth while working in the laboratory.
- Do NOT wear sandals or other open-toed shoes in the lab.
- Remove jewelry on hands and wrists before doing lab work. Loose jewelry, such as chains and long necklaces, should be removed to prevent them from getting caught in equipment.
- Do NOT wear clothing that is loose enough to catch on anything. If clothing is loose, tape or tie it down.
- Tie back long hair to keep it away from flames and equipment.
- Do NOT use hair spray, mousse, or other flammable hair products just before or during laboratory work where an open flame is used. These products ignite easily.
- Eating, drinking, chewing gum, applying makeup, and smoking are prohibited in the laboratory.
- Students are expected to behave properly in the laboratory. Practical jokes and fooling around can lead to accidents and injury.
- Students should notify their teacher about allergies or other health conditions that they have which can affect their participation in a lab.

Making Wise Choices

- When obtaining consumable laboratory materials, carefully dispense only the amount you will use. If you dispense more than you will use, check with your teacher to determine if another student can use the excess.
- If you have consumable materials left over after completing an investigation, check with your teacher to determine the best choice for either recycling or disposing of the materials.

Working in the Laboratory and the Field

- Study all procedures before you begin a laboratory or field investigation. Ask questions if you do not understand any part of the procedure.
- Do NOT begin any activity until directed to do so by your teacher.
- Work ONLY on procedures assigned by your teacher. NEVER work alone in the laboratory.
- Do NOT handle equipment without permission. Use all lab equipment for its intended use only.
- Collect and carry all equipment and materials to your work area before beginning the lab.
- Remain in your own work area unless given permission by your teacher to leave it. Keep your work area uncluttered.
- Learn and follow procedures for using specific laboratory equipment such as balances, microscopes, hot plates, and burners. Do not hesitate to ask for instructions about how to use any lab equipment.
- When heating or rinsing a container such as a test tube or flask, point it away from yourself and others.
- Do NOT taste, touch, or smell any chemical or substance unless instructed to do so by your teacher.
- If instructed to smell a substance in a container, hold the container a short distance away and fan vapors towards your nose.
- Do NOT substitute other chemicals/substances for those in the materials list unless instructed to do so by your teacher.
- Do NOT take any materials or chemicals outside of the laboratory.
- Stay out of storage areas unless you are instructed to be there and are supervised by your teacher.

Laboratory Cleanup

- Turn off all burners, gas valves, and water faucets before leaving the laboratory. Disconnect electrical devices.
- Clean all equipment as instructed by your teacher and return everything to the proper storage places.
- Dispose of all materials properly. Place disposable items in containers specifically marked for that type of item. Do not pour liquids down a drain unless your teacher instructs you to do so.
- Clean up your work and sink area.
- Wash your hands thoroughly with soap and warm water after each activity and BEFORE removing your goggles.

Emergencies

- Inform the teacher immediately of any mishap, such as fire, bodily injuries or burns, electrical shock, glassware breakage, and chemical or other spills.
- In most instances, your teacher will clean up spills. Do NOT attempt to clean up spills unless you are given permission and instructions on how to do so.
- Know the location of the fire extinguisher, safety shower, eyewash, fire blanket, and first-aid kit. After receiving instruction, you can use the safety shower, eyewash, and fire blanket in an emergency without your teacher's permission. However, the fire extinguisher and first-aid kit should only be used by your teacher or, in an extreme emergency, with your teacher's permission.
- If chemicals come into contact with your eyes or skin, notify your teacher immediately, then flush your skin or eyes with large quantities of water.
- If someone is injured or becomes ill, only a professional medical provider or someone certified in first aid should perform first-aid procedures.

METRIC SYSTEM AND SI UNITS

The International System of Measurement, or SI, is accepted as the standard for measurement throughout most of the world. The SI is a modernized version of the metric system, which is a system of measurement based on units of ten. In the United States, both the metric system and the standard system are used.

The SI system contains seven base units. All other units of measurement can be derived from these base units by multiplying or dividing the units by a factor of ten or by combining units.

- When you change from a smaller unit to a larger unit, you divide.
- When you change from a larger unit to a smaller unit, you multiply.

Prefixes are added to the base unit to identify the new unit created by multiplying or dividing by a factor of ten.

SI Base Units

Measurement	Unit	Symbol
length	meter	m
mass	kilogram	kg
time	second	s
electric current	ampere	A
temperature	Kelvin	K
amount of substance	mole	mol
intensity of light	candela	cd

Frequently Used Non-SI Base Units

Measurement	Unit	Symbol
volume	liter, cubic centimeter	L, cm³
density	grams/cubic centimeter, grams/liter	g/cm³, g/L

Common SI Prefixes

Prefix	Symbol	Equivalents
mega-	M	1,000,000
kilo-	k	1000
hecto-	h	100
deka-	da	10
deci-	d	0.1 or 1/10
centi-	c	0.01 or 1/100
milli-	m	0.001 or 1/1000
micro-	μ	0.000001 or 1/1,000,000
nano	n	0.000000001 or 1/100.000,000,000
pico-	p	0.000000000001 or 1/100.000,000,000,000

THE MICROSCOPE

Microscope Care and Use

1. Always carry the microscope by holding the arm of the microscope with one hand and supporting the base with the other hand.
2. Place the microscope on a flat surface. The arm should be positioned toward you.
3. Look through the eyepieces. Adjust the diaphragm so that light comes through the opening in the stage.
4. Place a slide on the stage so that the specimen is in the field of view. Hold it firmly in place by using the stage clips.
5. Always focus first with the coarse adjustment and the low-power objective lens. Once the object is in focus on low power, the high-power objective can be used. Use ONLY the fine adjustment to focus the high-power lens.
6. Store the microscope covered.

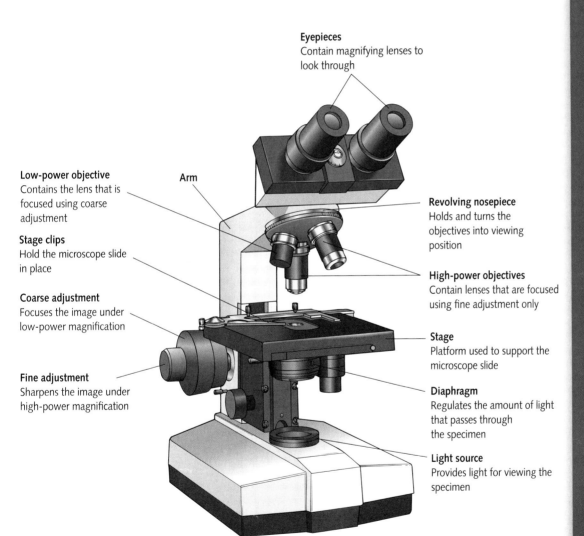

Eyepieces
Contain magnifying lenses to look through

Low-power objective
Contains the lens that is focused using coarse adjustment

Arm

Stage clips
Hold the microscope slide in place

Coarse adjustment
Focuses the image under low-power magnification

Fine adjustment
Sharpens the image under high-power magnification

Revolving nosepiece
Holds and turns the objectives into viewing position

High-power objectives
Contain lenses that are focused using fine adjustment only

Stage
Platform used to support the microscope slide

Diaphragm
Regulates the amount of light that passes through the specimen

Light source
Provides light for viewing the specimen

THE PERIODIC TABLE OF ELEMENTS

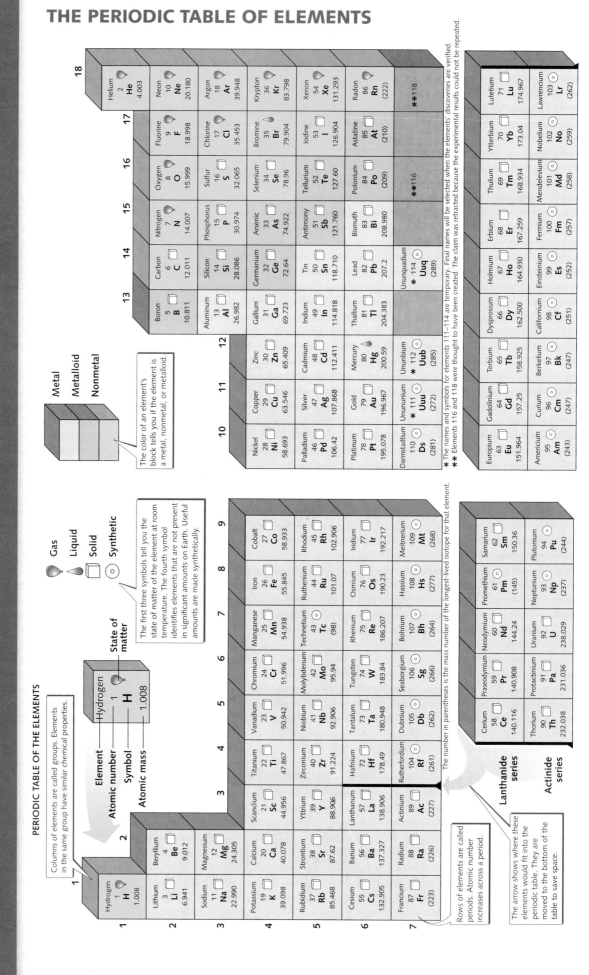

PERIODIC TABLE OF THE ELEMENTS

Element ——— Hydrogen
Atomic number ——— 1
Symbol ——— H
Atomic mass ——— 1.008

State of matter

The first three symbols tell you the state of matter of the element at room temperature. The fourth symbol identifies elements that are not present in significant amounts on Earth. Useful amounts are made synthetically.

Gas
Liquid
Solid
Synthetic

Metal
Metalloid
Nonmetal

The color of an element's block tells you if the element is a metal, nonmetal, or metalloid.

Columns of elements are called groups. Elements in the same group have similar chemical properties.

Rows of elements are called periods. Atomic number increases across a period.

The arrow shows where these elements would fit into the periodic table. They are moved to the bottom of the table to save space.

The number in parentheses is the mass number of the longest-lived isotope for that element.

* The names and symbols for elements 111–114 are temporary. Final names will be selected when the elements' discoveries are verified.
** Elements 116 and 118 were thought to have been created. The claim was retracted because the experimental results could not be repeated.

Glossary/Index

E

Glossary/Index

G

Galactose, 29

Galápagos Islands, Darwin and, 101

Gamete a sex cell, 57, 169, 419

alleles in, 61

chromosome numbers in, 66

haploid number of, 66

in meiosis, 68

in mitosis, 68

Gametophyte, 171

of club mosses, 173

fern, 175

of horsetails, 173

of liverworts and mosses, 170

pine, 179

Gametophyte generation the stage of a life cycle that produces sex cells, 169

Gas exchange, 352

Gaseous wastes, 390

carbon dioxide as, 392

Gases

in air, 28

movement into and out of body, 352

Gastropods, 219

GAVI Alliance, 443

Gecko, 255

Geese, 261

Gender, determination of, 69

Gene a heritable trait, 60

on chromosomes, 65–70

jumping, 66

Gene therapy, viruses in, 127

Gene, as code in genetic engineering, 121

Generations

alternation of, 169

gametophyte, 169

sporophyte, 169

Genetic code, 73

Genetic diversity the variation in genes among members of the same species, 87

Genetic engineering a variety of techniques that directly change the hereditary material of an organism, 76

benefits and risks of, 77

as career, 120–121

economic consequences of, 77

Genetic information, storage in DNA, 31

Genetic material

DNA in, 94

in virus, 125

Genetics the field of biology devoted to studying heredity, 6, 57. *See also* Chromosomes; Heredity

Genome, human, 76

Genotype a combination of alleles, 61, 62

Genus a group that helps make up a family, 112

dogs in, 113

species and, 114

Geobacter sulfurreducens, 152

Geologic time scale a record of living organisms' history, divided into geologic eras, periods, and epochs, 91

Germ theory, 436

Germination the process by which a seed embryo develops into a new plant, 203

Germs, Semmelweis and, 9

Gestation the period of time during which a developing mammal stays in the uterus, 268

Ghost orchid, 186, 189

Giant barrel sponge, 212

Giant desert centipede, 229

Giant kelp, 135, 141

Giant panda, bears and, 109

Giardia lamblia, 435

Gills

of crustaceans, 228

of fishes, 241, 244, 247

ventilation through, 351

Ginkgo a gymnosperm with leaves that closely resemble the leaves on more familiar trees, 177

Girls, in puberty, 428

Gland an organ that produces and releases a substance, 412. *See also* specific types

endocrine, 412–414

salivary, 381

sweat, 395

Glass sponge, 213

Gliding joint, 342, 343

Global warming the rise in the average temperature of Earth's air and oceans over the past few decades, 321, 327

Glomerulus a ball of capillaries in the kidney surrounded by the cup of the nephron, 393

Glucose, 25, 386, 415

in cellular respiration, 162

Glycerol, 30

Glycogen, 30

Gnawing mammals, 268

Gnetophytes, 177

Golden algae, 140

Golden dart frog, 251

Golden plover, migration by, 265

Golden toad, 248

Golgi apparatus an organelle that modifies and packages proteins and other molecules so they can be stored in a cell or sent outside the cell, 41, 331

Goodall, Jane, 5

"Goose bumps," 339

Gorillas, 95

Graduated cylinder, 16

Grafting, 206

Grand Canyon, stream communities in, 291

Graphs, 11

Grasshoppers, 230

damage by, 234

Grassland a community in which grasses are the most important plants, 300

Gravitropism, 191

Gravity, pollination and, 183

Grazing animals, energy for, 25

Great Barrier Reef, 215

Great blue heron, 263

Great Potato Famine (Ireland), parasitic fungus-like protists in, 145

Great white shark, 245

Green algae, 140, 141

Green color, in plants, 161

Greenhouse gases, 287

Growth hormone (hGH), 413, 415

Growth rate, of human infants, 427

Growth rings, of trees, 194

Glossary/Index

agents of, 183
of angiosperms, 182
by birds, 261
by honeybees, 234
Pollinators, 183

Pollution the release of harmful substances into the environment, **321**

antipollution measures and, 322
Pollution-fighting microbes, 152
Polyhedral virus, 125

Polyp the body part of a cnidarian that is shaped like a vase and usually does not move from place to place, **214**, 215

Polysaccharides, 30
Polyunsaturated fatty acids, 30
Pond scum, 134
Ponds, 303
Pons, 404

Population a group of individuals from the same species living in a specific area at the same time, **281**

of animals, 308
changes from natural events, 312
density of, 310
equation for, 312
growth and sizes of, 309–312
growth of human, 319
limit factors for, 310–311
relationships among, 313–315
Pores, 39
in stem, 193
Portuguese man-of-war, 214
Posture, 452, 453
Potato blight, 145
Potatoes, as tubers, 193, 195
Pouched mammals, 267
Power plants, smoke from, 321
Prairies, 300

Precipitation heavy drops of water that fall back to Earth as rain, snow, sleet, or hail, **286**, 296

Predation when one creature kills and eats another creature, **315**, 317

Predator an organism that kills and eats another organism for food, 84, **85**

adaptations by, 86
in ecosystems, 315
prey as part of, 318

Prediction a statement that suggests what a person thinks will happen in the future based on past experience and evidence, **12**

Preening a process in which birds apply oil to their feathers to keep them from drying out, **259**

Prefixes, metric, 14

Pregnant carrying a developing fertilized egg, **423**, 424

Prescription drug, 454
Preserves, nature, 322

Prey an organism that is eaten by other organisms, 84, **85**

adaptations by, 85
in ecosystems, 315
as part of predator, 318

Primary succession the process of living things invading previously lifeless places on Earth, **289**

Primate a mammal with a rounded head, flat face, opposable thumbs and a large, complex brain, **269**

ancestors of, 96
evolution of, 96
human-like, 98–99
opposable thumb of, 95

Probability the likelihood that a certain event will occur, **64**

Producer, (1) an organism that conducts photosynthesis, 48 (2) a living thing that can capture matter and energy from abiotic sources, **283**

Progesterone a female hormone that helps the uterine lining grow thick during the menstrual cycle, 414, **422**

Prokaryotes, bacteria as, 129

Prokaryotic cell a cell that does not contain membrane-enclosed organelles, **43**, 128

cell division by, 53
Prolactin, 413
Prop roots, 188
Propagation, vegetative, 189
Prophase, of mitosis, 67

Prosthetics, 460
Protected areas, 322
Protected species, alligators as, 276
Protection, blood as, 358
Protective covering, 85

Protein, (1) an organic molecule made up of carbon, hydrogen, oxygen, and nitrogen, 31 (2) a long chain of amino acids, **376**

nitrogen and, 288
as organic molecules, 29
on ribosomes, 73
Protein channels, 46
Protein coat, over virus, 125

Protein synthesis the process by which proteins are made, **73**

Protist an organism that is a eukaryote but is not a plant, animal, or fungus, 117, 134, **135**

animal-like protozoans, 136–138
fungus-like, 143–145

Protista the most diverse kingdom that includes organisms such as fungi and slime mold, **117**, 135

Protoceratops, 88

Protozoans animal-like protists that are unicellular, **136–138**

with cilia, 137
as decomposers, 283
disease spread by, 435
with flagella, 137
roles of, 138
with spores, 138
types of, 136–138

Pseudopod extension of the cell membrane that fills with cytoplasm, **136**

Psychological dependence a condition in which the user has an overwhelming, emotional desire to continue using a drug, **455**

Puberty the beginning of adolescence when a child's body turns into an adult body, **428**

Puerperal fever, Semmelweis on, 9
Puffballs, reproduction by, 147

Glossary/Index

South American cane toad, in
Australia, 321
Space science, 5
Space, for living things, 28
Spallanzani, Lazzaro, 37
Spanish moss, as living thing, 21

Species, (1) a group of organisms
that have similar physical
characteristics and can reproduce
offspring capable of producing their
own offspring, 23 (2) the smallest
and most specific group into which
organisms are classified, 112

animal, in Sequoia National Park,
80–81
of bryophytes and tracheophytes,
174
dogs in, 112, 113
extinction of, 90
genus and, 114
human killing of, 320
interspecific relationships
between, 315
introduced, 321
preservation of, 323
reproduction and, 23–24
succession by, 289
tracking and naming of, 120
variation in, 102

Sperm a male sex cell, 169, 170,
171, **419,** 420

fertilization by, 423
male gametes as, 68
in sponges, 213
Sperm cell, 51

Spicule a sharp, small structure
that helps make up some sponges'
bones, 213

Spiders, 226–227
adaptation of, 317
in California national parks, 80
innate behavior of, 271
structure of, 226
Spinal cord, 405
in central nervous system, 402
reflexes and, 406
severing of, 405
Spine, 405
Spiny anteater, 267

Spirilla bacteria that are shaped
like corkscrews or spirals, 129

Spitting cobra, 255
Spleen, 364

Sponges, 209, 212–213
asymmetry of, 211
characteristics of, 212
reproduction in, 213
structure of, 213

Spongy bone a strong structure of
bony tissue that makes up the inside
of a bone, 341

Spontaneous generation the
belief that living things come from
nonliving matter, 36–38

Sporangia, (1) special reproductive
hyphae that have grown up and
away from the mycelium, 147 (2)
structures located at the bottom
of a frond in which fern spores are
produced, **174**

Spore production
in fungus-like protists, 143
in plasmodium, 144

Spore a reproductive cell produced
by sporozoans that develops into
new organisms when environmental
conditions are favorable, 138, 169,
171

distribution by fungi, 147
fern, 175
in meiosis, 170
Plasmodium, 138
Sporophyte, 170, 171
of ferns, 174

Sporophyte generation the stage
of a life cycle that produces spores,
169

Spruce trees, 178, 299
Squash bug, 234
Squids, 220, 221

Stamen the male reproductive
organ of a plant, composed of the
anthers and filaments, 180

Starches, 30, 375
in photosynthesis, 162
in tubers, 193
Starfish. *See* Sea star
Steam engine, invention of, 38

Stem part of a plant's shoot system
that supports its leaves, cones, fruit,
and flowers, 192–195

herbaceous, 192
monocot, 195

specialized, 195
structure of, 193–195
types of, 192
woody, 192, 194
Steppes, 300

Sternum a bird's breastbone, 260

Stigma a structure on top of the
flower's pistil that holds onto any
pollen grains that land on it, 181

Stimulants, 455

Stimulus, (1) a condition or event
in your surroundings, 23, 399 (2) a
signal that causes an animal to react
in a certain way, 23, 165, 271, 399

plant responses to environmental,
165
Stinging cells, in cnidarians, 214
Stingray, 245
Stinkbugs, 233
Stirrup, in ear, 410

Stomach a large, muscular bag that
digests, breaks down, and stores
food, 382

human, 382
of sea star, 236

Stomata tiny pores that control
gas exchange between the leaf and
the environment, 198

Stonecrab, 228
Stonewashed jeans, fungus and, 153
Streams, 303
Strepsirrhines, 96

Stress the body's response to the
demands of everyday life, 452

management of, 449, 452
Striated muscle, 345

Stroke a condition that occurs
when an artery in the brain is
blocked, cutting off the blood supply
to part of the brain, 371

deaths from, 369
Structure
cell, 115
of flowers, 180–181
of fungi, 146
similarities as basis for
classification, 109
Structures (body), 92–94
analogous, 93
homologous, 93

Glossary/Index

Art Credits

McGraw-Hill, Garry Nichols, and Howard Friedman

Photo Credits

Cover © Art Wolfe/Getty Images; **2** (t) ©Kennan Ward/Corbis, (m) ©Tom Brakefield/Corbis, ©hybrid medical animation/Photo Researchers, (b) ©Stephen Dalton/Photo Researchers; **3** ©kristian; **4** ©Vasilev Ivan Mihaylovich; **5** (t) ©Michael Nichols/National Geographic Image Collection, (b, l-r) ©Yo Nayaya/Getty Images, ©David Sutherland/Getty Images, ©StockTrek/Getty Images and ©Chee-Onn Leong; **7** ©Tammy Wolfe; **9** Wikipedia; **12** ©ABN Stock Images/Alamy; **13** ©Eugene Buchko; **14** ©Christine Balderas; **18** ©ABN Stock Images/Alamy; **20** ©Ronald Sherwood; **21** ©S. Greg Panosian, (t) ©Digital Vision/PunchStock, (m) ©Mark Grenier, (b) ©Ivan; **22** (tl) ©Purestock/Getty Images, (tr) ©Emilia Stasiak, (bl) ©Gschmeissner/Photo Researchers, (br) ©Frank Leung; **23** (t) ©Charles D. Winters/Photo Researchers, (b) ©Elena Sherengovskaya; **24** ©Mark D. Phillips/Photo Researchers; **25** (t) ©iStockphoto.com/ Grafissimo, (b) ©Anita Elder; **26** (t) ©Jose Manuel Gelpi Diaz, (i) ©Tomasz Pietryszek, (b) ©Stockbyte/Getty Images; **27** ©Ted Kinsman/ Photo Researchers; **28** (l) ©U.S. Fish and Wildlife Services, (r) ©Andrew Syred/Photo Researchers; **30** (t, l-r) ©Jovan Nikolic, ©Sebastian Kaulitzki, ©Frank Boellmann, (b) ©Diane Diederich, ©iStockphoto.com/EddWestmacott, and ©Darryl Brooks; **31** ©Johan Swanepoel; **32** ©Tomasz Pietryszek; **34** ©Roy Morsch/zefa/Corbis; **35** (t) ©Omikron/Photo Researchers, (b) ©The Granger Collection, New York; **36** (t) ©Roman Krochuk, (b, l-r) ©Rachel Dewis, ©webartworks.de/Fotolia, ©Maria Veras; **42** (t) ©Keith R. Porter/Photo Researchers, (bl) ©Biophoto Associates/Photo Researchers, (br) ©LSHTM/Photo Researchers; **43** ©Dr. Jeremy Burgess /Photo Researchers, **44** ©Bloomimage/Corbis; **48** ©Elena Elisseeva; **49** ©Eric Wong; **50** ©Wing Tang; **51** (l-r) ©Alfred Pasieka/Photo Researchers, ©Sebastian Kaulitzki, ©Eye of Science/Photo Researchers; **56** ©Frans Lanting/Corbis; **57** ©Image 100/Corbis; **58** ©James King-Holmes/Photo Researchers; **62** ©Liga Gabrane; **66** (l) ©Digital Food Shots, (r) ©Kevin Snair; **70** ©David Nicholls/Photo Researchers; **72** ©Science Source; **74** ©Tom Bean/Corbis; **75** (t) ©John S. Sfondilias; (b) ©Derek Dammann; **77** ©Seoul National University/Handout/Reuters/ Corbis; **80** (t) ©Jean Krejca, Ph.D, (b) ©Kim Karpeles/Alamy; **81** (t) ©iStock International, (b) ©Michael Abbey/Photo Researchers; **82** (t) ©Falk Kienas, (b) ©Susan McKenzie; **83** ©Royalty-Free/Corbis; **84** ©Manoj Shah/Getty Images; **85** (t) ©PunchStock, (l-r) ©Claudia Adams/Alamy, ©Gary Buss/Getty Images, ©Joel Sartore/Getty Images; **86** (tl) ©PunchStock, (m) ©Johan Swanepoel/Shutterstock, Inc., (bl) ©Nik Niklz, (br) ©Lane Erickson; **87** (r) ©Matt Meadows, (b) ©J. McPhail; **88** (t, l-r) ©Marlene DeGrood, ©Ismael Montero Verdu, ©www.handini.com/iStock; **89** ©Richard T. Nowitz/Photo Researchers; **93** (t) ©Mark William Penny, (b) ©Creatas/PunchStock; **95** (l) © Christian Riedel, (r) ©Steven Tilston; **97** ©John Reader/Photo Researchers; **98** (t) ©Des Bartlett/Photo Researchers, (b) ©Marion Kaplan/ Alamy; **99** (l) ©The Field Museum, (r) ©Volker Steger/Nordstar-"4 Million Years of Man"/Photo Researchers; **101** (t) ©Marius Hainal, (m) ©Nancy Nehring, (b) ©Science Source; **105** ©Falk Kienas; **106** ©Blasius Erlinger/Getty Images; **107** (l-r) ©Tom Brakefield/Getty Images, ©Bettmann/Corbis, ©GK Hart/Vikki Hart/Getty Images; **109** (tr) ©Marina Cano Trueba, (m) ©Getty Images, (bl) ©Digital Vision/PunchStock, (br) ©Lynsey Allan; **110** ©Alex Balako; **111** (l-r) ©PhotoLink/Getty Images, ©Dr. Dennis Kunkel/Getty Images, (r) ©PhotoLink/Getty Images; **112** (l-r) ©Getty Images, ©Erik Lam, ©Getty Images; **113** (1) ©Getty Images, (2) ©Getty Images, (3) ©Corbis, (4) ©PhotoLink/Getty Images, (5) ©Hakan Karlsson, (6) Andrew Manley, (7) ©Stanislav Khrapov; **114** (t-b) ©Getty Images, ©Nancy Nehring, ©Pat Powers and Cherryl Schafer/Getty Images; **116** (t) ©Ines Gesell, (b) ©Michael Abbey/Science Photo Library; **117** (t, l-r) ©Sanamyan/Alamy, ©Paul Whitted, ©John Walsh/Photo Researchers, (bl) ©Corbis Royalty Free, (br) ©PhotoLink/Getty Images; **119** ©Getty Images; **120** (t) ©Pierre Perrin/Corbis Sygma, (b) ©KCNA/epa/Corbis; **121** Bryan L. Stuart; **122** (l) ©Stefan Glebowski, (r) ©Lee D. Simon/Photo Researchers; **123** ©Joseph Van Os, (i) ©Eye of Science/Photo Researchers; **124** (l-r) ©Chris Bjornberg/Photo Researchers, ©Omikron/Photo Researchers, ©Lee. D. Simon/Photo Researchers, ©Hans Gelderblom/Getty Images; **125** ©Lee D. Simon/Photo Researchers; **127** (t) ©Keystone/Getty Images, (m) ©Geogre/Wikipedia; **129** (t, l-r) ©Scimat/Photo Researchers, ©Eric V. Grave/Photo Researchers, ©James Cavallini/Photo Researchers; **130** (l) ©M. I. Walker/Photo Researchers, (r) ©Dr. Linda Stannard, Uct/Photo Researchers; **131** (l-r) ©Ben Osborne/Getty Images, ©Natalia Klenova, ©Nadezda Firsova, ©Scimat/Photo Researchers, ©Aleksander Bolbot; **132** ©Lee D. Simon/Photo Researchers; **133** ©Dr. L. Caro/Photo Researchers; **134** ©Jamie Steffens; **135** (l-r) ©M. I. Walker/Photo Researchers, ©Jeff Rotman/Photo Researchers, ©Daniel Puleo/Wikipedia; **136** ©Wilm van Egmond/Getty Images; **137** (t) ©Michael Abbey/Photo Researchers, (br) ©Michael Abbey/Science Photo Library; **139** (t) ©Paul Whitted, (b) ©Dee Breger/Photo Researchers; **140** ©PHOTOTAKE Inc./Alamy; **141** (t, l-r) ©2005 GettyImages, ©Visual&Written SL/Alamy, ©Steven P. Lynch, (b, l-r) ©M. I. Walker/Photo Researchers, ©John Walsh/Photo Researchers, ©Andrew J. Martinez/Photo Researchers; **142** ©Jackson Vereen/Cole Group/Getty Images; **143** (t) ©Eric Guinther/Wikipedia, (b) ©Bill Banaszewski/Visuals Unlimited; **144** ©Nigel Cattlin/Holt Studios International/ Science Photo Library; **145** ©Courtesy of the General Research Division, the New York Public Library, Astor, Lenox and Tilden Foundations; **146** (t) ©Steve McWilliam, (b) ©Photolink/Getty Images; **147** (l) ©SPL/Photo Researchers, (r) ©Michael & Patricia Fogden/ Corbis; **148** (l-r) ©David Toase/Getty Images, ©Bryan Eastham, ©Aleksandra, ©Emily Keegin/fStop/Getty Images; **149** (t-b) ©Tomas Bogner, ©fotosav (Victor & Katya), ©Digital Vision/Getty Images; **151** ©Tomas Bogner; **152** ©Stephen Ausmus/www.ars.usda.gov; **153** ©Rubberball Productions; **154** (t-b) ©Ron Nichols, USDA Natural Resources Conservation Service, ©Jeff Foott/Getty Images, ©Leo; **155** (l) ©Sarah Scott, (r) ©Ei Katsumata/Alamy; **156** (t) ©Comstock/Jupiter Images, (b) ©Royalty-Free/Corbis; **157** (l) ©Maria & Bruno Petriglia/Photo Researchers, (r) ©Louie Schoeman; **158** (l-r) ©Steven P. Lynch, ©Dr. Jeremy Burgess/Photo Researchers, (r) ©Royalty-Free/Corbis; **159** (t-b) ©Dubrovskiy Sergey Vladimirovic, ©Bruce Heinemann/Getty Images, ©Bob Gibbons/Photo Researchers; **160** ©Stapleton Collection/Corbis; **161** ©Heather Barr; **163** (l) ©Maryann Frazier/Photo Researchers, (r) ©Nigel Cattlin/Photo Researchers; **164** (t-b) ©Ivaschenko Roman, ©Kate Tilmouth, ©Zastavkin; **165** ©Maxine Adcock/Photo Researchers; **166** ©Leo; **168** ©Aaron Beernaert; **170** ©Steven P. Lynch; **171** ©Bruce Coleman Inc./Alamy; **172** ©Michael P. Gadomski/Photo Researchers; **173** ©amana images/Getty Images (i)©John W. Bova/Photo Researchers; **174** (t) ©Dennis Purse/Photo Researchers, (b) ©iStockphoto.com/ chepatchet; **175** ©Biophoto Associates/Photo Researchers; **176** (t) ©Reg Morrison/Auscape/Minden Pictures, (m) ©Frank Lane Picture Agency/Corbis, (bl)Lisa McDonald, (br) ©Wayne Atkinson c/o p.taihaku@googlemail.com; **177** (t) ©Joseph Malcolm Smith/Photo Researchers, (i) ©Charlotte Erpenbeck/Shutterstock, (bl) ©Michael Grube, (br) ©Rod Planck/Photo Researchers; **178** ©Charlie Ott/ Photo Researchers; **180** ©InsideOutPix/Corbis; **182** (l) ©Hannamariah, (r) ©PhotoLink/Getty Images; **183** (t-b) ©Dr. Jeremy Burgess/ Photo Researchers, ©Dr. Merlin D. Tuttle/Photo Researchers, ©Michael_Patricia Fogden/Minden Pictures; **184** ©Charlotte Erpenbeck/ Shutterstock; **186** ©Adeline Lim, (i) ©Craig Churchill/Alamy; **187** (l) ©John Kaprielian/Photo Researchers, (r) ©suravid; **188** (t) ©Frances Twitty, (b) ©E. R. Degginger/Photo Researchers; **189** (t) ©paradoks_blizanaca, (b) ©Nigel Cattlin/Photo Researchers; **191** (l) ©The McGraw-Hill Companies/Al Telser, Photographer, (r) ©Biophoto Associates/Photo Researchers; **192** (l) ©Muriel Lasure, (r) zastavkin; **193** (tl) ©BSIP/Photo Researchers, (tr) ©Alan L. Detrick/Photo Researchers, (m) ©Norman Tomalin/Alamy, (b) ©The McGraw-Hill Companies/Al Telser, photographer; **195** (t) ©Steven P. Lynch, (b) Brand X Pictures/PunchStock; **196** (l, b-t) ©Mark A. Schneider/Photo Researchers, ©Andrzej Tokarski, ©Melba Photo Agency/PunchStock, (tr) ©Sheila Terry/Photo Researchers, (br) ©Ilya D. Gridnev; **197** (tl) ©Royalty-Free/Corbis, (tr) ©Stockdisc/PunchStock, (bl) ©Science Source/Photo Researchers; **199** (r, t-b) ©Brand X Pictures/ PunchStock, ©Jacques Jangoux/Alamy, ©Comstock/PunchStock, (bl) ©iStockphoto.com/Vickie Sichau, (br) ©chai kian shin; **200** (t) ©Dr. Jeremy Burgess/Photo Researchers, (b) ©blickwinkel/Alamy; **201** (t) ©The McGraw-Hill Companies/Ken Cavanagh, (b) ©Siede Preis/ Getty Images; **202** ©Jean D'Alembert/Wikipedia; **205** ©Brand X Pictures/PunchStock; **206** (t) ©Royalty-Free/Corbis, (b) ©Scott Bauer; **207** (t) ©Martin Rogers/Getty Images, (b) ©Danny E. Hooks; **208** (t) ©Comstock Images/PictureQuest, (l) ©Travis Klein, (r) ©Creatas/ PunchStock, (b) ©Jeremy Woodhouse/Getty Images; **209** (l-r) ©Comstock Images/PictureQuest, ©Medioimages/PunchStock, ©Chee-